Artificial Intelligence (AI) for Smart and Sustainable Urban Transportation

IEEE Press

445 Hoes Lane

Piscataway, NJ 08854

Artificial Intelligence (AI) for Smart and Sustainable Urban Transportation

Edited by

Sathyan Munirathinam
ASML Corporation, San Jose, CA, USA

Pethuru R. Chelliah
Infocion Inc., Bangalore, India

Peter Augustine
Department of Computer Science, Christ University, Bangalore, India

Beaulah Soundarabai
Department of Computer Science, Christ University, Bangalore, India

IEEE*Press*
WILEY

Published by John Wiley & Sons, Inc., Hoboken, New Jersey.

For general information on our other products and services or for technical support, please contact our Customer Care Department within the United States at (800) 762-2974, outside the United States at (317) 572-3993 or fax (317) 572-4002.

Wiley also publishes its books in a variety of electronic formats. Some content that appears in print may not be available in electronic formats. For more information about Wiley products, visit our web site at www.wiley.com.

Library of Congress Cataloging-in-Publication Data has been applied for:

Print ISBN: 9781394351060
ePDF ISBN: 9781394351084
ePub ISBN: 9781394351077
oBook ISBN: 9781394351091

Cover Design: Wiley
Cover Image: © filo/Getty Images

Contents

List of Contributors

Prerna Ajmani
School of Information Technology
Vivekananda Institute of Professional Studies-Technical Campus
Delhi
India

S. Ashmad
Audisankara Group of Institutions
Nellore
Andhra Pradesh
India

Anoop Bhardwaj
School of Civil Engineering
Lovely Professional University
Phagwara
Punjab
India

Shantanu Bindewari
School of Computer Science Engineering
IILM University
Greater Noida
India

S. Cyciliya Pearline Christy
Department of Computer Applications and Research Centre
Sarah Tucker College (Autonomous)
Affiliated to Manonmaniam Sundaranar University
Tirunelveli
Tamil Nadu
India

Prakhar Consul
School of Computer Science Engineering
IILM University
Greater Noida
India

Sneha Das
School of Civil Engineering
Lovely Professional University
Phagwara
Punjab
India

Sudeepa Roy Dey
Computer Science and Engineering
PES University
Bengaluru
Karnataka
India

Kathi Durgesh
School of Civil Engineering
Lovely Professional University
Phagwara
Punjab
India

Siddharth Garia
School of Civil Engineering
Lovely Professional University
Phagwara
Punjab
India

Nancy Jasmine Goldena
Department of Computer Applications and Research Centre
Sarah Tucker College (Autonomous)
Affiliated to Manonmaniam Sundaranar University
Tirunelveli
Tamil Nadu
India

Kirti Goyal
School of Civil Engineering
Lovely Professional University
Phagwara
Punjab
India

Hriday Gupta
School of Information Technology
Vivekananda Institute of Professional Studies-Technical Campus
Delhi
India

Ishika Gupta
School of Information Technology
Vivekananda Institute of Professional Studies-Technical Campus
Delhi
India

Siddharth Gupta
School of Information Technology
Vivekananda Institute of Professional Studies-Technical Campus
Delhi
India

A. Immanuel
Audisankara Group of Institutions
Nellore
Andhra Pradesh
India

Ashik Shah Jahangeer
Department of Computer Science and Applications
The Gandhigram Rural Institute
Dindigul
Tamil Nadu
India

Vishal Jain
Department of Computer Science and Engineering
Sharda School of Engineering and Technology
Sharda University
Greater Noida
India

A. Jane
Department of Computer Applications and Research Centre
Sarah Tucker College (Autonomous)
Affiliated to Manonmaniam Sundaranar University
Tirunelveli
Tamil Nadu
India

M. Robinson Joel
Department of Computer Science and Engineering
KCG College of Technology
Karapakkam, Chennai, Tamil Nadu
India

M. Kannan
Department of Computer Science with Data Science
SRM Arts and Science College
Affiliated to the University of Madras
Chengalpattu
Tamil Nadu
India

A. Jothi Kumar
Thinkcove Technologies
Chennai
Tamil Nadu
India

Mary Immaculate Sheela Lourdusamy
Department of Computing and Engineering
School of Engineering Technology and Applied Science (SETAS)
Heritage Christian University College (HCUC)
Accra
Ghana

K. Merriliance
Department of Computer Applications and Research Centre
Sarah Tucker College (Autonomous)
Affiliated to Manonmaniam Sundaranar University
Tirunelveli
Tamil Nadu
India

Archan Mitra
Department of Mass Communication and Journalism
Amity School of Communication
Amity University
Mumbai
India

Vishal Kumar Narnoli
School of Civil Engineering
Lovely Professional University
Phagwara
Punjab
India

Basab Nath
School of Computer Science Engineering
IILM University
Greater Noida
India

Animesh Nayak
School of Civil Engineering
Lovely Professional University
Phagwara
Punjab
India

Sanchita Paul
Neil Patel Digital
Bangalore
India

S. Rakshika
Computer Science and Engineering
PES University
Bengaluru
Karnataka
India

Barath Ramesh
Department of Computer Science with Data Science
SRM Arts and Science College
Affiliated to the University of Madras
Chengalpattu
Tamil Nadu
India

Raviprakash R Salagame
Autobotik Pte Ltd.
Singapore

R. Saradha
Department of Computer Science
SDNB Vaishnav College for Women
Chennai
Tamil Nadu
India

M. Saravanan
Sreenivasa Institute of Technology and Management Studies
Chittoor
Andhra Pradesh
India

Hilal Ahmed Shah
School of Computer Science Engineering
IILM University
Greater Noida
India

P. Shanmugavadivu
Department of Computer Science and Applications
The Gandhigram Rural Institute
Dindigul
Tamil Nadu
India

Amol Sharma
CEPT University
Ahmedabad
Gujarat
India

L. Ameer Shohail
Saudi Telecom Company (STC)
Riyadh
Saudi Arabia

J. Steffi
Department of Computer Applications and Research Centre
Sarah Tucker College (Autonomous)
Affiliated to Manonmaniam Sundaranar University
Tirunelveli
Tamil Nadu
India

R. Rashia Subashree
Department of Computer Applications and Research Centre
Sarah Tucker College (Autonomous)
Affiliated to Manonmaniam Sundaranar University
Tirunelveli
Tamil Nadu
India

P. Sudheer
Sreenivasa Institute of Technology and Management Studies
Chittoor
Andhra Pradesh
India

A. Jasmine Sugil
Department of Computer Applications and Research Centre
Sarah Tucker College (Autonomous)
Affiliated to Manonmaniam Sundaranar University
Tirunelveli
Tamil Nadu
India

S. Thangapriya
Department of Computer Science with Data Science
SRM Arts and Science College
Affiliated to the University of Madras
Chengalpattu
Tamil Nadu
India

Mansi Trivedi
School of Computer Science Engineering
IILM University
Greater Noida
India

T. S. Vasughi
Department of Computer Science with Data Science
SRM Arts and Science College
Affiliated to the University of Madras
Chengalpattu
Tamil Nadu
India

About the Editors

Dr. Sathyan Munirathinam is working as a cloud solution architect manager for ASML Corporation. He is responsible for developing a cloud strategy and roadmap for addressing dynamic cloud solutions, as well as working on machine learning, deep learning, and IoT issues. He has a PhD in Machine Learning and more than 20 years of experience in working with business intelligence and 15 years of experience in manufacturing. His research interests include artificial intelligence, equipment health monitoring, IoT and big data analysis, statistical machine learning and data mining, ubiquitous computing, and human–computer interaction. He holds a Master of Science from Illinois Institute of Technology and is a Certified Business Intelligence Professional.

Dr. Pethuru Raj is currently serving as principal AI architect at Infocion Inc., AKR Tech Park, Bangalore. He has over 17 years of extensive experience in the IT industry and more than eight years of research experience, with deep expertise spanning artificial intelligence, cloud computing, enterprise architecture, and site reliability engineering. Prior to his current role, Dr. Pethuru Raj worked as chief architect in the Site Reliability Engineering (SRE) division at Reliance Jio Infocomm Ltd. (RJIL), Bangalore. He earlier served as a cloud infrastructure architect at the IBM Global Cloud Center of Excellence (CoE), IBM India, Bangalore, where he worked for four years. Before joining IBM, he had a long and impactful tenure as a TOGAF-certified Enterprise Architecture Consultant with Wipro Consulting Services (WCS). In addition, he has held the position of lead architect in the Corporate Research Division of Robert Bosch, Bangalore.

Dr. Peter Augustine has been working as a professor in the Department of Computer Science, CHRIST University, Bangalore. He has a PhD in Medical Image Processing in Cloud Environment, with more than eight years of experience in cloud computing and five years in Big Data Analytics. He has collaborated with St. John's Medical Research Institute in a research project to diagnose lung diseases using cutting-edge AI and Machine Learning. His research interests include artificial intelligence, IoT, and big data analysis in the areas of healthcare, data mining, and human–computer interaction.

P. Beaulah Soundarabai is a professor in the Department of Computer Science, Christ University, Bangalore, India. She has 20 years of teaching experience. She has been associated with Christ University for the past 14 years. Prior to this, she has teaching experience in SFR College for Women, Sivakasi, and as a lecturer, in AGCS, Kolkata, for three years respectively. She has 10 years of research experience in the areas of Distributed Computing, Computer Networks, IoT, Edge and Cloud computing, and Data Analytics.

1

Recent Trends in Intelligent Transportation Systems

*Kathi Durgesh, Siddharth Garia, and Vishal Kumar Narnoli**

School of Civil Engineering, Lovely Professional University, Phagwara, Punjab, India

*Corresponding author: vishal.ce.imps@gmail.com

1.1 Introduction

The transportation sector is undergoing a profound transformation in advanced technologies and is experiencing a demand for more efficient, safe, and sustainable mobility solutions. In the era of transportation, the intelligent transportation system (ITS) is widely used. ITS been used in airways, roadways, seaways, and railways. Today, the transportation sector is at the forefront of innovation and development. To address all the problems in transportation, there are some systems that have been allocated to ITS. ITS has been divided into some subsections based on their application of many transportation management concepts, i.e., Advanced traveler information system (ATIS), Advanced traffic management system (ATMS), Advanced public transportation system (APTS), Emergency management system (EMS) [1]. Parallel traffic management system (PTMS) [2]. Advanced rural transportation system (ARTS) [3]. Connected and Automated vehicle (CAV) [4]. Dynamic intelligent traffic light control system (DITLCS) [5]. Over the last few years, there has been a significant change in the transportation infrastructure, which plays a major role in Traffic problems. To sustain that situation, we are utilizing the ITS. In this, we know some of the trends and changes in the ITS. ITS has been addressing various problems in development by taking the help of machine learning, sensors, networks, and IoT. ITS provides optimized fuel consumption, traffic congestion, navigation assistance, blind spot monitoring, an automatic collision alert system, adaptive cruise control, and a collision-avoiding system, and more.

ITS been working on wireless sensors network technology and traditional acquisition systems [6]. In a case study, dynamic route planning by VANETs demonstrates that 116.8% of vehicles reach their location, 54% journey duration, and 125.8% wait time [7] traffic, travel and safety information to traveler by robust tracking algorithms by video and multimedia technology analysis based on IoT and Ai technology ITS proposed high speed traffic management [8] and Adopting LEDs to Lds to create a functional automotive VLC network will allow wireless connectivity in ITS using diode-based VLC technology [9] a interface system based on traffic behavior by DITLCS dynamic and intelligent traffic light control system by three phases named as fair mode FM, priority mode PM and emergency mode EM [5].

In this chapter, ITS can be used for more solutions to get better results, approach should be taken. V2I communication can be elucidated in ITS. Revamp logistics and freight management system, autonomous navigation system, enhanced vision, and incident detection management system to get more trends for development of transportation systems, in this generation, by using supplementary smart technologies and systems.

Reviewing several forms of ITS and obtaining the best traveling information for a traveler to shorten their traveling time, fuel consumption, weather notification, road clearance data, development of EV charging infrastructure in urban and rural areas, safety precautions, and road monitoring assistance by AI in Transportation. Concern development of AMR sensors to identify the vehicle's impact on the earth's magnetic field during its travel. Implementing the VANET system in transportation can reduce the delay time. Hazardous material monitoring and tracking could be done by the development of ITS.

The Intelligent Transport System (ITS) works in the framework of the combination of various components, which are interdependent, such as vehicles, road infrastructure, communication networks, urban environments, and users. These components are dynamic to support real-time traffic monitoring, exchange of data, and making decisions. The bilateral interactions between roads, cities, vehicles, networks, and people form the structural basis of the operations of the ITS. The entire elements and their connection are illustrated in Figure 1.1.

ITS is much more than the traditional vehicle-to-road interactions and includes a wide range of urban and environmental elements, such as positioning systems with satellites, traffic signals cameras, traffic signal arrays, communication towers, pedestrian surveillance systems, and embedded intelligibility of road infrastructure in urban environments. By combining the environmental factors, including weather conditions and urban infrastructure characteristics, one will be encouraged to consider more advanced real-time decision-making procedures and build upon the overall control of traffic safety. The overall interconnectivity of the ITS elements with the elements taking place around the city is outlined in Figure 1.2.

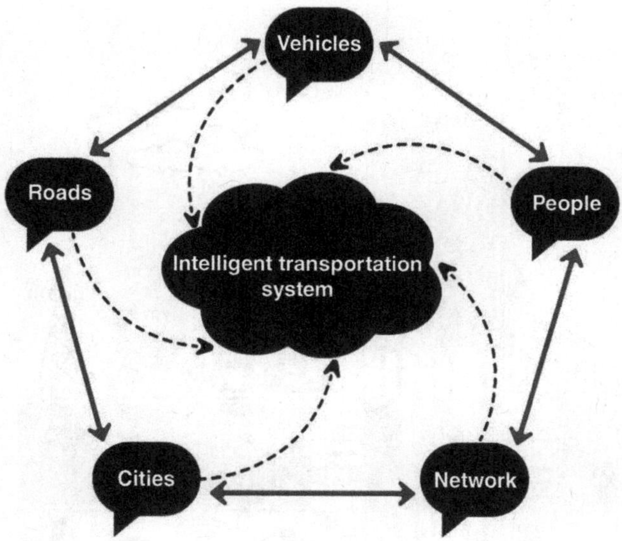

Figure 1.1 Components and interconnections of an ITS.

1.2 Methodology

ITS cited to the comprehensive application of information processing, communication, and control technology in transportation infrastructure. ITS provides many solutions for some type of problems in transportation. There are some types like ATIS, ATMS, APTS, EMS, PTMS, ARTS, CAV, and DITLTCs under these systems.

ATIS provides a broad spectrum of communication technologies, which includes internet services, television, radio, mobile phones, traditional telephones, and SMS-based information systems. Their offerings also include in-vehicle information systems, variable message signs, and multiple communication methods tailored for individuals and large groups at every stage of travel—before, during, and after. This system enables travelers and drivers to make informed decisions regarding departure times, route optimization, and available travel options. Furthermore, it includes an advanced vehicle control system that supports driver-assisted vehicle operation and an automated highway system designed for platooning [1].

ATMS system is primarily used by traffic departments to monitor and manage the traffic network, facilitating the quick creation of effective traffic plans. By leveraging real-time data, it improves vehicle movement through the adjustment of measures to improve traffic flow, such as traffic signals. This system incorporates VANET technology, enabling the detection of cargo vehicles, the monitoring

Figure 1.2 Integrated overview of ITS and their connectivity to diverse urban and environmental elements.

of hazardous materials, and automatic vehicle identification. It features capabilities for gathering traffic information, assigning traffic, optimizing traffic, and predicting traffic patterns.

APTS is a public transportation system designed to enhance the operational efficiency of various public transit modes. The main areas of focus include planning, Scheduling systems, signal prioritization, traffic management, automated payment options, dynamic ride-sharing, arrival predictions, and coordination of emergency services, which are all categorized as subcomponents. The system employs IoT sensors, AI technology, and advanced sensing techniques in electric vehicles to enhance public transportation services [8, 3].

EMS has developed a transportation system specifically designed for emergencies, with the goal of reducing accident rates within its network. This system features an Emergency Accident Rescue System (EARS) for highways, which is based on Geographic Information Systems (GIS). It comprises two key components: one focuses on early alerts and identifying accident locations, while the other automatically generates rescue plans [10].

Parallel transportation management systems (PTMS) describe the concurrent interaction between real-world transportation systems and their artificial or virtual counterparts. In this framework, tasks and computations are performed simultaneously by ACP-based PTMS. The system is composed of five essential components: actual transportation systems (ATS), operator and administrator training systems (OTSt), decision evaluation and validation systems (DynaCAS), and traffic sensing control and management systems (aDAPTS). The PTMS process depends on the co-evolution of real systems and their associated ATS [2].

ARTS is for the development of rural areas connection for travelers, and it has sub-categories like self-guided navigation system, improved visibility, emergency distress alerts, and traveler safety and protection. These can be done by using GPS and GIS we can help through this [3].

Based on the goals of the operation and the realms of application, Intelligent Transportation Systems (ITS) can be divided into a number of functional subsystems. These are sophisticated rural transportation systems, traveler information systems, traffic management systems, public transportation systems, connected autonomous vehicles, emergency management systems, and parallel transportation management systems. Every subsystem will help in improving mobility, safety, and operational efficiency in transportation networks. Figure 1.3 represents the key categories of Intelligent Transportation Systems.

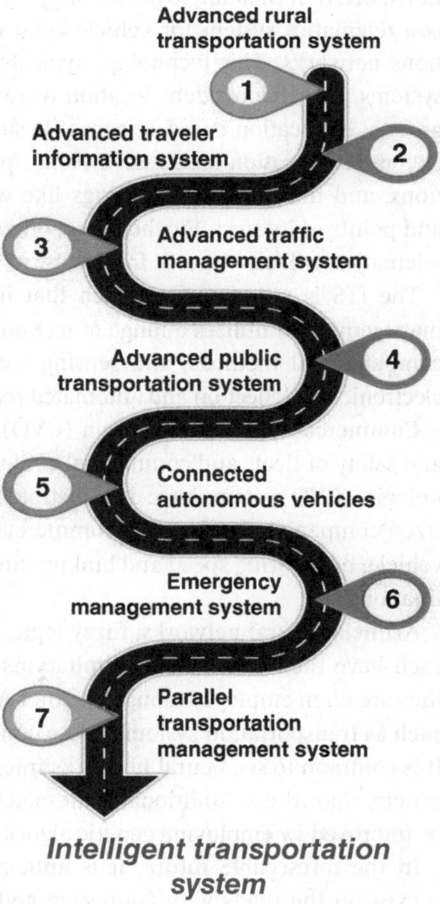

Figure 1.3 Types of Intelligent transportation system.

DITLTCS method is designed specifically for varied vehicular traffic networks, taking into account different types of vehicles, including emergency and priority ones. It operates in three stages: fair mode (FM), priority mode (PM), and emergency mode (EM), ensuring every vehicle is given the same level of significance. This approach combines deep reinforcement learning with traffic signal generation and utilizes a fuzzy inference system [5].

AMR sensors use anisotropic magnetoresistance to detect variations inside the Earth's magnetic field, generated by the movement of vehicles above. These signals they capture vary depending on the type of vehicle, like cars, buses, or trucks. To aid in this detection process, a detection state machine (DSM) has been implemented to recognize vehicles as they pass by Mishra et al. [11].

The automatic vehicle location system (AVLS) is a modern approach to public transportation that employs automatic vehicle location (AVL) technology, which is a telematics system for vehicles that generally operates via telecommunications networks. This technology typically depends on terrestrial radio, satellite systems, and independent location techniques. Consequently, there is expected growth in location-based services. Besides wireless E911, these services may encompass location-based advertising, position-sensitive billing, tracking solutions, and informational offerings like weather updates, navigation assistance, and points of interest. Furthermore, other ITS, such as public transit and vehicle telematics, will also benefit from these advancements [12].

The ITS is an area of research that integrates electronics with global traffic management. It utilizes a range of technologies, such as wireless communication, computational methods, and sensing technologies, to support applications like electronic toll collection and automated road enforcement, among others [13].

Commercial vehicles operation (CVO), these systems enhance the efficiency and safety of fleets and commercial vehicles through the use of various ITS technologies. CVO systems have demonstrated their value for both large and medium-sized companies that operate commercial fleets, as they aid in managing each vehicle, monitoring speed and braking times, and ensuring timely arrivals at destinations [14].

Artificial neural networks, fuzzy logic, genetic algorithms, and expert systems each have their own distinct limitations role at ITS-related methods; however, they are often employed alongside other AI techniques. In complex environments such as transportation systems, these combinations show considerable potential. It is common to see neural networks integrated with fuzzy logic or trained using genetic algorithms. Additionally, the effectiveness of fuzzy rule-based systems can be improved by employing genetic algorithms [15].

In the foreseeable future, it is anticipated that human-driven vehicles will coexist on the roads with connected and automated vehicles (CAVs). This scenario necessitates collaboration between human operators and autonomous systems to efficiently oversee CAVs on roadways. The erratic behavior of human drivers, combined with their inclination to overlook traffic rules, could interfere with traffic movement. Therefore, it is essential to consider the interactions between manually-driven cars and CAVs when studying traffic dynamics. Additionally, CAVs have the ability to communicate with infrastructure, traffic management systems, and other vehicles using established protocols. By utilizing this information, we can improve traffic flow and enhance road safety [4].

The great potential for improving road usage efficiency while also decreasing fuel consumption, alleviating traffic congestion, and minimizing the environmental impacts of vehicle operations. Investigations have examined conventional wireless sensor network technology, as well as in-vehicle and road infrastructure sensing systems. Ultra-low power sensing and communication capabilities, lower installation and maintenance costs, improved flexibility and adaptability, scalable mesh networking, and more are all provided by the wireless sensor network solution. the capability to function on battery power for several years [16].

ITS solutions for individual drivers will play a vital role in our daily routines and commutes, making the study of driving behavior both engaging and promising. Until self-driving cars emerge and transform the driving environment and responsibilities, it is essential to carefully craft solutions that tackle the challenges of mixed traffic and the interactions between other vehicles and powered two-wheelers. The objective will be to seamlessly integrate all drivers to improve the efficiency of traffic networks [17].

Integrating technology into the infrastructure of smart cities is crucial for improving mobility, alleviating congestion, managing traffic efficiently, and reducing environmental impact. Recent advancements, including the Internet of Things (IoT), vehicle-to-everything (V2X) communication, and 5G networks, offer new opportunities for innovative solutions such as intelligent traffic lights (ITL) and vehicle traffic lights (VTL). By utilizing the real-time communication capabilities of 5G and combining various sensor technologies—such as AI cameras, roadside units (RSUs), and onboard units (OBUs)—with ITL and VTL systems, traffic flow can be enhanced, resulting in a significant reduction in CO2 emissions [18].

Key strategic initiatives, including collaboration, research and development investment, and the integration of emerging technologies like edge AI and advanced 5G, can significantly boost the growth and economic impact of the V2X industry [19].

The progress of ITS and the understanding of system dynamics in smart cities can be ascribed to several important factors, such as: (1) the adoption of hybrid methodologies, (2) enhanced adaptability and efficiency, (3) cost-effectiveness and sustainable environmental practices, and (4) valuable insights for policymakers [20].

As a component of a broader ITS and the Internet of Things (IoT), its objective is to improve reliability while reducing average wait times. Furthermore, the system enables lane priority adjustments when identical conditions are encountered more than four times, ensuring that vehicles previously unrecognized are eventually granted priority. This capability underscores the system's ability to withstand failures. As smart cities continue to emerge, this system provides an economical approach to upgrading traffic management, rendering it both adaptive and intelligent. Being connected to the cloud, traffic controllers and developers can oversee real-time traffic data from afar. The adaptive traffic management system is flexible in terms of the types of sensors used for measuring traffic density. The strength, adaptability, and capability for real-time data processing render it ideal for effective implementation [21].

Efforts to track the location of mobile phones have intensified in response to the deadline of October 1, 2001. Organizations that set telecommunications standards are collaborating to incorporate new location technologies into their frameworks, which include GSM, UMTS, CDMAOne, CDMA2000, W-CDMA, TDMA, and UWC-136. Among these technologies, TOA, TDOA, and assisted GPS are the leading candidates for existing communication networks. Once these technologies undergo standardization, phones equipped with location capabilities will be launched, enabling services based on location, such as wireless E911, location-based billing, tracking, advertising, and informational features like navigation and weather updates. Additionally, automotive telematics and public transportation will gain advantages, as will various smart transportation systems. These innovations are expected to streamline operations while improving service ease, ultimately resulting in safer, more efficient transport with less congestion, reduced pollution, and minimized environmental effects [12].

The technological progress is paving the way for smart transportation systems. It presents significant opportunities for optimizing roadway usage while curbing fuel consumption, reducing traffic congestion, and lessening the environmental effects of vehicle movement. The paper examines conventional wireless sensor network technologies and sensing systems seen in automobiles and roadside infrastructure. Extremely low-power sensing and communication capabilities, reduced installation and maintenance costs, more flexibility, improved mesh networking adaptability and scalability, and years of battery-powered operation are all provided by the wireless sensor network solution. We anticipate more advancements in intelligent transportation technologies and vehicle communication systems, along with broader implementation in the years ahead [16].

The Intelligent Transportation Systems are divided into separate functional categories referred to as Connected Autonomous Vehicles (CAV), Advanced Traveler

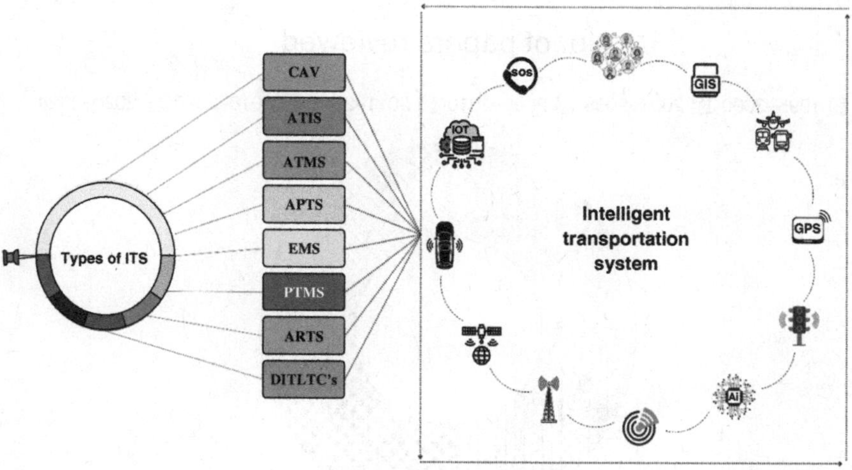

Figure 1.4 Classification of ITS and their interconnection with key functional components.

Information Systems (ATIS), Advanced Traffic Management Systems (ATMS), Advanced Public Transportation Systems (APTS), Emergency Management Systems (EMS), Parallel Transportation Management Systems (PTMS), and Advanced Rural Transportation Systems (ARTS). These subsystems are based on the basic technologies such as the Internet of Things (IoT), Geographic Information systems (GIS), Global Positioning System (GPS), artificial intelligence, communication networks, and real-time monitoring systems. Figure 1.4 illustrates that the relationships between ITS typologies are interrelated with each other, and through the use of its functional units.

1.3 Results

ITS are revolutionizing the future of mobility by employing advanced technologies to tackle key issues in transportation. These systems seek to enhance safety, boost efficiency, and foster environmental sustainability through advancements like real-time traffic control, interconnected vehicle networks, and intelligent infrastructure. However, despite their considerable promise, the implementation of ITS faces obstacles concerning compatibility, cybersecurity, and expenses. By promoting collaboration among governmental bodies, companies, and researchers, ITS can change how people and goods are moved, leading to smarter, safer, and more environmentally friendly transportation systems that serve the greater good of society.

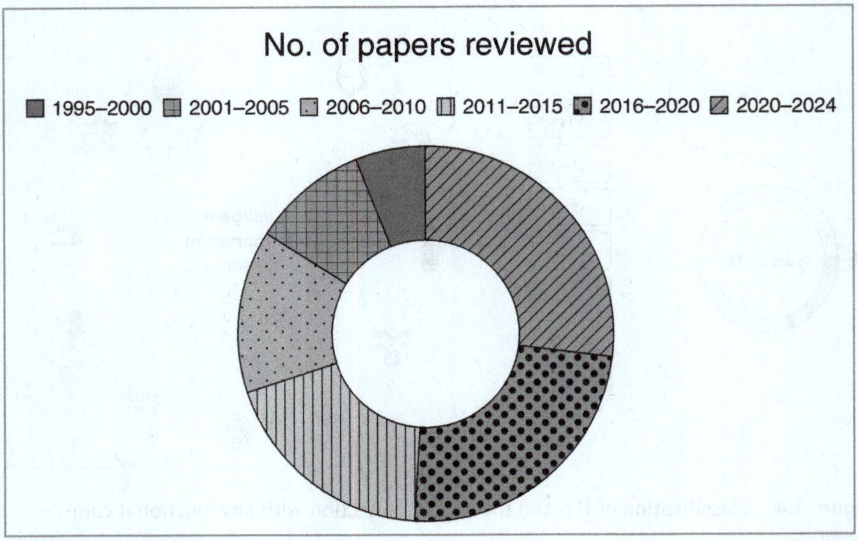

1.4 Discussion

This chapter shows that recent trends in intelligent transportation system how AI and IoT been utilized to form infrastructure supporting transportation, big data real time analytic, 5G and its role in transportation, block chains of ITS, shared mobility and ride hailing services, VANETs services more over we discussed in this paper. A brief summary of ITS and its subtopics regarding Transportation has been done in this paper. EMS, ATMS, and traffic related ai services systems have seen greater development in this chapter. It shows that improvement in this system, which shows a proper vision and more helping to the people in the transportation system. This ITS has a great scope of development in many segments to get an information system regarding ITS, which helps save many resources in transportation.

References

1 Singh, B. and Gupta, A. Recent trends in intelligent transportation systems: A review. *Journal of Transport Literature*, 9(2): 30–34, 2015.

2 Wang, F. Y. Parallel control and management for intelligent transportation systems: Concepts, architectures, and applications. *IEEE Transactions on Intelligent Transportation Systems*, 11(3): 630–638, 2010.

3 Wootton, J. R., Garcia-Ortiz, A., and Amin, S. M. Intelligent transportation systems: A global perspective. *Mathematical and Computer Modelling*, 22 (4–7): 259–268, 1995.

4 Patil, P. A review of connected and automated vehicle traffic flow models for next-generation intelligent transportation systems. *Applied Research in Artificial Intelligence and Cloud Computing*, 1(1): 10–22, 2018.

5 Kumar, N., Rahman, S. S., and Dhakad, N. Fuzzy inference enabled deep reinforcement learning-based traffic light control for intelligent transportation system. *IEEE Transactions on Intelligent Transportation Systems*, 22(8): 4919–4928, 2020.

6 Riaz, M. T., Aaqib, S. M., Ahmad, S., Amin, S., Ali, H., Husnain, S., and Riaz, S. The intelligent transportation systems with advanced technology of sensor and network. In *2021 International Conference on Computing, Electronic and Electrical Engineering (ICE Cube)*, pages 1–6. IEEE, 2021.

7 Yang, Y., and Bagrodia, R. Evaluation of VANET-based advanced intelligent transportation systems. In *Proceedings of the sixth ACM international workshop on VehiculAr InterNETworking*, 2009.

8 Cui, Y. and Lei, D. Design of highway intelligent transportation system based on the internet of things and artificial intelligence. *IEEE Access*, 11: 46653–46664, 2023.

9 Shaaban, K., Shamim, M. H. M., and Abdur-Rouf, K. Visible light communication for intelligent transportation systems: A review of the latest technologies. *Journal of Traffic and Transportation Engineering (English Edition)*, 8(4): 483–492, 2021.

10 Long, K., Liu, Y., and Luo, X. Emergency accident rescue system in freeway based on GIS. In *2008 International Conference on Intelligent Computation Technology and Automation (ICICTA)*. IEEE, 2008.

11 Mishra, D. P., Asutkar, G. M., and Dorale, S. S. An application of wireless sensor network in intelligent transportation system. In *2013 6th International Conference on Emerging Trends in Engineering and Technology*, pages 90–91. IEEE, 2013.

12 Zhao, Y. Mobile phone location determination and its impact on intelligent transportation systems. *IEEE Transactions on Intelligent Transportation Systems*, 1(1): 55–64, 2000.

13 Qi, L. Research on intelligent transportation system technologies and applications. In *2008 Workshop on Power Electronics and Intelligent Transportation System*, pages 529–531. IEEE, 2015.

14 Figueiredo, L., Jesus, I., Machado, J. T., Ferreira, J. R., and De Carvalho, J. M. Towards the development of intelligent transportation systems. In *ITSC 2001. 2001 IEEE Intelligent Transportation Systems. Proceedings (Cat. No. 01TH8585)*, pages 1206–1211. IEEE, 2001.

15 Machin, M., Sanguesa, J. A., Garrido, P., and Martinez, F. J. On the use of artificial intelligence techniques in intelligent transportation systems. In *2018 IEEE Wireless Communications and Networking Conference Workshops (WCNCW)*, pages 332–337. IEEE, 2018.

16 Tewolde, G. S. Sensor and network technology for intelligent transportation systems. In *2012 IEEE International Conference on Electro/Information Technology*. IEEE, 2012.

17 Barmpounakis, E. N., Vlahogianni, E. I., and Golias, J. C. Intelligent transportation systems and powered two wheelers traffic. *IEEE Transactions on Intelligent Transportation Systems*, 17(4): 908–916, 2015.

18 Elassy, M., Al-Hattab, M., Takruri, M., and Badawi, S. Intelligent transportation systems for sustainable smart cities. *Transportation Engineering*, 16: 100252, 2024.

19 Rammohan, A. Revolutionizing intelligent transportation systems with cellular vehicle-to-everything (C-V2X) technology: Current trends, use cases, emerging technologies, standardization bodies, industry analytics and future directions. *Vehicular Communications*, 43: 100638, 2023.

20 Chen, G. and wan Zhang, J. Intelligent transportation systems: Machine learning approaches for urban mobility in smart cities. *Sustainable Cities and Society*, 107: 105369, 2024.

21 Dubey, A., Lakhani, M., Dave, S., and Patoliya, J. J. Internet of Things based adaptive traffic management system as a part of Intelligent Transportation System (ITS). In *2017 International Conference on Soft Computing and its Engineering Applications (icSoftComp)*, pages 1–6. IEEE, 2017.

Further Reading

Bojan, T. M., Kumar, U. R., and Bojan, V. M. An internet of things based intelligent transportation system. In *2014 IEEE International Conference on Vehicular Electronics and Safety*, pages 174–179. IEEE, 2014.

Cheng, H. Y., Gau, V., Huang, C. W., and Hwang, J. N. Advanced formation and delivery of traffic information in intelligent transportation systems. *Expert Systems with Applications*, 39(9): 8356–8368, 2012.

Ganin, A. A., Mersky, A. C., Jin, A. S., Kitsak, M., Keisler, J. M., and Linkov, I. Resilience in intelligent transportation systems (ITS). *Transportation Research Part C: Emerging Technologies*, 100: 318–329, 2019.

Gosman, C., Cornea, T., Dobre, C., Pop, F., and Castiglione, A. Controlling and filtering users data in intelligent transportation system. *Future Generation Computer Systems*, 78: 807–816, 2018.

Jain, S. and Jain, S. S. Development of intelligent transportation system and its applications for an urban corridor during COVID-19. *Journal of The Institution of Engineers (India): Series B*, 102(6): 1191–1200, 2021.

Lin, Y., Wang, P., and Ma, M. Intelligent transportation system (ITS): Concept, challenge and opportunity. In *2017 IEEE 3rd International Conference on Big Data Security on Cloud (Bigdatasecurity), IEEE International Conference on High Performance and Smart Computing (HPSC), and IEEE International Conference on Intelligent Data and Security (IDS)*, pages 167–172. IEEE, 2017.

Maimaris, A. and Papageorgiou, G. A review of intelligent transportation systems from a communications technology perspective. In *2016 IEEE 19th International Conference on Intelligent Transportation Systems (ITSC)*, pages 54–59. IEEE, 2016.

Mfenjou, M. L., Ari, A. A. A., Abdou, W., and Spies, F. Methodology and trends for an intelligent transport system in developing countries. *Sustainable Computing: Informatics and Systems*, 19: 96–111, 2018.

Nkoro, A. B., and Vershinin, Y. A. Current and future trends in applications of Intelligent Transport Systems on cars and infrastructure. In *17th International IEEE Conference on Intelligent Transportation Systems (ITSC)*, pages 514–519. IEEE, 2014.

Nuzzolo, A. and Comi, A. Advanced public transport and intelligent transport systems: new modelling challenges. *Transportmetrica A: Transport Science*, 12(8): 674–699, 2016.

Qu, F., Wang, F. Y., and Yang, L. Intelligent transportation spaces: Vehicles, traffic, communications, and beyond. *IEEE Communications Magazine*, 48(11): 136–142, 2010.

Rammohan, A. Revolutionizing intelligent transportation systems with cellular vehicle-to-everything (C-V2X) technology: Current trends, use cases, emerging technologies, standardization bodies, industry analytics and future directions. *Vehicular Communications*, 43: 100638, 2023.

Ravi, S. and Mamdikar, M. R. A review on its (intelligent transportation systems) technology. In *2022 International Conference on Applied Artificial Intelligence and Computing (ICAAIC)*, pages 155–159. IEEE, 2022.

Tewolde, G. S. Sensor and network technology for intelligent transportation systems. In *2012 IEEE International Conference on Electro/Information Technology*, pages 1–7. IEEE, 2012.

Tsu-Tian, L. Research on intelligent transportation systems in Taiwan. In *2008 27th Chinese Control Conference*, pages 18–23. IEEE, 2008.

Yan, X., Zhang, H., and Wu, C. Research and development of intelligent transportation systems. In *2012 11th International Symposium on Distributed Computing and Applications to Business, Engineering & Science*, pages 321–327. IEEE, 2012.

Zhang, J., Wang, F. Y., Wang, K., Lin, W. H., Xu, X., and Chen, C. Data-driven intelligent transportation systems: A survey. *IEEE Transactions on Intelligent Transportation Systems*, 12(4): 1624–1639, 2011.

2

Artificial Intelligence and IoT Applications Transforming the Automotive Industry

Raviprakash R Salagame

Autobotik Pte Ltd., Singapore

Corresponding author: ravirs2912@gmail.com

2.1 Introduction

The mobility industry is undergoing a phenomenal transformation with the adoption of new technologies. Over the past decade, foundational shifts have emerged in the form of electrified powertrains, autonomous driving with guided safety features, and vehicles that are connected to each other and to the infrastructure network. These megatrends are fueled by advances in artificial intelligence (AI), large language models (LLMs), IoT, and deep learning (DL) methods. Many of these applications are already enhancing unique customer experiences. For smart cities, these technologies will play a significant role in providing safe, seamless, and efficient transportation solutions. This chapter will explore the adoption of these technologies and their impact on the future of the mobility industry.

Consider a scenario where you are driving at high speed on a highway and suddenly encounter a parked vehicle in your lane, leaving insufficient time to brake and stop the car. However, your vehicle, equipped with advanced sensors, would have "seen" the parked vehicle from a distance and begun decelerating. Consequently, the car automatically stops a few meters from the stationary vehicle on its own. This is an example application called automated emergency braking (AEB). Various technologies are implemented in this feature such as computer vision, path planning, machine learning (ML), and adaptive control. Input to the system is fed from sensors in the car such as cameras, radars, and lidars. This illustrates the growing number of digital features in modern vehicles. Over the next decade or so, these transformations will accelerate, bringing a host of safety and internet connectivity features to the users built on green transportation platforms such as electric vehicles (EVs). Similar features are implemented across transportation industries such as off highway vehicles, railroads, large trucks, and commercial aviation. In general, the mobility industry encompasses a broad range of businesses and technologies focused on the movement of people and goods. The purpose of this chapter is to introduce these technologies with applications and use cases to provide an overview and references to incite the readers to pursue them further.

First, we begin with a basic overview and description of the foundational digital technologies such as AI and IoT, and examples of their application in the mobility industry. We will then delve deeper into a detailed discussion of four of the key application verticals namely, autonomous, connected, electric and shared (sometimes called ACES). Discussions assume no particular prerequisite in the field. While these technologies have broad applicability across domains like commercial aerospace and drones, the examples and use cases presented here will primarily focus on the automotive industry, particularly urban transportation.

2.1.1 Overview of Key Digital Technologies and Applications

Since the advent of the first petrol-powered car in the late 1880s, the automotive industry has focused mostly on power, speed, and fuel efficiency. Over the last 50 years, the industry has developed multiple technologies that have increased average fuel efficiency by over three times while reducing emissions by over two times [1]. Fossil fuel availability, transportation demands, and substantial urban migration [2] have fueled these improvements. Until the internet's emergence in the 1990s, the focus on in-cabin experience, car safety, and personal features had been secondary goals. Although the first car radio was introduced in the 1930s and evolved over the decades with audio features and personal music options, the focus on personalization of car experience using digital technology has been a phenomenon of the last two to three decades. Current transportation technology is mainly driven by urban traffic, the necessity for enhanced safety and reduced pollution, and the desire for a better personal experience. Thus, today's automobile may appear to be a smart mobile device rather than a simple transportation vehicle. Key technologies driving this transformation are discussed in the following section.

The mobility industry has broadly classified technology applications into four key verticals (or megatrends) expected to drive future growth.

- **Autonomous**: Driven by the vision of reducing accidents and increasing road safety by enabling vehicles to perceive their surroundings to assist the driver or operate on their own.

- **Connected**: Supported by the growth of 5G, IoT, and internet technologies to create seamless connections between vehicles, infrastructure, and resources.
- **Electrification**: Motivated by the imperative to lower carbon footprint and promote environmentally friendly transportation.
- **Shared Mobility**: Focused on comprehensive mobility solutions utilizing multimodal transportation options like ride-hailing, shared cabs, bike sharing, and public transit

These interconnected verticals are facilitated by four fundamental digital technologies, which are discussed in more detail, and are anticipated to converge, creating powerful new transportation alternatives.

2.1.2 IoT (Internet of Things) and Sensors

The term "Internet of Things (IoT)" was coined by Kevin Ashton, a Britain-born technologist, to describe the concept of connecting devices over the internet. IoT involves a network of physical objects or "things" embedded with electronics, software, sensors, and connectivity to enable objects to exchange data with the manufacturer, operator, and/or other connected devices. In contemporary vehicles, this manifests into a host of electronic devices with embedded software and sensors mounted on them and connecting them to other components within the car and externally.

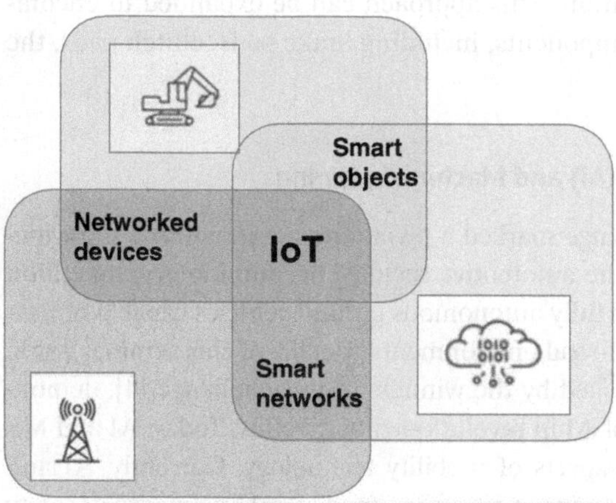

A modern car has between 60 and 200 sensors depending on the sophistication [3]. Over half of them would be IoT-based sensors. Sensor data is typically processed and periodically uploaded to a central cloud server. This server analyzes the data, which can then be used to enhance drivers' personal experience.

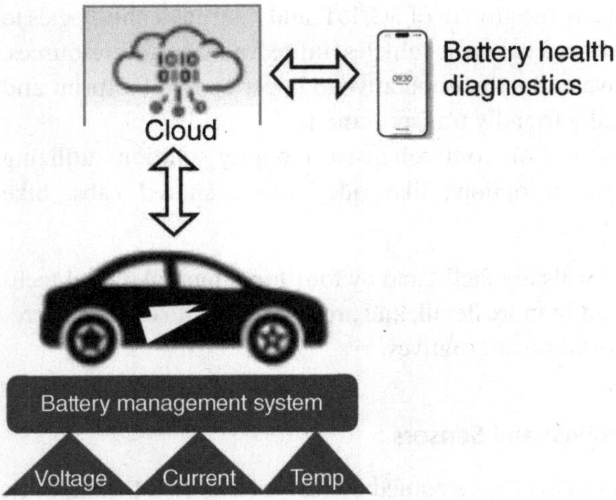

As an example, consider an EV battery management system (BMS). It is mounted with sensors to measure battery parameters like energy level, temperature, and other parameters. EV batteries go through charge and discharge cycles where they store energy during charging and deliver it to power the vehicle. During battery cycling, sensors within the BMS collect battery parameters. This data undergoes local processing before being transmitted to a cloud server. Drivers can access a web portal via a smartphone dashboard to monitor battery health and diagnostic information. This approach can be expanded to encompass other essential vehicle components, including brake pads, clutch pads, the engine, and tire pressure.

2.1.3 Artificial Intelligence (AI) and Machine Learning

The 2003 DARPA Grand Challenge marked a pivotal moment, igniting the widespread adoption of AI within the automotive sector. This ambitious competition challenged researchers to create fully autonomous ground vehicles capable of navigating complex, unprepared off-road environments. Details of this seminal work, documented in the paper published by the winners of this challenge [4], demonstrated the immense potential of AI in revolutionizing mobility. Today, AI and ML are penetrating practically all aspects of mobility technology. Currently, AI and ML algorithms power a diverse range of features aimed at enhancing driver safety and convenience. Advanced driver-assistance systems (ADAS), utilizing sophisticated sensor data and AI-driven analysis, can detect potential hazards, assist with parking, maintain lane discipline, and even adapt vehicle speed to traffic conditions. These capabilities represent a significant step toward semi-autonomous and eventually fully autonomous driving. Beyond driver assistance, AI plays a crucial role in improving personalization of the driver experience. Natural language processing (NLP) enables sophisticated voice control systems, allowing drivers to interact with hands-free vehicle functions. The continuous advancements in AI and ML promise even more transformative applications in the automotive industry, paving the way for safer, more efficient, and more personalized mobility solutions.

AI is the ability of a computer or a robot to perform tasks done by humans. ML, a key approach to realizing AI, enables systems to learn automatically from past data. This makes big data crucial for many AI applications in the vehicle. Modern connected cars generate an average of 20–30 gigabytes of data per hour, sourced from drive data, entertainment, apps, performance data, and traffic data. The automotive sector implements AI through diverse methodologies, including ML, heuristic and rule-based systems, statistical learning, and integrated hybrid algorithms [5].

2.1.3.1 Machine Learning

ML techniques leverage existing data to train programs, known as ML models. These trained models can then be used for learning and making predictions. AI and IoT methods can be implemented using four main approaches: supervised learning, unsupervised learning, semi-supervised learning, and reinforcement learning.

Supervised learning involves training the model with labeled datasets where each input is paired with a corresponding output label. Common supervised learning algorithms include linear regression, logistic regression, support vector machines (SVMs), decision trees, random forests, and neural networks.

Unsupervised learning focuses on discovering hidden patterns, structures, and relationships within unlabeled data. Techniques like k-means clustering, hierarchical clustering, principal component analysis (PCA), and autoencoders fall under unsupervised learning.

Semi supervised learning bridges the gap between supervised and unsupervised learning by employing a combination of both labeled and unlabeled data for training. Various techniques, such as self-training, co-training, and graph-based methods, are employed in semi-supervised learning.

Reinforcement learning (RL) employs a reward and penalty system to iteratively enhance the performance of a model through feedback on its actions. Examples of RL algorithms are Q-learning, deep Q-networks (DQNs), and policy gradient methods.

Each of these methods perform well for specific applications. For example, supervised learning is applied to failure prediction based on historical data and unsupervised learning for anomaly detection problems. Deep learning, a ML subset employing multi-layered neural networks and extensive datasets, enables applications like efficient autonomous navigation in complex scenarios. Further applications and case studies will be explored later in this chapter.

2.1.3.2 AI and ML Applications in Automotive

AI and ML applications in the automotive industry vary from product development to launch in various forms. Some of the main application areas are discussed in the following section.

Product Development: During vehicle development and testing, AI is used to improve productivity, efficiency and innovation during the complete life cycle of product development. For instance, vehicle requirements are managed using AI based tools, capable of automatically converting customer requirements into test cases. Using a vast library of past design solutions, new and innovative concepts are generated using Gen AI methods. During simulation of design solutions, ML techniques are used to simplify predictive analytics by building intelligent models [6]. This is useful in time consuming analysis such as crash simulation. The increasing adoption of software-driven architectures in vehicles is driving the widespread use of LLMs and AI in the automated verification process.

Personalization of Consumer Experience: Modern vehicles increasingly integrate AI applications to enhance the in-cabin experience for both drivers and passengers. These applications encompass a wide range of functionalities, including sophisticated voice recognition systems that allow occupants to control various vehicle features through natural language commands. Intelligent navigation systems leverage real-time data and AI algorithms to provide optimal routing, traffic updates, and points of interest recommendations. Consider a scenario where a driver is on a journey and expresses a desire to visit a nearby grocery store. An AI-based recommendation system, having learned the driver's past preferences for specific stores or types of products, can proactively identify and suggest the most suitable options in the vicinity.

Autonomous Driving and Safety: This is one of the most promising areas of automotive technology with wide application of AI [7]. These systems represent a suite of technologies designed to aid drivers and improve road safety. These include a range of features, such as adaptive cruise control, lane departure warning, automatic emergency braking, blind-spot detection, and parking assist. These functionalities rely heavily on data collected from various sensors embedded in the vehicle (an aspect of IoT), which is then processed by AI algorithms to make real-time decisions and provide assistance or take corrective actions eventually leading to full self-driving (FSD) capabilities. Tesla has emerged as a prominent early adopter and innovator in the field of FSD.

The company has pursued a strategy of building proprietary software within its vehicles, leveraging deep learning techniques.

Manufacturing and Supply Chain: Product quality, accuracy, and efficiency are significantly boosted by using AI in manufacturing along with IoT [8]. Broadly labeled as digital manufacturing and encapsulated by the terms Industry 4.0 and smart factories, it represents a fundamental shift in industrial operations. One key application of this integration lies in predictive maintenance. By continuously monitoring machine yield data and critical health parameters through a network of IoT sensors, sophisticated AI algorithms can be trained to identify patterns and anomalies that precede potential equipment failures. This proactive approach to maintenance scheduling allows manufacturers to move away from reactive repairs and time-based maintenance cycles. Instead, maintenance interventions are triggered only when necessary, minimizing downtime, reducing maintenance costs, and extending the lifespan of valuable equipment. The ability to predict and prevent failures not only ensures smoother production flows but also enhances safety within the manufacturing facility. Similarly, combining supply chain optimization and manufacturing demand using predictive models leads to significant productivity improvements. By gathering real-time data from various points in the supply chain, including supplier lead times, transportation logistics, and inventory levels, AI-powered models can predict future demand with greater accuracy, optimize inventory levels across the supply chain and prevent potential disruptions.

Vehicle Health Monitoring: Modern vehicles are equipped with hundreds of sensors to measure critical vehicle parameters in real time. This real-time stream of information provides a comprehensive view of the vehicle's condition across numerous parameters. Leveraging this wealth of sensor data with ML models has emerged as a rapidly expanding and highly impactful application within the automotive sector [9]. This approach enables sophisticated vehicle health monitoring, moving beyond traditional reactive maintenance to proactive and predictive strategies. Consider the example of brake pad wear. Instead of relying solely on mileage or visual inspections, which can be inaccurate and inefficient, ML models can be trained on a diverse range of influencing factors. These factors can include individual driving habits (e.g., frequency and intensity of braking), intrinsic wear parameters of the brake pad material, the typical drive profile (e.g., urban vs. highway driving), and temperature profiles experienced by the brakes. By analyzing the complex interplay of these variables, a highly accurate model of brake pad wear can be developed. Using this model, brake pad replacement schedule can be predicted, and preventive maintenance can be customized to improve brake pad performance.

2.1.4 Cloud Computing

Cloud computing can be described as on-demand access to computational resources such as hardware, storage, applications, development tools, networking, and more. These resources are hosted at a remote data center managed by a cloud service provider and accessed over the internet. Automotive cloud is essentially a cloud infrastructure with all the vehicle related data. Cloud computing is a very attractive proposition in the automotive industry and vital to the connected car experience. Cloud services can be one of the following—infrastructure-as-a-service or platform-as-a-service or software-as-a-service depending on resources under the service contract. A popular application of the automotive cloud is the ability to perform software updates. In the automotive world, without cloud connectivity, a software feature update or a bug fix needed a visit to the service center. If a formal recall was involved with thousands of vehicles, a critical software fix to the control unit in the vehicle would have to be done for each vehicle over several weeks. In the world of connected vehicles, such updates are easily done through the automotive cloud directly to the vehicle control unit without human intervention. This technology is called over-the-air (OTA).

Several AI applications leverage cloud infrastructure in a connected car. For example, autonomous vehicles would send incidents and driving patterns to the cloud for analysis and model tuning. Remote diagnostics applications take data from the vehicle diagnostics and predict maintenance related issues. An illustration of automotive cloud infrastructure is shown in Figure 2.1.

2.1.5 Cybersecurity

Growing reliance on software-driven architecture in the vehicle (called software driven vehicle or SDV) has amplified the vulnerability of vehicles to cyberattacks. Autonomous vehicles, designed to operate without human intervention, present

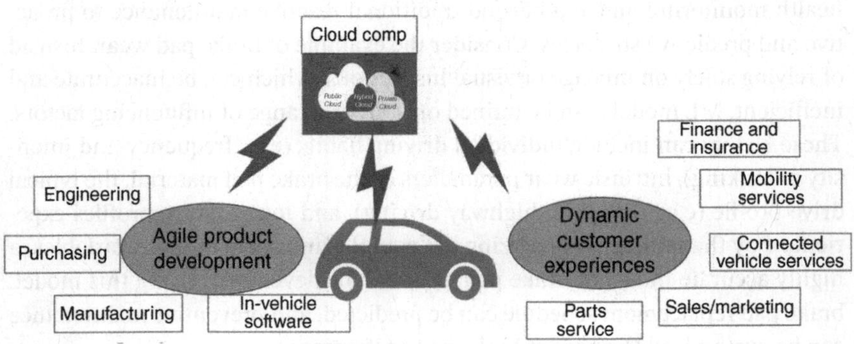

Figure 2.1 Automotive cloud infrastructure.

a particularly concerning scenario. This looming threat has propelled cybersecurity from a peripheral concern to a fundamental digital pillar of the mobility industry. As future urban landscapes evolve into interconnected smart cities, the seamless integration of software and digital platforms with physical infrastructure will become increasingly pervasive. Consequently, robust cybersecurity will be a critical glue between physical and digital infrastructure. ML and AI offer powerful solutions for enhancing cybersecurity to improve threat detection and other applications [10]. By leveraging vast datasets and sophisticated algorithms, ML and AI can identify patterns and anomalies that traditional rule-based security systems might miss. Beyond threat detection, ML and AI contribute to automated vulnerability analysis, intelligent security orchestration and automation (SOAR), and enhanced user behavior analytics.

2.2 Automotive AI Applications and Urban Use Cases

Urbanization and city growth create significant transportation challenges, including increasing populations, aging infrastructure, and the need to support all citizens. AI offers potential solutions, such as intelligent traffic control, efficient last-mile delivery, and smart EV solutions for logistics, to name a few. AI can adapt to various transportation and communication problems using the vast data collected from vehicles and city infrastructure [11]. Automotive megatrends such as autonomous, connected, and electrification fit nicely into these solution frameworks. These solutions, ranging from urban transit to highway systems, integrate various technologies in addition to AI. Our focus here would be AI-based applications, while providing an overview of all the technologies and how they bind together to address the larger problem. The subsequent sections are structured according to the four technology application verticals (autonomous, connected, electrified, and shared) or megatrends outlined previously.

2.2.1 Autonomous Vehicles

The development of autonomous vehicles, also called self-driving vehicles, promises to revolutionize transportation by potentially improving safety, reducing traffic congestion, and offering greater mobility to a wider range of people. Although the mass market for fully autonomous cars is a few years away, many profitable applications of autonomous technology already exist. These applications can be found in land based vehicles, railroad vehicles, aerial vehicles, and in some cases, water-based vehicles. Before understanding autonomous technology, it is useful to know the industry's definition of "driving automation."

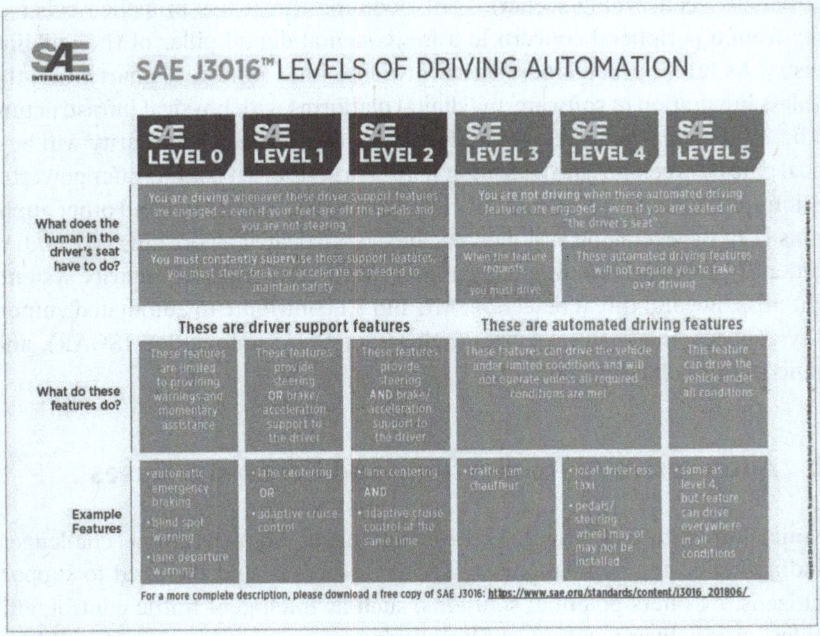

Figure 2.2 Levels of autonomy based on SAE J3016 standard.

Land based autonomous vehicles are classified based on SAE defined levels [12]. These standards were initially released in 2014 and updated recently in 2021. Figure 2.2 shows various levels of autonomy based on these standards.

Level 0: *No Driving Automation*: The driver will handle all aspects of driving.

Level 1: *Driver Assistance*: Driver's primarily responsible, though the vehicle can assist with one or more driving functions such as braking or acceleration.

Level 2: *Partial Driving Automation*: Drivers must be alert and be prepared to take over while the vehicle can handle multiple driving functions at the same time.

Level 3: *Conditional Driving Automation*: The vehicle can handle most driving tasks in specific operating domains called ODD, like on highways, but the driver must take over if the system fails or in the event of sudden changes.

Level 4: *High Driving Automation*: Though the vehicle can handle all driving tasks in designated environments, like city roads, the driver still needs to be able to take over if the system fails.

Level 5: *Full Driving Automation*: The vehicle handles all situations, and the driver is not required to intervene.

Modern vehicles feature ADAS with automation primarily up to Level 2 and occasionally Level 3. Full automation, or Level 5 applications, are currently in pilot programs and are being introduced in various countries.

2.2.1.1 How Autonomous Vehicles (AV) Work

Autonomous vehicles navigate by continuously perceiving their surroundings through onboard sensors. These vehicles primarily employ three types of object sensors: cameras (for vision), radars (for radio wave-based position and distance detection), and lidars (for light reflection-based object and position detection). Each sensor type has unique strengths and weaknesses, including cost considerations. Automakers may utilize one, two, or all three sensor types, often integrating their outputs before further processing. To determine the vehicle's relative and absolute position, GPS, inertial measurement units (IMUs), and inertial navigation systems (INS) are also integrated. The process of an autonomous vehicle moving from a starting point A to a destination B involves several steps as shown in Figure 2.3.

Step 1 Perception: Data obtained from the sensors is fed into an AI based software, which detects the surrounding environment and classifies it as various objects such as vehicles, lane markings, static objects, pedestrians, road signs, and so on. The information includes the type of object, location, distance, speed, and all the relevant information to interpret the environment. Once objects are detected and classified, their movement is tracked over time. If multiple sensors are used, data from cameras, lidars, and radars are combined using sensor fusion algorithms to create a more comprehensive and clear view of the environment. This step answers the question "what's around me" at this instance in time.

Step 2 Localization: Using this information, the software can build a localized 3D map of the environment with all the objects, their distances, and speeds relative to the current position of the vehicle. This step answers the question "where am I and what is the environment around me." The process of sensing and building

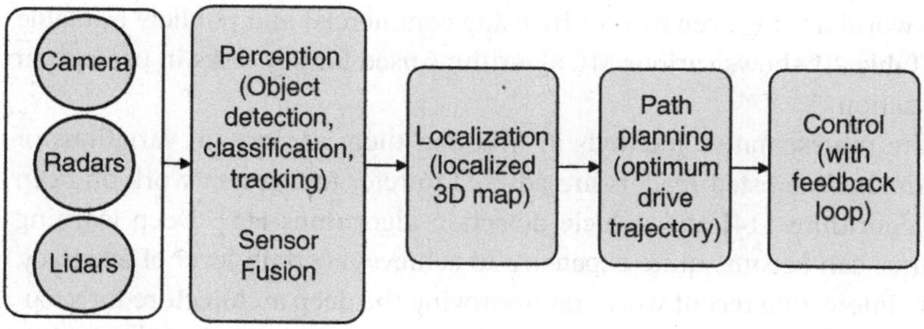

Figure 2.3 Illustration of autonomous function.

high-definition 3D maps of the local environment is called simultaneous localization and mapping (SLAM).

Step 3 Path Planning: In the next step, the vehicle determines how to navigate from the current position to the next position in the current environment. This is called "path planning," which determines the optimal path to follow. Path planning can be at high level to determine safe and optimal routes to follow and low level or local trajectory to follow safely. These algorithms use behavior arbitration to resolve conflicts and determine appropriate path strategies.

Step 4 Control: Finally, the vehicle control system executes motion while reactively correcting for errors.

With the knowledge of the route map, current location, and application of SLAM, autonomous vehicles navigate from source to destination.

2.2.1.2 How AI Is Used in Autonomous Systems

The core of autonomous technology is the AI based software that delivers autonomous functionality. However, AI is applied in building autonomous systems from end to end, commencing with the creation and curation of training datasets to rigorous verification process and safe commercial deployment. This discussion assumes a foundational understanding of ML concepts, principles, and methodologies. Numerous resources, including publicly available materials and massive open online courses (MOOCs), offer comprehensive introductions to these essential topics [13]. These examples will demonstrate how various ML algorithms and approaches are leveraged to achieve specific functionalities and address the unique challenges associated with building intelligent, self-driving systems.

2.2.1.3 Perception and Localization

Sensors mounted on the vehicle send input data in the form of visual images or point clouds. AV software interprets this input data using trained ML models to detect objects, classify them, and build the environment around the vehicle. These models are trained with large datasets. The datasets are either obtained from real-world driving scenarios or by using commercial and publicly available datasets. Table 2.1 shows various ML algorithms used for key tasks in perception and localization

These are representative methods in ML, and there are several variations of these methods. Interested readers are advised to refer to a recent work on deep learning algorithms [14] and vehicle detection algorithms [15]. Deep learning architectures can become quite expensive to achieve a certain level of accuracy. One of the interesting recent works on improving the deep architecture for accuracy and efficiency is on a special architecture called " HydraNet" [16], which is also implemented by Tesla.

Table 2.1 A few ML algorithms are used in perception and localization.

Tasks	Algorithms
Object detection and classification	Supervised learning, deep learning (convolution neural networks), unsupervised learning
Behavior detection of surroundings (e.g., pedestrians)	Recurrent neural networks
Anomaly or hazard detection	Unsupervised learning
Localization (SLAM)	Deep learning

2.2.1.4 Path Planning

Continuous determination of safe and efficient trajectory to move from the current position is a fairly complex task in autonomous driving. Path planning involves both local trajectory planning and high level path planning to align with the route map. Local trajectory planning determines steering parameters to safely navigate in the local environment, while the global path ensures the vehicle is on its planned route.

Path planning algorithms may be divided into three categories [17]—traditional methods using optimization, machine and deep learning, and meta-heuristic optimization. Path prediction algorithms sometimes use supervised learning based on convolution neural networks (CNN) or reinforced neural networks (RNN). High level path planning also uses deep learning and RNN.

2.2.1.5 Synthetic Data Generation

The success of autonomous systems depends on a comprehensive dataset with as much coverage of various scenarios as possible. However, in real world driving situations, potential failure modes are infinite, and it is difficult to obtain large amounts of good nominal and edge case scenarios to train ML models. As a consequence, synthetic data is increasingly gaining acceptance as a viable means to test autonomous systems. By definition, synthetic data is the data generated in the digital world as opposed to live data generated in the real world (by recording a video, for example). This may be done either by morphing real world data or by using scenario generation methods using abstract language inputs.

There are three broad methods of synthetic data generation [18]—rules-based methods, which have predefined data structures, agent-based methods which simulate environments and deep learning methods which use neural network based methods. There are three main families of deep-learning techniques used in synthetic data generation: generative adversarial networks (GANs), variational autoencoders (VAEs), and diffusion models.

Many commercial tools (Applied Intuition, Foretellix, to name a few) generate synthetic data (scenarios) besides providing a platform for simulation and verification of AV systems

2.2.1.6 Verification and Validation of AV Systems

ML is widely used to validate autonomous systems on simulation platforms. A sample application could be creating virtual environments to simulate real world driving scenarios. Another area of application is the automatic generation of test cases from requirements using abstract language inputs using LLMs. Similarly, when large amounts of simulation results are available, ML is used to identify vulnerabilities and missed edge cases.

2.2.1.7 Safe Deployment of AV Systems

The route to safe commercial deployment of autonomous systems involves a process called "monitored deployment." Here, a validated autonomous vehicle is deployed on a pilot basis along with a "safety driver." Vehicle behavior and AV system performance in handling various situations are recorded and closely studied. Feedback from this pilot is used to retrain the model and improve its performance. ML methods help in identifying edge or near-edge cases and in analyzing data to determine vulnerabilities.

2.2.1.8 Autonomous Urban Use Cases

Autonomous technology driven by AI has numerous commercial applications. The most well-known example is autonomous ride-hailing services, often called robotaxis. Companies like Lyft, Waymo, Zoox, and WeRide are popular providers of commercial autonomous rides. Beyond robotaxis, there are other compelling commercial applications of autonomous technology.

2.2.1.8.1 Delivery Drones and Last Mile Logistics

The land-based autonomous vehicles are complex because of infinite possibilities for accidents and surrounding uncertainties. To mitigate these challenges, the concept of an operational design domain (ODD) has been introduced. An ODD defines a specific set of operational conditions, including geographic boundaries, within which an autonomous vehicle is designed to function safely and effectively. By focusing on deep learning application development within a well-defined ODD, the design and validation processes become considerably more tractable.

Delivery drones are one such compelling application area for autonomous systems for logistics, and it is a growing area in the field [19]. Logistics applications can be either on land or in the form of unmanned aerial vehicle (UAV) carrying a payload to the delivery destination. Several prominent companies are actively operating in the UAV delivery space, including Amazon Prime Air,

Zipline, Manna, Wing Drone Delivery, and UPS Flight Forward. These sophisticated drones leverage a combination of global positioning system (GPS) technology, onboard sensors (such as cameras, lidar, and radar), and advanced AI algorithms to autonomously navigate optimal delivery routes. Their navigation capabilities take into account dynamic factors such as prevailing weather conditions and the terrain's topography, ensuring efficient and safe delivery operations. The controlled nature of last-mile delivery within specific geographic areas makes it a particularly suitable domain for the initial deployment and refinement of autonomous vehicle technologies.

2.2.1.8.2 Self-parking Vehicles

These are autonomously maneuvered vehicles that move from traffic lanes to parking spaces and perform angular or parallel parking [20]. Early versions of this technology appeared in the late 1990s but eventually improved using ML algorithms. Autonomous parking is now a prevalent feature in numerous vehicles, enhancing driver convenience and reducing the stress associated with parking in challenging conditions.

In short, vehicles which park autonomously use sensor data from cameras, radars and ultrasonic sensors to sense the surroundings. The algorithms analyze the sensor data to identify available parking spaces, assess their suitability based on the vehicle's dimensions and maneuvering capabilities, and plan an optimal parking trajectory. The system calculates the precise steering angles, acceleration, and braking inputs required to execute the parking maneuver safely and efficiently.

Autonomous parking use cases extend to more comprehensive parking solutions, such as robotic garages and fully automated parking facilities. Vehicles equipped with autonomous navigation capabilities can independently navigate through parking structures, locate vacant parking spots, and park themselves without any driver intervention.

2.2.1.8.3 Robotic Street Sweepers

Handling vast amounts of waste generated in cities is a formidable challenge. Autonomous technology is used to solve this problem by deploying robotic systems for waste management. Specifically, autonomous robotic street sweepers are deployed to navigate city streets utilizing an array of cameras, lidars, and ultrasonic sensors to locate waste on the street. Once identified, robotic arms automatically pick up and segregate the waste. By employing ML algorithms and object recognition capabilities, the robots can differentiate between various types of waste, such as recyclable plastics, paper, organic matter, and nonrecyclable materials. These systems use AI for route optimization for efficient coverage, dynamic adaptation for urban environments, and refining waste recognition and manipulation skills. Many urban municipalities are piloting these solutions [21].

Trombia Technologies® is a representative company that develops and deploys such solutions.

2.3 Connected Vehicles

Connected technology is ubiquitous in modern vehicles and is central to most AI-centric applications for urban transportation. A connected vehicle, in simple terms, is a vehicle networked to the outside world with the ability to communicate with other vehicles and to external devices such as cloud applications, road infrastructure, and devices such as tablets and smartphones. Early ideas on connected vehicles came from General Motors in 1966 with the introduction of the driver aid information and routing (DAIR) system. This system provided assistance to drivers with information for navigation, road conditions, and potential hazards. In 1996, GM introduced the OnStar® telematics system on Cadillac models, which is seen in the industry as pioneering early work on connected cars. The system integrated a cellular network with GPS to activate emergency calls in the event of a crash [22]. Eventually, these systems evolved into today's sophisticated connected cars. The connected car ecosystem is illustrated in Figure 2.4. It enables communication between vehicles, vehicle to infrastructure and cloud, and vehicle to external services and devices. This is accomplished with cellular network (4G/5G), in-vehicle network (WiFi, Bluetooth, Ethernet) and dedicated short range communication (DSRC).

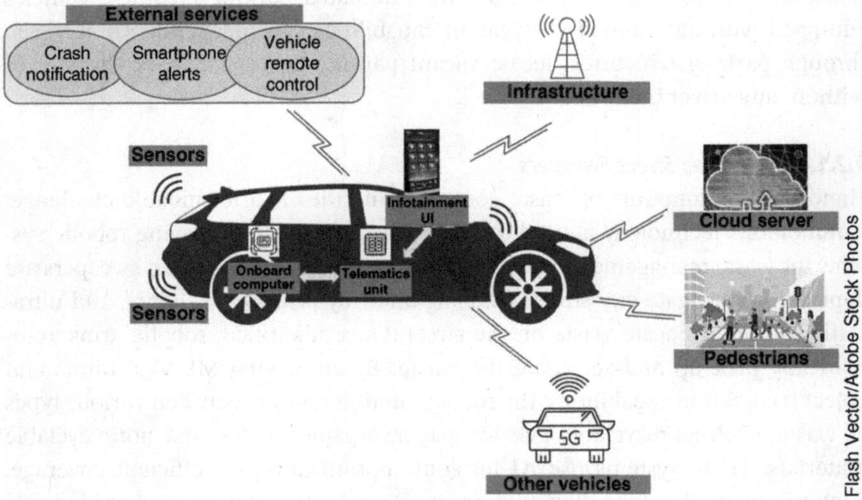

Figure 2.4 Connected vehicle ecosystem.

2.3.1 How AI Is Used in Connected Vehicles

Several applications are built using this infrastructure. These applications are meant to enhance user experience, increase the safety of the driver/pedestrians, and streamline urban traffic.

2.3.1.1 Personalized User experience

AI transforms user experience inside a vehicle from a generic interaction to a highly personalized interaction. Based on user preferences and choices, AI algorithms adjust vehicle settings to provide personalized experiences such as music choices, news and traffic updates, and streaming services. Recommendation systems, classification, and clustering algorithms are used to provide these services. Similarly, natural language processing (NLP) is used for voice recognition functions. Aptiv® and BMW® implemented the first gesture recognition system in cars in 2015, where drivers could interact with the infotainment system using hand gestures. Systems such as this use sensors and ML algorithms like CNN [23] to implement gesture recognition.

2.3.1.2 Navigation Guidance

Route optimization is a fairly common application leveraging connected car features. Using inputs such as traffic, road closures, and constructions, intelligent navigation and route planning are done using methods such as graph analysis and reinforcement learning. This is a widely studied area with applications in logistics solutions. An example is a research paper [24] that proposes a deep reinforcement learning (DRL) method to build a real-time intelligent vehicle routing and navigation system by formulating the task as a sequence of decisions.

2.3.1.3 Vehicle Status Monitoring

Remote control of the vehicle functions such as door lock, ignition, and keyless entry is one of the popular features implemented in connected vehicles. This capability allows users to interact with their vehicles from a distance through smartphone applications or other connected devices. Functions such as locking and unlocking doors, initiating the engine ignition, and enabling keyless entry systems can now be managed remotely, enhancing convenience and security for vehicle owners. This remote access not only provides ease of use in everyday scenarios but also opens up possibilities for innovative services and features, such as remote diagnostics, over-the-air software updates, and integration with smart home ecosystems. The reliability and security of these remote control functionalities are paramount, demanding robust communication protocols and stringent cybersecurity measures to protect against unauthorized access and potential vulnerabilities.

2.3.1.4 Connected Vehicle Urban Use Cases

2.3.1.4.1 Fleet Monitoring Systems

Fleet operators own a large fleet of commercial vehicles such as LCVs and trucks. Using GPS sensors. With telematics and in-vehicle cameras, fleet management companies get real-time operational data from each vehicle and the driver in the fleet. Using AI and real time analytics, fleet management efficiency, safety, and performance of vehicles, plus drivers, can be improved. This is a rapidly growing use case of AI in the automotive domain. In-vehicle cameras collect real time driver data for monitoring driving patterns and driver alertness during driving. ML methods enable the detection of alertness levels based on camera images. Data from the vehicle sensors provide driving parameters. These data are combined to analyze driver performance and trends. The feedback is used to train drivers and reward positive behaviors. Similarly, traffic safety analytics from real time data can be used to predict potential hazards and alert drivers. Monitoring logistics vehicles with smart solutions is a key piece of smart city solutions in urban transportation.

By combining fleet monitoring with autonomous technology (CAV or connected autonomous vehicle), vehicle manufacturers are able to offer unique solutions that deliver steep improvements in operational efficiency. For example, Arora (US-based autonomous truck company) launched long haul autonomous trucks for transportation in Texas [25] recently.

2.3.1.4.2 Uber's DeepETA

Uber uses the physical model and combines it with deep learning methods to predict the estimated time of arrival (ETA) for its services such as rides, freight, and eats [26]. The physical model includes real time traffic data and maps. Uber needed to reduce latency and increase the accuracy of ETA estimates. This was accomplished using a deep learning model called DeepETA, a low-latency deep neural network architecture designed for global ETA prediction.

2.4 Electric Vehicles (EV)

Urban transportation is witnessing the growth of EVs in recent years. Although EV technology was developed decades ago, technological advancements in battery physics and the downward trend of battery cost have led to an accelerated pace of EV adoption in cities. Densely populated countries like India and China are absorbing significant growth in electric two-wheelers and three-wheelers in the last decade. Besides, the integration of electric, connected, and autonomous technologies is driving a convergence toward a comprehensive and sustainable transportation solution.

EV systems are powered by a "battery pack," which is a collection of interconnected cells delivering DC power. Typically, individual cells are based on lithium ion chemistry and have high energy density. Battery pack thermal management is an important feature in the EV system. The battery pack is housed in a metallic enclosure that also contains a cooling system for temperature regulation, a system of IoT sensors to monitor crucial parameters, and a BMS. The DC power from the battery pack is converted to AC power by an inverter and fed to a vehicle's electric motor, typically an AC induction motor, which drives the vehicle. The sensors continuously monitor various parameters of the battery pack, such as voltage, current, temperature, and state of charge. BMS controls the charging and discharging of the cells. A vehicle control unit performs power management functions.

EV systems offer many opportunities for the application of AI and IoT sensors to optimize their performance, safety, and user experience. In the previous section on IoT and sensors, we discussed an example of remote access to battery parameters.

The sensor data is usually locally processed in the BMS system for immediate actions and frequently transmitted to a central computer for analysis for remote monitoring. By deploying AI algorithms, health status and current battery performance are constantly assessed and optimized. BMS acts as a controlling system to adjust battery parameters based on insights from AI analysis. AI algorithms can identify subtle patterns and anomalies in the sensor data that might be indicative of potential issues, such as degradation, thermal runaway risks, or inefficiencies in charging and discharging cycles. This proactive assessment allows for timely interventions and predictive maintenance strategies, ultimately extending the lifespan of the battery and improving its safety.

2.4.1 AI-based Urban Use Cases of EVs

AI finds applications in various aspects of EVs, such as remote battery assessment, charging strategies, cost optimization of charging stations, and route optimization. A few representative examples are provided in the following section.

2.4.1.1 Remote Monitoring of EV Fleets

Battery cell capacity diminishes with charge-discharge cycles, eventually leading to failure. Predicting capacity fade is crucial for estimating the lifespan of individual cells and the entire battery pack. Real-time battery parameters, combined with the vehicle's energy consumption rate, allow for estimations of the remaining driving range of an EV at any point of time. ML techniques, such as neural networks and regression algorithms, are commonly employed for these estimations. Remote monitoring of EV batteries is applied to EV fleets used by the logistics industry and ride-hailing industry.

Table 2.2 AI-based battery parameter estimation examples.

Battery parameter	Neural network model
Battery aging	Feed-forward neural networks and RNNs with long short-term memory [27]
State of charge (SoC)	Feed-forward neural networks and RNNs [28]
State of health (SoH)	Long short-term memory neural network models [29]
Battery temperature field	Convolutional neural network model, long Short-term memory [30]
Battery range	Deep neural networks
Remaining useful life (RUL)	Deep neural networks, recurrent neural networks

IoT sensor data is collected in real time and sent to a central computer after a local preprocessing step. AI algorithms will then assess the condition of battery packs and recognize patterns in the data. Using predictive models, these algorithms can provide estimates for various parameters such as State of Charge (SoC), range of the battery, and remaining useful life (RUL). Table 2.2 illustrates a few examples from recent literature on using neural network architectures to build predictive models for battery remote diagnostics.

2.4.1.2 EV Charging Infrastructure Management

AI methods can be leveraged to plan and maintain EV charging infrastructure networks and optimize efficiency, utilization, and availability. By employing sophisticated AI methodologies, it is possible to optimize charging station utilization, overall network efficiency, and consistent availability of charging points for EV users. Data sources include prevailing weather, energy generation and capacity details, load demand on the grid, and dynamic cost of electricity. AI-driven optimization can lead to significant reductions in operational costs by strategically managing energy consumption and scheduling maintenance proactively. By combining these insights with EV vehicle data, charging locations of individual EVs can be intelligently managed to optimize overall cost and efficiency.

2.4.1.3 Enhancing User Experience

AI algorithms improve EV user experience by enabling features like predictive maintenance, intelligent navigation, and personalized settings. Smart navigation is achieved by AI using route maps, traffic, weather, and battery charge to recommend routes and nearby charging stations. Moreover, AI techniques incorporating personal preferences and voice recognition further enhance the user experience.

2.5 Shared Mobility

The increasing congestion within urban centers, coupled with the escalating costs and inefficient transportation solutions, has propelled shared mobility to the forefront of urban transportation planning. Shared mobility is a concept built on collaborative use of multiple transport services such as vehicle, bicycle, and other alternative modes of transportation. This paradigm shift is further integrated under the concept of mobility as a service (MaaS) offering users a unified platform to manage their commute needs. Carpooling, ride-sharing, micro-transit, and bike-sharing services fall in this category. These services can supplement fixed-route bus and rail services. Early adoption of shared solutions was driven by peer-to-peer platforms, connecting individuals with shared transportation needs.

The proliferation of digital technology, including AI, has played a key role in widespread development and optimization of modern shared mobility solutions. AI-powered systems are instrumental in complex logistics challenges such as dynamic route planning. ML can enhance the management of shared mobility systems by improving the accuracy of user journey assessments [31].

2.6 Future Trends Toward Smart Transportation

The trajectory of urban transportation is shifting toward future cities where infrastructure is intricately designed to facilitate seamless urban mobility, powered by the presence of data and sophisticated AI applications. Complementing this evolution, the automotive industry is witnessing a significant shift with the increasing integration of LLMs across the entire product life cycle, from initial research and development phases to the final launch and post-market analysis. As foundational mobility technologies encompassing electrification, autonomous driving capabilities, enhanced safety features, and interconnected vehicles mature in their availability and demonstrate greater reliability, the landscape of smart transportation solutions within urban environments will be fundamentally reshaped by the adoption of advanced technologies aimed at optimizing efficiency, bolstering safety protocols, and drastically reducing environmental emissions. Several key AI-driven trends would exert influence on the future of smart transportation, and we have listed five key trends here:

a. *Convergence* will supercharge transportation efficiency with the integration of electric, connected, and autonomous technologies. EVs, which offer a sustainable transportation solution when combined with the power of connected technology and autonomous features, can create powerful shared mobility solutions that can be accessed on demand. Robotaxi pilot programs are underway in numerous locations globally, indicating a trend toward this mode of transportation.

b. *Integration of IoT sensors* across urban infrastructure, coupled with data acquisition mechanisms and robust cloud computing infrastructure, will serve as the foundation for deployment of AI-based methodologies for technology-driven traffic control. Real-time data streams from traffic cameras, vehicle sensors, and pedestrian movement trackers will feed into AI algorithms capable of analyzing traffic patterns and predicting congestion hotspots, and assisting in dynamic routing options. Furthermore, the integration of abstract language inputs and the power of LLMs will enable more intuitive and natural interfaces for traffic management systems, allowing for more flexible and adaptive control strategies based on real-time needs and events.
c. *Smart public and Shared multimodal mobility solutions* are likely to pave the way for the development of coordinated services to solve last-mile transportation problems. Smart public transport will leverage real-time information, predictive analytics, and dynamic scheduling to optimize routes, reduce waiting times, and enhance the overall passenger experience. AI-powered algorithms will play a crucial role in optimizing fleet management, predicting demand fluctuations, and ensuring smooth transitions between different transportation modes
d. *Connected vehicles* equipped with V2X infrastructure will be integral to the smart city transportation fabric, providing a continuous flow of real time information enabling intelligent traffic control solutions.
e. *Autonomous last mile solutions* for the delivery of packages, food, medicine, and other supplies will be increasingly driven by land based autonomous vehicles and UAVs. Autonomous ground vehicles offer potential for cost-effective and efficient transportation and delivery services, particularly in dense urban environments with well-defined operational domains such as large campuses, residential communities, and airports.

Thus, AI and IoT sensor-based solutions driven by big data as a backbone will drive the industry forward with innovative urban transportation solutions.

2.7 Conclusion

This chapter explores the impact of AI, ML, and IoT systems on the transportation and mobility sector. While the examples and use cases primarily focus on urban transportation, their applicability often extends to other areas. For instance, autonomous drone technology delivers essential supplies in remote regions. The chapter offers a selection of representative examples and use cases with limited references, hoping to spark the reader's interest and encourage further exploration in these domains.

References

1 EPA, USA, The 2024 EPA automotive trends report. *EPA-420-S-24-001*, November 2024, https://nepis.epa.gov/Exe/ZyPDF.cgi?Dockey=P101CUZD.pdf

2 UN, 2018 Revision of world urbanization prospects, *UN Department of Economic and Social Affairs*, 2025, https://www.un.org/uk/desa/68-world-population-projected-live-urban-areas-2050-says-un#:~:text=68%25%20of%20the%20world%20population,of%20Economic%20and%20Social%20Affairs

3 Scott Shriber, S. The important role of sensors in modern vehicles, *Counterman*, November 13, 2024, *https://www.counterman.com/role-sensors-modern-vehicles/#:~:text=The%20average%20modern%20vehicle%20contains%20anywhere%20from,the%20vehicle's%20systems%20make%20split%20second%20decisions.*

4 Thrun, S., Montemerlo, M., Dahlkamp, H., Stavens, D., Aron, A., Diebel, J., Fong, P., Gale, J., Halpenny, M., Hoffmann, G., Lau, K., Oakley, C., Palatucci, M., Pratt, V., Stang, P., Strohband, S., Dupont, C., Jendrossek, L., Koelen, C., Markey, C., Rummel, C., van Niekerk, J., Jensen, E., Alessandrini, P., Bradski, G., Davies, B., Ettinger, S., Kaehler, A., Nefian, A., and Mahoney, P. Stanley: The robot that won the DARPA grand challenge. *Journal of Field Robotics* 23(9): 661–692, 2006.

5 Hossain, M. N., Rahim, M. A., Rahman, M. M., and Ramasamy. D. Artificial intelligence revolutionising the automotive sector: A comprehensive review of current insights, challenges, and future scope. *Computers, Materials and Continua*, 82(3): 3643–3692, 2025. https://doi.org/10.32604/cmc.2025.061749.

6 Despot, K., Srebrenkoska, S., and Sandeva, V. The role of artificial intelligence in automotive design. *Knowledge - International Journal*, 61(3): 423, 2023.

7 Maurer, M., Gerdes, J. C., Lenz, B., and Winner, H., editors. *Autonomous Driving Technical, Legal and Social Aspects*, Springer Open, 2016

8 Rakholia, R., Suárez-Cetrulo, A. L., Singh, M., and Carbajo, R. S. Advancing Manufacturing Through Artificial Intelligence: Current Landscape, Perspectives, Best Practices, Challenges, and Future Direction, *IEEE Access*, 12: 131621, 2024.

9 Theissler, A., Pérez-Velázquez, J., Kettelgerdes, M., and Elger, G. Predictive maintenance enabled by machine learning: Use cases and challenges in the automotive industry, *Reliability Engineering and System Safety*, 215: 107864, 2021.

10 Refat, R. U. D., Elkhail, A. A., Malik, H. Machine learning for automotive cybersecurity: Challenges, opportunities and future directions. In: Murphey, Y.L., Kolmanovsky, I., Watta, P. editors, *AI-enabled Technologies for Autonomous and Connected Vehicles. Lecture Notes in Intelligent Transportation and Infrastructure*. Springer, Cham., 2023. https://doi.org/10.1007/978-3-031-06780-8_20.

11 Abduljabbar, R., Dia, H., Liyanage, S., and Bagloee, S. Applications of artificial intelligence in transport: An overview. *Sustainability*, 11: 189, 2019. https://doi.org/10.3390/su11010189.

12 SAE International, *Taxonomy and Definitions for Terms Related to Driving Automation Systems for On-road Motor Vehicles*, SAE J3016_202104, April 30, 2021. https://doi.org/10.4271/J3016_202104.

13 Andrew Ng, "Machine learning specialization," 2024, DeepLearning.Ai https://www.coursera.org/specializations/machine-learning-introduction

14 Grigorescu, S., Trasnea, B., Cocias, T., and Macesanu, G. A survey of deep learning techniques for autonomous driving. *Journal of Field Robotics*, 37(3): 362–386, 2019.

15 Karangwa, J., Liu, J., and Zeng, Z. Vehicle detection for autonomous driving: A review of algorithms and datasets. *IEEE Transactions on Intelligent Transportation Systems*, 24(11): 11568–11594, 2023.

16 Shazeer, N., Fatahalian, K., Mark, W. R., and Mullapudi, R. T. HydraNets: Specialized dynamic architectures for efficient inference. In *2018 IEEE/CVF Conference on Computer Vision and Pattern Recognition*, Salt Lake City, UT, USA, pages 8080–8089. 2018.

17 Reda, M., Onsy, A., Haikal, A. Y., and Ghanbari, A. Path planning algorithms in the autonomous driving system: A comprehensive review. *Robotics and Autonomous Systems*, 174: 104630, 2024.

18 Deng, H. Exploring synthetic data for artificial intelligence and autonomous systems: A primer. *UNIDIR Report*, November 2023. *https://unidir.org/publication/exploring-synthetic-data-for-artificial-intelligence-and-autonomous-systems-a-primer.*

19 Jazairy, A., Persson, E., Brho, M., von Haartman, R., and Hilletofth, P. Drones in last-mile delivery: A systematic literature review from a logistics management perspective. *The International Journal of Logistics Management*, 36(7): 1–62, 2025.

20 Paromtchik, I. and Laugier, C. Motion generation and control for parking an autonomous vehicle" (PDF). In *Proceedings of the IEEE International Conference on Robotics and Automation* Vol. 4. Minneapolis, MN, USA, pages 3117–3122, 1996.

21 CiTTi. Helsinki pilots world's first high-power autonomous street sweeper, September 2020, *https://www.cittimagazine.co.uk/videos/helsinki-pilots-worlds-first-high-power-autonomous-street-sweeper.html*

22 Zhao, Y. Efficient and reliable data transmission for cellular-and-GPS-based mayday systems. In *Proceedings of Conference on Intelligent Transportation Systems*, Boston, MA, pages 555–559, 1997. https://doi.org/10.1109/ITSC.1997.660534.

23 Tateno, S., Zhu, Y., and Meng, F. Hand gesture recognition system for in-car device control based on infrared array sensor. In *2019 58th Annual Conference of the Society of Instrument and Control Engineers of Japan (SICE)*, Hiroshima, Japan, pages 701–706, 2019. https://doi.org/10.23919/SICE.2019.8859832.

24 Koh, S., Zhou, B., Fang, H., Yang, P., Yang, Z., Yang, Q., Guan, L., and Ji, Z. Real-time deep reinforcement learning based vehicle navigation. *Applied Soft Computing*, 96: 106694, 2020.

25 Aurora Press Release. Aurora begins commercial driverless trucking in Texas, ushering in a new era of freight, May 2025. https://ir.aurora.tech/news-events/press-releases/detail/119/aurora-begins-commercial-driverless-trucking-in-texas.

26 Uber blog. DeepETA: How uber predicts arrival times using deep learning, February 2022, *https://www.uber.com/en-IN/blog/deepeta-how-uber-predicts-arrival-times/?ref=louisbouchard.ai.*

27 Hoque, M. A., Hassan, M. K., Hajjo, A., and Tokhi, M. O. Neural network-based Li-ion battery aging model at accelerated C-rate. *Batteries*, 9: 93, 2023. *https://doi.org/10.3390/batteries9020093.*

28 Dini, P. and Paolini, D. Exploiting artificial neural networks for the state of charge estimation in EV/HV battery systems: A review. *Batteries*, 11: 107, 2025.

29 Min, H., Yan, Y., Sun, W., Yu, Y., Jiang, R., and Meng, F. Construction and estimation of battery state of health using a De-LSTM model based on real driving data. *Energies*, 16: 8088, 2023. https://doi.org/10.3390/en16248088

30 Wang, B., Chen, Z., Zhang, P., Deng, Y., and Li, B. The lithium-ion battery temperature field prediction model based on CNN-Bi-LSTM-AM. *Sustainability* 17: 2125, 2025. *https://doi.org/10.3390/su17052125.*

31 Kubik, A. The use of artificial intelligence in the assessment of user routes in shared mobility systems in smart cities. *Smart Cities*, 6: 1858–1878, 2023. https://doi.org/10.3390/smartcities6040086.

3

Artificial Intelligence in Transportation Using Automated Grading, Adaptive Learning, and Predictive Maintenance to Increase Efficiency

S. Cyciliya Pearline Christy[1,], K. Merriliance[1], and Mary Immaculate Sheela Lourdusamy[2]*

[1] Department of Computer Applications and Research Centre, Sarah Tucker College (Autonomous), Affiliated to Manonmaniam Sundaranar University, Tirunelveli, Tamil Nadu, India
[2] Department of Computing and Engineering, School of Engineering Technology and Applied Science (SETAS), Heritage Christian University College (HCUC), Accra, Ghana

*Corresponding author: cyciliyapearlinechristy@gmail.com

3.1 Introduction

3.1.1 Background of AI in Transportation

AI has emerged as a transformative force across multiple industries, with transportation standing out as one of the most significantly impacted sectors. Over the past decade, the integration of AI technologies in transportation has expanded rapidly, ranging from AVs and smart traffic management systems to intelligent logistics and infrastructure monitoring. These advancements are driven by the growing demands of urban mobility, environmental concerns, and the need for safer, more reliable systems.

Key areas where AI is making a mark include real-time decision-making, data-driven route optimization, dynamic traffic predictions, and vehicle diagnostics. Technologies such as ML, computer vision, natural language processing (NLP), and neural networks are now embedded in everyday transportation tools. Governments, private companies, and researchers alike are investing heavily in AI to revolutionize how goods and people move.

3.1.2 Purpose and Scope of the Chapter

The purpose of this chapter is to explore how AI-driven solutions—specifically **automated grading**, **adaptive learning**, and **predictive maintenance**—can be strategically employed to improve efficiency in transportation systems. While AI offers many capabilities, the focus here will be on these three interconnected applications and how they collectively contribute to operational improvements, cost reduction, and enhanced safety.

This chapter will:

- Define each of the three AI methods within a transportation context.
- Present case studies and real-world examples.
- Discuss implementation challenges and technological requirements.
- Highlight future trends and opportunities for growth.

3.1.3 Importance of Efficiency in Modern Transportation Systems

Efficiency is a cornerstone of modern transportation systems, directly affecting economic productivity, environmental sustainability, and public satisfaction. Inefficient systems contribute to traffic congestion, increased emissions, delayed deliveries, and elevated operational costs. As populations grow and urbanization intensifies, the pressure on transportation networks continues to rise.

In this context, AI technologies offer a promising pathway toward achieving higher levels of efficiency [1]. By leveraging automated systems that can learn, adapt, and predict, transportation networks can become more responsive, resilient, and sustainable. The integration of automated grading (for infrastructure assessment), adaptive learning (for traffic and behavior modeling), and predictive maintenance (for vehicles and infrastructure) represents a shift from reactive to proactive management—ensuring transportation systems are not only faster but also smarter.

3.2 Overview of Artificial Intelligence in Transportation

3.2.1 What is AI and How It Applies to Transportation

AI refers to the simulation of human intelligence processes by machines, especially computer systems. These processes include learning (acquiring data and rules for using it), reasoning (using rules to reach conclusions), and self-correction. In transportation, AI applications are leveraged to improve decision-making, optimize operations, and enhance safety.

AI in transportation can take many forms:

- **Machine learning** enables systems to learn from data and improve over time without being explicitly programmed, such as in dynamic traffic light control systems.
- **Computer vision** is used for object detection and lane tracking in AVs.
- **NLP** powers voice-controlled navigation systems and customer service bots.
- **Reinforcement learning** optimizes complex tasks like multi-agent traffic management or fleet logistics.

From intelligent traffic systems and AVs to real-time route planning and infrastructure monitoring, AI is embedded across various transportation layers. It acts as the brain behind smarter, more adaptable systems that react to changing environments and user demands.

3.2.2 Current Trends in AI Adoption

AI adoption in transportation is growing rapidly due to advances in computing power, big data availability, and a push for sustainable and efficient mobility. Some of the leading trends include:

- **Autonomous and Connected Vehicles**: Self-driving technology is maturing, with AI playing a central role in perception, decision-making, and navigation.
- **Smart Traffic Management**: Cities are using AI to reduce congestion through real-time traffic flow analysis and adaptive signal control.
- **Predictive Maintenance**: Transport agencies are increasingly using AI models to forecast equipment failures before they occur, reducing downtime and repair costs.
- **AI in Logistics**: Companies use AI for route optimization, demand forecasting, and warehouse automation to improve delivery efficiency.
- **AI-powered Public Transit Systems**: Real-time tracking, rider behavior analysis, and personalized services are enhancing public transport experiences.

Government initiatives, such as smart city frameworks and infrastructure digitization, are further accelerating AI integration into transportation.

3.2.3 Challenges in Implementing AI Technologies

Despite its promise, implementing AI in transportation comes with several challenges:

- **Data Quality and Availability**: AI systems depend on vast amounts of high-quality data, which is often fragmented, inconsistent, or proprietary.

- **Integration with Legacy Systems**: Many transportation infrastructures rely on outdated systems that are not easily compatible with modern AI solutions.
- **Regulatory and Ethical Concerns**: Issues around privacy, accountability, and safety raise questions about how AI should be deployed and governed.
- **High Implementation Costs**: Advanced AI systems often require substantial investments in hardware, software, and skilled personnel.
- **Public Trust and Acceptance**: Resistance to change, concerns about automation, and lack of transparency in AI decision-making can hinder adoption.

Overcoming these challenges will require cross-sector collaboration, strong policy frameworks, and continuous innovation to ensure AI systems are reliable, secure, and beneficial for all stakeholders.

3.3 Automated Grading Systems in Transportation

3.3.1 Definition and Role in Infrastructure Evaluation

Automated grading systems in transportation refer to AI-powered technologies used to assess and evaluate the condition of infrastructure such as roads, bridges, tunnels, and railways. These systems automate the process of grading—assigning condition scores or performance metrics—based on visual, structural, or operational data. This reduces the reliance on manual inspections and subjective assessments.

The role of automated grading in infrastructure evaluation is critical, especially as transportation networks age and expand. Regular, data-driven evaluations help transportation agencies prioritize maintenance, allocate budgets efficiently, and enhance public safety. These systems can detect cracks, potholes, material wear, and other structural anomalies early, allowing for proactive intervention.

3.3.2 AI Techniques Used (e.g., Computer Vision, ML)

Automated grading systems leverage a combination of advanced AI technologies [2]:

- **Computer Vision**: Analyzes images or video data from drones, mobile units, or fixed sensors to detect surface defects, structural deformations, and anomalies in road and bridge conditions.
- **Machine Learning**: Trains models on historical data to identify patterns and predict infrastructure degradation. Supervised and unsupervised learning models can both be used to classify damage severity.

- **Deep Learning**: Especially convolutional neural networks (CNNs), is used for image recognition tasks such as identifying cracks or corrosion in pavement and bridges.
- **Sensor Fusion**: Combines data from multiple sensors (LIDAR, infrared, accelerometers) for a more accurate and holistic evaluation.

These techniques make grading systems not only faster but also more consistent and scalable compared to traditional methods.

3.3.3 Real-world Applications (e.g., Road Surface Assessment, Bridge Safety)

AI-driven grading systems are already being used in a variety of infrastructure monitoring and evaluation tasks:

- **Road Surface Assessment**: Mounted cameras on vehicles scan and classify surface issues like potholes, rutting, and cracks. Cities like Amsterdam and New York have adopted such systems for efficient pavement management.
- **Bridge Safety Monitoring**: Drones equipped with AI vision systems can inspect difficult-to-access bridge components, detecting corrosion, fatigue, or deformation.
- **Railway Track Inspection**: Automated grading systems identify track misalignments, ballast issues, and wear using computer vision and vibration sensors.
- **Tunnel Inspection**: AI tools analyze footage from robotic crawlers or drones for detecting leaks, lining wear, and structural cracks in tunnels.

These real-world applications reduce human error, increase the frequency of evaluations, and help transportation authorities manage infrastructure assets more effectively.

3.3.4 Benefits and Limitations

Benefits:

- **Efficiency**: Significantly reduces the time and labor involved in infrastructure inspection.
- **Consistency**: Provides objective, repeatable evaluations across different locations and times.
- **Cost Savings**: Minimizes unnecessary repairs and extends infrastructure lifespan through early intervention.
- **Safety**: Reduces the need for inspectors to physically access dangerous or hard-to-reach areas.

Limitations:

- **Data Dependency**: Performance is heavily reliant on the quality and quantity of training data.
- **Environmental Constraints**: Poor lighting, weather, or surface debris can affect detection accuracy.
- **Initial Investment**: Requires significant upfront cost in AI model development, hardware (e.g., sensors, drones), and skilled personnel.
- **Integration Challenges**: Difficulty in aligning AI outputs with existing asset management systems or inspection standards.

Despite these challenges, ongoing advancements in AI and sensor technologies continue to improve the reliability and affordability of automated grading systems in transportation.

3.4 Adaptive Learning in Traffic Management and Logistics

3.4.1 What Is Adaptive Learning?

Adaptive learning is a subset of AI that enables systems to continuously improve their performance by learning from new data and adjusting their behavior accordingly. Unlike traditional rule-based systems, adaptive learning models can respond in real-time to changes in the environment, user behavior, or system performance [3].

In transportation, adaptive learning algorithms play a vital role in dynamically optimizing operations, predicting outcomes, and personalizing services. These systems evolve with usage—becoming smarter and more efficient as they process more traffic data, logistics patterns, or commuter behavior.

3.4.2 Traffic Signal Optimization and Route Planning

Adaptive learning is transforming urban mobility through smarter traffic management. Traditional traffic signals often operate on fixed cycles, which don't adapt well to real-time conditions like accidents, surges in traffic, or weather disruptions. Adaptive systems powered by machine learning (ML) can change this:

- **Real-time Signal Adjustment**: AI systems analyze live traffic flows and adjust signal timing to reduce congestion and improve throughput.
- **Predictive Modeling**: Using historical data and current traffic conditions, these systems can forecast congestion points and reroute vehicles before problems arise.

- **Vehicle-to-infrastructure (V2I) Communication**: Vehicles equipped with sensors can share real-time data with traffic lights, helping to optimize light changes for smoother flow.

Cities like Pittsburgh, Los Angeles, and Singapore have deployed adaptive traffic control systems that have significantly reduced commute times and carbon emissions [4].

3.4.3 AI-driven Logistics and Fleet Management

In logistics, adaptive learning enables systems to respond in real-time to variables such as fuel prices, weather, road conditions, and delivery demand. Applications include:

- **Dynamic Route Optimization**: AI continually learns from delivery patterns, road closures, and time-of-day data to find the fastest, most cost-effective routes.
- **Demand Forecasting**: Machine learning models analyze historical orders, seasons, and trends to predict future demand and optimize resource allocation.
- **Fleet Monitoring and Optimization**: Adaptive systems track vehicle health, driver behavior, and delivery status to enhance fleet performance and reduce fuel consumption.
- **Autonomous Dispatch Systems**: AI determines the best vehicle-driver-resource pairing based on real-time availability and constraints.

Companies like UPS, FedEx, and Amazon are already leveraging adaptive AI to streamline last-mile delivery and manage complex logistics networks.

3.4.4 Case Studies and Pilot Projects

1. **Surtrac (Pittsburgh, USA)**
   Pittsburgh's smart traffic signal system uses decentralized AI agents that adapt to real-time conditions at each intersection. Studies report up to a **25% reduction in travel times** and a **40% cut in idle time** at lights.
2. **DHL SmartTruck (Germany)**
   DHL implemented a smart truck project that uses adaptive algorithms to reroute deliveries throughout the day based on changing traffic and parcel data. This led to **improved on-time delivery rates** and better fuel efficiency.
3. **Singapore Land Transport Authority (LTA)**
   Singapore's LTA uses adaptive AI models in its Expressway Monitoring and Advisory System (EMAS), allowing it to manage traffic flow dynamically

during peak hours and special events, reducing congestion and enhancing commuter experience.

4. **Amazon's Adaptive Logistics Network**
   Amazon uses AI to dynamically assign packages to fulfillment centers and determine optimal delivery routes based on order location, inventory, and delivery time commitments. Their use of real-time adaptive learning has significantly improved scalability and delivery speed.

These adaptive systems are reshaping how traffic and logistics are managed—shifting from reactive to proactive control. The benefits include not only improved operational efficiency but also enhanced environmental sustainability and user satisfaction [5].

3.5 Predictive Maintenance in Transportation Systems

3.5.1 Introduction to Predictive Maintenance

Predictive maintenance (PdM) is an AI-driven strategy that uses real-time data, historical trends, and statistical modeling to anticipate equipment failures before they occur. In transportation systems, predictive maintenance represents a major shift from traditional reactive or scheduled maintenance practices.

Instead of adhering to fixed maintenance intervals (which can be inefficient or too late), PdM leverages AI to forecast when and where maintenance is actually needed. This approach reduces unplanned downtime, extends asset life, enhances safety, and lowers maintenance costs across vehicles, infrastructure, and mechanical systems.

3.5.2 Sensors and Data Collection Techniques

Predictive maintenance depends on continuous data collection from various physical components using advanced sensor technologies. Some of the most common include [6]:

- **Vibration Sensors**: Detect unusual movement patterns in engines, rail tracks, or bridge structures.
- **Temperature Sensors**: Monitor overheating in components like brakes, engines, or electronics.
- **Acoustic Emission Sensors**: Capture high-frequency sounds indicative of material stress or cracks.
- **Pressure Sensors**: Useful in monitoring fluid systems in vehicles, pipelines, and hydraulics.

- **Infrared and Thermal Cameras**: Identify overheating or component wear not visible to the naked eye.
- **LIDAR and Ultrasonic Sensors**: Inspect structural conditions or detect misalignments in rail and road systems.

Data from these sensors is transmitted to centralized platforms or edge computing units, where it is analyzed in real time or stored for longer-term trend analysis.

3.5.3 Machine Learning Models for Failure Prediction

ML plays a pivotal role in transforming raw sensor data into actionable insights. Several ML models and techniques are commonly used for failure prediction:

- **Supervised Learning Models**: Algorithms like random forests, support vector machines (SVMs), and neural networks are trained on labeled data to classify equipment as "healthy" or "at risk."
- **Unsupervised Learning**: Clustering algorithms detect abnormal behavior without labeling failure data—useful when failure examples are scarce.
- **Time-series Forecasting**: Recurrent neural networks and LSTMs analyze trends over time to forecast when specific components are likely to fail.
- **Anomaly Detection**: Algorithms flag deviations from normal operating patterns, serving as early warnings.

For instance, in railway systems, ML can predict track defects days before a failure, allowing operators to intervene proactively [7].

3.5.4 Impact on Cost, Downtime, and Safety

Cost Efficiency

- Predictive maintenance significantly reduces **maintenance costs** by avoiding unnecessary servicing and minimizing emergency repairs.
- It allows for **better inventory management**, as parts are replaced only when needed.

Reduced Downtime

- By predicting failures before they happen, systems can be serviced during off-peak hours or in scheduled windows, minimizing disruptions.
- This is particularly beneficial for **public transit**, where unexpected delays can affect thousands of commuters.

Improved Safety

- Early detection of wear or defects in critical components (e.g., aircraft engines, bridge joints, brake systems) prevents catastrophic failures.
- Enhances **compliance** with safety regulations and builds public trust in the reliability of transportation systems.

Environmental Impact

- Reduces emissions by maintaining optimal performance of engines and reducing fuel-wasting breakdowns or idle times.
- Limits unnecessary resource use through more efficient maintenance cycles.

3.6 Ethical, Regulatory, and Security Considerations

As AI becomes increasingly embedded in transportation systems, ethical, regulatory, and security dimensions must be critically assessed to ensure responsible development and deployment. These considerations are essential not only to protect public interests but also to sustain long-term adoption and trust in AI technologies [8].

3.6.1 Data Privacy in AI Systems

AI-powered transportation systems rely on vast amounts of data—ranging from vehicle telemetry and sensor inputs to passenger behavior and location information. Ensuring the privacy of this data is paramount.

- **Challenges**:
  - Collection of sensitive information (e.g., GPS data, biometric authentication in smart vehicles).
  - Risks of unauthorized access and data misuse.
- **Solutions**:
  - Implementing robust anonymization and encryption techniques.
  - Enforcing strict access controls and audit trails.
  - Adoption of privacy-by-design principles during system development.

3.6.2 AI Bias and Fairness

Bias in AI models can lead to discriminatory outcomes in decision-making processes, particularly in adaptive learning systems and automated grading of performance.

- **Sources of Bias**:
  - Historical data that reflects existing inequalities.
  - Nonrepresentative training datasets.

- **Mitigation Strategies**:
  - Continuous auditing of AI models.
  - Incorporation of fairness-aware algorithms.
  - Stakeholder engagement to define fairness metrics specific to transportation contexts.

3.6.3 Regulatory Frameworks and Standards

The deployment of AI in transportation is governed by emerging but fragmented regulatory landscapes.

- **Current Landscape**:
  - Global efforts like the EU's AI Act aim to establish comprehensive frameworks.
  - In transportation, regulatory bodies such as the U.S. Department of Transportation (DOT) and National Highway Traffic Safety Administration (NHTSA) play key roles.
- **Recommendations**:
  - Harmonization of international standards for AI in transport.
  - Development of sector-specific guidelines (e.g., for predictive maintenance in rail or autonomous trucks).
  - Mandatory transparency and documentation practices, such as model cards and audit logs.

3.6.4 Cybersecurity in AI-driven Infrastructure

AI integration introduces new vulnerabilities, particularly when systems are connected through IoT networks and cloud platforms [9].

- **Risks**:
  - Hacking of AVs.
  - Manipulation of predictive maintenance data leading to false diagnostics.
- **Defense Measures**:
  - Deployment of AI-based anomaly detection systems.
  - Real-time threat intelligence and response protocols.
  - Regular penetration testing and cybersecurity audits.

3.7 Future Outlook and Emerging Technologies

The future of AI in transportation is rapidly evolving, driven by breakthroughs in generative models, digital infrastructure, and autonomous systems. These innovations promise to revolutionize how transportation networks operate—enhancing efficiency, safety, and user experience—while also introducing new challenges and research opportunities.

3.7.1 Role of Generative AI and Digital Twins

Generative AI and digital twins are reshaping the design, simulation, and optimization of transportation systems.

- **Generative AI**:
  - Used to simulate traffic conditions, model driver behaviors, and generate synthetic data for model training.
  - Enhances system adaptability by generating real-time route optimization or maintenance predictions.
- **Digital Twins**:
  - Real-time, virtual replicas of physical transportation systems (vehicles, infrastructure, networks).
  - Allow scenario testing (e.g., simulating bridge wear over decades or stress-testing new AI models for AVs).

3.7.2 Integration with Smart Cities

Smart cities provide the ideal ecosystem for AI-enhanced transportation systems to thrive.

- **Synergies**:
  - Integration with IoT-enabled infrastructure (e.g., smart traffic lights, pedestrian sensors).
  - Seamless intermodal travel experiences (e.g., AI coordinating between buses, trains, and ride-shares).
- **Benefits**:
  - Reduced congestion, lower emissions, and improved accessibility.
  - Real-time adaptive control of urban mobility systems.

3.7.3 Autonomous Vehicles and AI Coordination

As AVs proliferate, AI coordination becomes vital for ensuring safe and efficient operation across complex environments [10].

- **Advances in AVs**:
  - Enhanced perception, decision-making, and control systems using deep learning and reinforcement learning.

  - Swarm intelligence techniques for coordinating fleets of autonomous delivery drones or vehicles.
- **Coordination with Infrastructure**:
  - Vehicle-to-everything (V2X) communication enables AVs to interact with traffic systems and other road users.
  - Central AI platforms can orchestrate traffic flow, reroute AVs in real-time, and prevent congestion.

3.8 Conclusion

AI is reshaping the transportation landscape, offering tools and techniques that significantly enhance efficiency, safety, and sustainability. From automated grading systems to predictive maintenance and adaptive learning, the integration of AI is helping create smarter, more resilient transportation networks.

3.8.1 Summary of Key Points

- **Automated Grading** streamlines the assessment of transportation infrastructure, enabling rapid and consistent evaluation of roads, bridges, and systems.
- **Adaptive Learning** empowers transportation systems to evolve in real-time, adjusting to dynamic conditions and user behaviors for optimal performance.
- **Predictive Maintenance** reduces downtime and costs by identifying potential failures before they occur, leading to more reliable operations.
- Ethical and security challenges—such as **data privacy, AI bias, and cybersecurity**—must be addressed with robust safeguards and transparent policies.
- The future is being shaped by **generative AI, digital twins, AV coordination**, and integration with **smart cities**, which together point to a more intelligent and interconnected transportation ecosystem.

3.8.2 Implications for Policymakers and Practitioners

- **For Policymakers**:
  - There is a pressing need to establish clear **regulatory frameworks** that govern the ethical use and accountability of AI systems in transport.
  - Governments must also invest in **digital infrastructure** and promote **interoperability standards** across regions and platforms.

- **For Practitioners**:
  - Transportation planners and engineers should adopt **AI tools** not only for operational gains but also to advance sustainability and equity goals.
  - Emphasis must be placed on **interdisciplinary collaboration**, combining data science, engineering, urban planning, and social sciences to build robust AI solutions.

3.8.3 Final Thoughts on the Future of AI in Transportation

AI's role in transportation is not just about efficiency—it's about **transforming how societies move, connect, and live**. While the technology offers tremendous potential, its successful integration depends on balancing innovation with **responsibility, inclusiveness, and foresight**.

As AI continues to evolve, so must our frameworks for deploying it—ensuring that the benefits are **shared equitably** and that our systems remain **secure, fair, and future-ready**.

References

1 Vasudevan, M., Townsend, H., Dang, T. N., O'Hara, A., Burnier, C., and Ozbay, K. Identifying real-world transportation applications using artificial intelligence (AI): Summary of potential application of AI in transportation, 2020.

2 Narayanan, L., Subbiah, P., Muralidharan, R. A., Venkataraman, A. P., and Sandhiya, M. Future professions in agriculture, medicine, education, fitness, R&D, transport, and communication. In Tyagi, A. K., editor, *Digital Twin and Blockchain for Smart Cities*, pages 529–546. Wiley, 2024.

3 Saeed, M. M. A., Saeed, R. A., Ahmed, Z. E., Gaid, A. G. S. A., and Mokhtar, R. A. AI Technologies in Engineering Education. In *AI-Enhanced Teaching Methods*, pages 61–87. IGIGlobal, 2024.

4 Ran, Y., Zhou, X., Lin, P., Wen, Y., and Deng, R. A survey of predictive maintenance: Systems, purposes and approaches. *arXiv preprint arXiv:1912.07383*, 2019.

5 Yu, H., Wang, J., Murugesan, M., and Rahman, A. B. M. S. Artificial intelligence-basedqualitymanagementanddetectionsystemforpersonalizedlearning. *Journal of Interconnection Networks*, 22(Supp02): 2143004, 2022.

6 Saranya, E., Shreenidhi, K. S., Anandaram, H., Upadhye, S., and Gukendran, R. Transforming hospitality, personalized medicine, and adaptive learning: Robotic innovations. In Shaik, A., editor, *AI-Powered Advances in Pharmacology*, pages 373–406. IGI Global, 2025.

7 Chen, X., Ryan, T., and Wang, H. Exploring AI in education: Personalized learning, automated grading, and classroom management. *MZ Computing Journal*, 3(1): 1–12, 2022.

8 Chen, L., Li, Y., Huang, C., Li, B., Xing, Y., Tian, D., Li, L., Hu, Z., Na, X., Li, Z., Teng, S., Lv, C., Wang, J., Cao, D., Zheng, N., and Wang, F.-Y. Milestones in autonomous driving and intelligent vehicles: survey of surveys. *IEEE Transactions on Intelligent Vehicles*, 8(2): 1046–1056, 2023.

9 Rayhan, A. The future of work: how AI and automation will transform industries. *Technical Report*, 2023. https://doi.org/10.13140/RG.2.2.36092.51848.

10 Okem, E. S., Iluyomade, T. D., and Akande, D. O. Revolutionizing US pavement infrastructure: A pathway to sustainability and resilience through nanotechnology and AI Innovations. *World Journal of Advanced Engineering Technology and Sciences*, 11(1): 411–428, 2024.

4

Autonomous Vehicles and Smart Mobility

P. Sudheer[1], S. Ashmad[2,], M. Saravanan[1], and A. Immanuel[2]*

[1] *Sreenivasa Institute of Technology and Management Studies, Chittoor, Andhra Pradesh, India*
[2] *Audisankara Group of Institutions, Nellore, Andhra Pradesh, India*

*Corresponding author: ashmad.syed@gmail.com

4.1 Introduction

Smart cities are a new idea of urban development where technology is a calculated decision aimed at augmenting the quality of life, squeezing the infrastructure efficiency, and sustainably developing environmentally friendly cities. These cities have used information and communication technologies (ICT) in order to guarantee that there is effective management of the resources, resolution of urban issues, and also that there is a provision of responsive, inclusive, and future oriented services to the citizens. In sum, it is expected to come up with the development of smart and effective urban environments that are also sustainable, sensible, and viable to all city residents [1].

The primary structure of a smart city is six-pillars system of smart economy, smart mobility, smart environment, smart people, smart living, and smart governance [1]. Each of these pillars applies to some feature of city development, and all of them offer an ecosystem. The key characteristics of a smart economy are innovation, entrepreneurship, and flexibility. It enhances styles of productivity through the means of having a new age technological development and also economic development through permitting variation and improved global interaction. A cognitive setting leads to green orthodoxy and hence leads to green initiatives such as energy conservation, preventing pollution, and renewable resource management to save the urban ecology [1].

The smart people element observes that human resources are important to smart cities. These citizens are meant to be educated, open, and culturally diverse, while also being active citizens. They are required to provide dynamism and flexibility to the city [1], therefore, their innovation and role in decision making. At the same time, smart governance theorizes open, user-input and online supported systems of administration which can leave a stronger impression on the service provision, the organization of the cities, as well as the responsibility [1]. Smart mobility is among the items [1]. reduces the most significant factors and discusses the establishment of intelligent transportation network on a sustainable level in order to enhance the reality of mobility in the city. It applies the ICT to improve the network of transport between the people and the public to provide some emphasis on relief on congestion and emissions. The development of real-time data on the traffic flows should be possible thanks to technologies such as GPS, WI-FI, mobile networks (3G/4G), and the internet of things (IoT), etc. that allow collecting data and controlling and tracking it [2].

Intelligent transportation systems (also known as ITS) would be vital to smart cities because they would offer the services of ICT in a bid to offer better mobility, safety, and urban sustainability. These are systems that are able to get real time data, which contains GPS, Wi-Fi, 3G/4G networks, and IoT devices in the process of monitoring the traffic conditions and making judgmental decisions [2]. Almost all applications vital to the city at large, like the optimization of traffic, implementation of emergency services, and management of resources, are captured in ITS, virtually eliminating the congestions, consumption of fuel, exhaust gases, and resulting economical losses [3]. Despite the benefits, the primary research problem is related to dealing with large quantity and variety of real-time data. As one of the most developed technologies defining smart cities, ITS is becoming popular in being applied in collaboration with autonomous vehicles, thus ensuring fully automated paths in the transport system [2, 3]. The other features of ITS, which help a commuter in decision-making, are the real-time warning and real-time traffic forecasts, which reduce the level of time wasted and emissions harmful to the atmosphere [4].

Intelligent cities hugely depend on the concept of synergy. The combination of the six pillars creates a value addition in a better way than by combining the parts. An intelligent system of governance, targeting of which is one of the examples of positive effects of smart mobility, is capable of multiplying positive effects. the organization of policies, and incorporation of photographs into the infrastructure [5]. In similar manner, smart people are able to make smart living projects more valuable by embracing technology in healthcare, education, and law enforcement.

Smart mobility in specific is regarded to be a significant attribute of the development of smart city. It links all city activities and also helps each resident, worker, and visitor to have convenient and permanent access to resources [6].

Due to the pressures associated with urban living, including traffic, the growth in the use of privately owned vehicles, and the number of accidents, smart mobility is an important tool in ensuring that urban areas remain livable and usable. The smart pillar of living involves an integration of technology in the daily lives to improve on health services, housing, educational amenities, cultural practices, and recreational opportunities. It concerned the issue of social cohesion and providing high-quality urban-services in an unbiased way [7, 8].

Autonomous vehicles (AVs) and smart vehicular networks have dawned in ITS systems by virtue of increasing technological advances. AVs with the technologies of artificial intelligence and sensor technology may be considered one of the realities of the future since transport is not only automatic but also safer, more acceptable, and eco-friendlier [9, 10]. The integration of AVs with ITS systems promotes a new paradigm of the mode of urban transport, the foundation of an environment of uninterrupted and intelligent mobility systems [11].

AVs may be categorized as an autonomous system, equipped with technology (that drives the vehicles) or rather a system that has no human driver, which is one of the most substantial updates of ITS [12]. The said ones operate within the framework of IoV, which enables them to intercommunicate with the infrastructure, the other vehicles along with the pedestrians through a variety of communication means, including V2V, V2I, and V2P [13]. SAE recognizes six levels of automation with the levels 3 being autonomous, which are capable of driving in some conditions or all of them [14]. Connected autonomous vehicles (CAVs) are an extension of AVs with the ability to communicate with its environment, improving real-time decision-making and traffic coordination [15].

ITS is used to assist AVs manage traffic, give routing commands, and respond in emergencies, receiving the necessary information in real-time both on vehicle and infrastructure levels [16]. Intelligent traffic systems are able to adjust timings of traffic signals depending on the traffic and make the signals more efficient and safe [17]. Even though AVs are developing and are still maturing, it should be noted that they must first co-exist with the current fleet of vehicles, and this necessitates a phased configuration and road infrastructure improvements. Taken together, AVs and ITS are the future pillars of smart and sustainable urban transport.

Smart roadside infrastructure (SRI) is classified as a network of smart systems along roads to assist in driverless cars that enhance the vehicle awareness, security, and chokehold the traffic. It involves roadside units (RSUs), V2X, sensors (e.g., cameras, LiDAR, and radar), and edge computing to power real time data processing. SRIs help autonomous vehicles by offering a long perception of the environment and cooperative driving, especially in complicated traffic situation. They provide dynamic traffic control and effective routing as well. Some issues affecting SRIs, which include interoperability and privacy, as well as costs and exorbitant costs, outweigh its benefits, such as end-to-end safety, improved traffic, and better traffic flow. However, they play a vital role in promoting automatization of vehicles and more intelligent transport systems.

The Soar communications framework is targeted toward high-fidelity data transfers between roadside infrastructure (I2I) and vehicles (I2V) which is essential to real time autonomous driving applications. Within the I2I deploy type, Soar applies a linear multi-hop network that consists of two 802.11ac-form interfaces in AP and STA modes, 100 Mbps was obtainable over 10 hops to 500 m. In this arrangement, deployment can scale and cost-effectively be achieved along roads, where the shared sensor data can be processed on a distributed basis. Soar combines batched sparse (BATS) codes, an effective network coding scheme that quickly enhances a wireless link to one that has greater throughput and has minimal computation overhead to support reliable wireless links that suffer significant losses.

In the case of I2V communication, Soar uses unidirectional broadcast with injector-sniffer model in the sense that roadside nodes broadcast information (such as LiDAR or detection results) without connection establishment, and vehicles passively listen to the information. The approach is scalable, and it enables real-time delivery to a large number of vehicles. In order to deal with packet loss, Soar uses error correction on the application level and adjusts the best modulation rates according to the environmental requirements but not car speed. Cars channel hop dynamically with reference to a control channel, thus avoiding the reassociation overhead. Lastly, malicious protection is achieved through protection against malicious data injection. Soar uses private key encryption and digital signatures to ensure secure, verifiable broadcasts [18].

Reinforcement learning via human feedback (RLHF) is an extension of traditional reinforcement learning, in which standard reinforcement learning is applied onto training a reward model using human preferences rather than designed reward functions. Direct frame-by-frame feedback are hard to obtain in self-driving. As an example, a car may not collide and reward itself but may engage in sudden movements that are uncomfortable to passengers. The response of the person in situations such as sudden braking or sudden laning is one of the critical signs of driving quality. RLHF takes into consideration these minor preferences, like comfort and safety, and applies them in training. Within the context, the sensor information (i.e., LiDAR, cameras, and physiological controls) drives choice in actions, and the reward structure indicates human-compatible feedback. The new policy is kept close to the initial supervised model using a KL divergence term, hence the stability of learning.

Large language models (LLMs) also aid this system in three major ways: as simulation agents, by augmenting human-simulation interaction, and by enhancing the RLHF loop. Being agents, LLMs have the capacity to portray human or environmental behavior, and they are even more complex in the case of a real absence of users. They also adapt to the levels of skill of the users and change the conditions of simulation. Most importantly, the physiological reactions (such as high rates of the heart) can be translated by the LLMs into decipherable preference messages. The insights serve to further the RL model in order to make decisions that consider the need for human comfort and usefulness in the real world, and this enhances the safety performance of the autonomous driving systems [19].

Energy-aware DMARL cooperative driving system of CAEVs gives a framework to energy-aware cooperative driving of connected and automated electric vehicles (CAEVs) utilizing decentralized multi-agent reinforcement learning (DMARL). The CAEVs (green) and HDVs (blue) are present within the traffic environment on a four-lane and on-ramp highway. Merging vehicles traveling on the ramp should be able to merge with other traffic in order to get onto the highway. Cars on the road need to cooperate in order to improve the efficiency of the traffic flow and to make it safe and energy-efficient.

The HDVs state is not fully observable and therefore, control of CAEVs in the mixed traffic scenario is more complicated than in full automation scenarios. In order to fight this problem, the suggested solution considers a model of the CAEV platoon management task as a cooperative multi-agent task where all the agents participate in one global goal of maximizing the overall mean reward.

Under the IoV infrastructure, every CAEV monitors the local traffic and also receives the state and learnt policies of other CAEVs. Multi-agent reinforcement learning (MARL) incorporates huge joint state and action areas, which are managed by using decentralized interaction-learning approach where the distinct CAsEVs act according to their local perceptions along with exchanging key information about the whole platoon.

There is a reward that is provided to the individual agents in a bid to attain a more efficient travel, the safe-driving enhancement, and the delivery of the State of Charge (SoC) at optimum levels. The information collected during the interactions, which were named DCMA-MDP transitions, is recorded to be played back as an experience.

A cooperative multi-agent actor-critic (CMAAC) structure is applied in the learning algorithm as it integrates the decentralized policy training and the centralized action-value functions. Some of the issues the design is responsive to our credit allocation and non-stationarity on environment. The joint advantageous policy and value network is also applied in the CMAAC framework to optimize the cooperation driving plan based on experience data stored in the memory [20].

System Architecture Digital Twin platform integrates cloud computing and edge computing to arrive at a cyber-physical model of traffic in the real time. It is divided into physical space (streets, constructions, cars, and people in the real world) and the cyberspace (static maps and data that is placed in the cloud). Sensor data were collected in LiDAR and camera, which were performed on the edge of the vehicle and the roadside. The utilization of units (RSUs) assists in completing activities like object identification, object location, and motion management. This minimizes latency since it offloads the computations that have been executed in the cloud. The cloud that is offloaded deals with computation-intensive activities like planning routes worldwide and traffic surveillance. It also aligns data from several sources to achieve collaborative perception and to reach high-level navigation decisions. In a bid to ensure stability in the connectivity, the platform employs a nonhomogeneous V2X communication network, and this incorporates the ability to switch between the cellular, DSRC, and the mm Wave technologies based on availability. This will accord continuous data transfer and adequate tracking even in the far-flung locations or facility-less territories; therefore, the system will be fully receptive and adjustable to self-driving [21].

4.2 Challenges in Smart Mobility and Autonomous Vehicles

Although autonomous vehicles (AVs) integration and the smart mobility system are promising ways for future urban transportation, the integration induces a range of challenges that should be adequately addressed. These issues range from the technical, infrastructural, legal, ethical, and socioeconomic aspects. To guarantee safe, efficient, and inclusive implementation of AV-based intelligent transport systems, a full cognizance of these barriers needs to be achieved.

Technically, AVs rely much on the use of sensors, including LiDAR, radar, cameras, and ultrasonic as well as detectors to sense out what is beyond them. Those sensors, however, perform poorly during adverse weather such as fog, snow, or heavy rain. An example is LiDAR, which interprets reflective surfaces like wet roads, thus wrongly detecting the objects. Real-time responsiveness is a major challenge in sensor fusion, combining data collected by more than one sensor to make better decisions, since the process involves a lot of processing power and low-latency systems. In addition, the artificial intelligence algorithms used to facilitate the autonomous decision-making process tend to perform poorly when encountered with new environments, edge cases, and unfamiliar objects that are not included in their training data [22]. Also, recent addition of unobstructed V2X (vehicle-to-everything) communications between AVs and the infrastructure or other vehicles presents a risk of cybersecurity. AVs can be susceptible to spamming of data, GPS jamming, or denial-of-service (DoS) attacks that may jeopardize the security of customers and the traffic system in general. These intelligent vehicles are vulnerable to malicious reactions unless they are effectively encrypted, shielded using authentication methods, and have intrusion detection measures.

The other barrier is infrastructural preparedness. There are many areas that are yet to be equipped with smart infrastructure that is needed to enable AVs: RSUs, high-bandwidth 5G network, and edge computing nodes. The AVs rely on high-definition (HD) maps, which act as the backbone of localization and path-planning, but it is costly to keep them up to date with the changes on the road (construction, traffic diversions, temporary closures), in real-time. AVs were not taken into consideration when opting to construct the kind of existing road infrastructure to support human drivers. When sharing roads with AVs, human-piloted vehicles, bicycles, and even pedestrians, autonomous systems are unable to overcome murky human behaviors. Moreover, AVs of the current era are mainly of the electric vehicle (EV) type, and the lack of a powerful fast-charging network restricts their range and field effectiveness [23]. AVs interoperability with legacy traffic and legacy traffic signals as well as traditional and conventional road signs is also a challenge, and this demands a phased and well-coordinated strategy of transition.

Legal and ethical aspects: Autonomous cars are bringing in new complexities on the legal and ethical front. Probably the most scandalous one is the aspect of moral decision-making in critical situations, which is commonly addressed using so-called trolley problem. In another case, in case of a crash that cannot be avoided, will an AV protect its internal passenger or a road pedestrian? Today, no international ethical model exists to help resolve such dilemma. Accountability in case of an accident related to an AV is also unclear—is it the responsibility of the car production company, the software developer, the government, as far as the implementation of the infrastructure, or the car owner? The majority of legal systems cannot currently cope with software-based negligence or algorithmic mistakes. In addition, AVs constantly acquire and multicast data such as video streams, GPS locations, and user actions, and these features increase privacy and data security concerns. In other jurisdictions like EU, there are strict laws governing the use, sharing or storage of personal data, such as GDPR. Providing compliance with respect to the well-functioning of the AV systems is another burden.

Human factors and social acceptance are also very significant issues. The development of the autonomous vehicle technology may challenge the old job market and especially the profession that involves driving a car, e.g., taxi services, trucking, and logistics, which may leave a million workers fall from the job. hurdle. Public distrust has been enhanced by high profile accidents involving AVs, including the Uber AV crash in Arizona, or the problems with the Autopilot system of Tesla. Individuals are yet to be willing to give way to machines when it comes to uncontrollable or complicated situations on the road. More questions entail access and fairness; there is a chance that AVs will be implemented only in wealthy, city localities with strong digital networks, and other communities will be underserved, including economically-deprived or rural territories.

Unless carefully planned, smart mobility may contribute to widening the digital gap instead of bringing it down. In addition, the expenses of AV system deployment and maintenance are still too high, with high-quality sensor systems and AI training and compliance being covered by too high of a price.

To sum up, although the concept of smart mobility and autonomous vehicles is revolutionary in terms of its possible transformations in the sphere of traffic movement, to achieve its full potential, it is essential to overcome several challenges. These are the consistent perception and real time processing in a varying environment, enhancement of the current infrastructure to make it V2x and high definition mapping ready, building strong cybersecurity networks, framing legally and ethically viable governance systems, and dealing with human and societal effects of automation. Every aspect of addressing these challenges in a global approach with the combination of governments, industry, academia, and civil society will be essential toward a future that is not only autonomous and intelligent, but also equitable, secure, and human-based.

4.3 Case Studies and Practical Applications of Smart Mobility and Self-driving Cars

The future of autonomous vehicles and intelligent mobility has long since transitioned into something that can be discussed in practice, rather than in terms of hypothetics or simulations; the scale of planned and implemented pilot projects nowadays provides tangible examples of what can, and cannot, be approached by self-driving technology, as well as how these technologies might be used. These case studies can be considered good guides in how to effect ITS, AVs, and connected mobility solutions into various urban systems. A world leader in smart city technology. In collaboration with the land transport authority (LTA), Singapore has created a comprehensive transportation vision under the Smart Mobility 2030 program that will incorporate on-demand mobility, autonomous shuttles, and intensive applications of real-time data analytics. The tests have also been autonomous in other regions, such as One-North and Sentosa, where the vehicles are driverless and are used in conveying and taking people around in a strictly measured setting. Such tests demonstrated that AVs operate predictably in geofenced and structured areas, but it is not so straightforward to expand to more difficult but no less important real-world urban streets, where challenges like dynamic obstacle avoidance and the ability to adapt to the unexpected behavior of pedestrians are raised. Singapore has been successful because of its top-down planning by the government, extensive investment in infrastructure, and public–private partnership between research institutions and major AV companies [24].

The Waymo company (Alphabet Inc.) is an example of a benchmark of the commercialization of AV in the United States. With a fleet of autonomous taxis in Phoenix, Arizona, Waymo has guaranteed millions of autonomous vehicle miles and is still honing its consistent autonomous ride-hailing plan. They have presented a solution that combines LiDAR sensors, radar, and high-definition mapping to enable perception of the environment, and in the background is continuous software upgrades. The implementation of Waymo also confirms the significance of developing AV systems that would be flexible in response to both regulatory standards and immediate data flows. Nevertheless, the resistance in the public eye is still present, and the safety, privacy, and liability remain the subject of concern, even despite the technical success. It is worth noting that Waymo works in the environment with favorable weather conditions and clearly marked roads, which restrict its applicability to other areas that have a harsher climate or a greater reliance on outdated infrastructure.

Horizon 2020 and urban mobility framework are some of the programs developed by the European Union that have aided in funding various AV research and implementation in pilot projects across the member countries. Other projects like AUTOPILOT (automated driving progressed by IoT) and L3Pilot (Level 3 pilot) have been conducted to test automated driving functions in highways and urban roads in various cities of Europe. Besides, the digital infrastructure preparedness, like 5G connectivity and vehicle-to-infrastructure (V2I) applications, are key element in the efforts that are pursued by Europe. One thing that is learned in such programs is that the legal and ethical framework should keep pace with advances in technology, especially when it comes to consideration data sharing, data transparency, and ethics of driving behavior under the European laws of data protection, such as GDPR [25].

Similar to the Apollo project established by companies such as Baidu in China, autonomous vehicles have been harshly tested. Baidu has already provided AV services in such cities as Beijing and Wuhan, such as autonomous taxis (robotaxis) with limited human control. The Chinese government also contributed to ensuring that the AV-friendly policy becomes a reality, including intelligent road construction and AV trial permits, which also impel mass adoption. The fact that China incorporates artificial intelligence, edge computing, and the use of urban digital twin in transportation planning is one of the differences in the China model. The centralized system of governance, however, creates concerns over the consent of the users and long term data governance.

India is one of those countries at an early phase of AV adoption, yet it has started to investigate smart mobility with projects such as smart cities mission and the National Automotive Testing and R&D Infrastructure Project (NATRiP). The cities such as Pune, Surat, and Bangalore have pilot programs on smart traffic lights, electric buses, and an intelligent transport management system. Although the fully autonomous vehicle deployment is not feasible yet because of the infrastructure limitations and sophisticated road dynamics, India has started focusing on the digital transformation, raising the efficiency rates of the current public transportation system and shaping mobility decisions on the basis of the data. These initiatives demonstrate that smart mobility does not have to be all about AVs, but it is rather a wide range of potential innovations, from as simple as GPS-based tracking of a fleet to all-in-one mobility-as-a-service (MaaS) solutions.

Besides national/regional implementations, there could be local testing, as in the example of the U.S. conducted by Uber/AVs and Navya-based in France. These are autonomous last-mile shuttles in gated communities, around industrial parks, pubs and campus grounds, and throughout airports. These launches indicate that AVs perform well when the operating environment is very structured: the route is fixed, the safety of pedestrians can be predicted, and infrastructure can be preprogrammed to operate with smart sensors.

These case studies, on the whole, revealed that the success of AV and smart mobility deployment relies not only on the technological maturity but also on the regulatory clarity, infrastructure accommodativeness, popular acceptance, and coordinated efforts of the stakeholders. As the adoption of AV has experienced a great leap forward in high-income countries, the developing ones are struggling with affordability, the legacies of road systems, and the heterogeneity of traffic conditions. However, every practical lesson in this world can be taken as a contribution to the theory and implemented or adjusted to regional priorities and socioeconomic realities. A combination of the three currents of the public policy, the private innovation, and community engagement will eventually determine the future changes of urban life all over the world due to smart mobility [26].

4.4 Policies, Ethics, and Governance in the Autonomous Vehicle Ecosystem

With the ADs (AVs) gradually moving out of experimental prototypes and into reality as a form of transportation, it is more important than ever that strong policy frameworks, moral decisions, and guiding principles should be developed. All these points are important in ensuring that AV technologies are not only safe and functional but also socially responsible, formally liable, and in tandem with the greater good of the people. In contrast to their traditional counterparts, AVs do not rely merely on advanced technology since they work through the use of intricate algorithms and data-driven decision-making activities, which raises the questions expanding beyond traditional engineering and into the contexts of law, ethics, civil policy, and human rights.

Among the most important challenges that governance has to face includes the ability to clarify regulatory structures on safe AV deployment. This fragmented approach to the regulation of AV is global. The United States is one of those countries; it replaced the federalist style that presupposes the voluntary nature of those guidelines which the National Highway Traffic Safety Administration (NHTSA) issues, as well as keeping the ability of the states to control the functioning of vehicles. On the contrary, Germany enacted the Autonomous Driving Act (2021) that allows Level 4 autonomous driving under certain conditions and does not allow autonomous driving without the presence of a technical supervisor. In the European Union, there is an enfolding coordinated approach toward vehicle safety standards that are coupled with cybersecurity measures and data governance regulations. Nevertheless, developing countries continue to encounter difficulties in regulatory regimes development of regulatory regimes because of the lack of infrastructure and the preparedness of institutions. One of the challenges in all jurisdictions is that there is no uniformity in liability laws, and it is not very clear who should bear the responsibility in the case of an accident related to AVs, whether it is the owner of the vehicle, the manufacturer, the person who developed the software, or the municipality in charge of vehicles.

Issues of ethics, like those in AV systems, are also urgent. Among the most controversial ones is the so-called trolley problem, which forces AVs to make one of two acts of harm (i.e., to save the passenger and cause harm to pedestrians or vice versa). Such situations require predetermined ethical judgment, and it is not agreed what moral should be in place regarding AVs. The problem is further complicated by cultural differences; say, surveys show that ethical preferences in Western societies and Eastern societies are very different, and therefore it is not easy to implement a universally ethical framework. In addition to that, ethical dilemmas are not confined to the cases of emergency. Value judgment also appears in our daily choices, such as whether a car will enter the traffic forcefully when it is its turn or whether it will make room for pedestrians. These micro-decisions affects the public satisfaction and credibility of autonomous systems. Therefore, it becomes increasingly clear that the consideration of ethics in machine learning must be integrated into such models with interdisciplinary cooperation between ethicists, engineers, psychologists, and legislators.

A further level of control entails the privacy and protection of information and surveillance. AVs have a constant stream of data being produced, streamed, and analyzed, e.g., images, their location, biometric input, and user behavior. This data is prone to misuse or exploitation, given that strict data protection laws are absent. In Europe, the general data protection regulation **(GDPR)** invokes very stringent regulations over data gathering and consent of the user. In our technology laws seeking a balance between innovation and individual privacy, in many countries, there is a push to protect individual privacy, such as in the proposed Digital

Personal Data Protection (DPDP) Bill in India, but enforcement is a problem. It is especially relevant since AVs perform their functions in the open territory and capture the passers-by and environment without their direct permission. In addition, cybersecurity risks of AV, including their control systems hacking, transmission jamming of V2X, or malicious code application, are a real threat to passengers and cargo security and the integrity of the traffic. These issues require secure software upgrading, encrypted processes, intrusion detection systems, and anomaly detection in real time, which is a part of the vehicle architecture [27].

As policy considerations, more focus should also be directed to long-term effects of AV adoption on the entire society, especially concerning job creation and equity as well as accessibility. The risk of some jobs being taken over by technology, especially in an industry such as trucking, taxi services, and logistics, and needs the proactive labor policies embracing reskilling initiatives. Reports on governance models have to answer the question of how the effects of automation have created economic disruptions that will be offset. Moreover, fair access to new services provided through AV should be provided, such that these latter technologies are restricted not just to well-off urban communities, but also to rural communities, older people, and disabled individuals. The policymakers should avoid the development of a mobility monopoly when AV fleets are under the control of only a few corporations, and digital exclusion and inequality are enhanced. Transparency and accountability in guiding AV pilot programs with publicly funded developments, requirements of open data, and inclusive design requirements are some ways smart mobility systems development can benefit the greater good.

Finally, international cooperation and standardization are also essential in the regulation of autonomous mobility. As car manufacturing chains and car standards, as well as car data streams, are cross-border, it is important to coordinate national policy with the international conventions. Safety standardization organizations such as the United Nations Economic Commission for Europe (UNECE) and International Organization for Standardization (ISO) are in the process of reaching consensus on an array of safety standards, testing paradigms, and definitions of automation levels. Countries such as the UK, Singapore, and South Korea are tempted to balance innovation and the safety of the masses through regulatory sandbox frameworks, where industry players are given the room to experiment with AVs in a controlled environment under the oversight of the government. This kind of global harmonization not only enables the diffusion of technology, but it also brings clarity to both manufacturers and users.

Finally, the regulation of autonomous vehicles and smart mobility systems requires a multilevel solution, which combines policy, ethics, data safety, and policymaking that is universal in nature. Although the technology is evolving at a very high speed, policy and law should also evolve at least at the same speed to manage the emerging risks and maintain responsible innovations. Governance in the future must be open, involving participation and flexible enough to involve partnership between governments, academia, industry, and civil society. It is only by means of such cooperative and foresighted governance that autonomous vehicles can be made a safe, trustworthy, and fair prospect of the new mobility world [28].

References

1 Kumar, T. M. and Dahiya, B. The key components of a smart city. In *Proceedings of the International Conference on Smart Cities*, 2017.

2 Menouar, H., Güvenç, I., Akkaya, K., Uluagac, A. S., Kadri, A., and Tuncer, A. UAV-enabled intelligent transportation systems for the smart city: Applications and challenges. *IEEE Communications Magazine*, 55(3): 22–28, 2017.

3 Meneguette, R. I., Delamaro, E., Boukerche, A., and Viana, A. C. Data summarization in the node by parameters (DSNP). In *Proceedings of the IEEE Symposium on Computers and Communications*, Natal, Brazil, pages 423–429, 2018.

4 Brčić, D., Marić, M., and Dragčević, S. The role of smart mobility in smart cities, In *Proceedings of the International Symposium on Novel and Intelligent Computing and Applications*, pages 112–121, 2018.

5 Vishnivetskaya, A. and Alexandrova, A. Smart cities and intelligent transportation systems: Integration of technologies. In *Proceedings of the International Conference on Urban Sustainability*, Moscow, Russia, pages 305–310, 2019.

6 Orlowski, K. and Romanowska, A. Smart cities concept – Smart mobility indicator, In *Proceeding of the Conference on Smart City Transitions*, Warsaw, Poland, pages 98–104, 2019.

7 Moura, F. and de Abreu e Silva, J. Smart cities: Definitions, evolution of the concept and examples of initiatives. In *Proceedings of the Urban Transportation and Planning Conference*, Lisbon, Portugal, pages 212–219, 2019.

8 Winkowska, J., Szpilko, A. and Pezarski, P. Smart city concept in the light of the literature review. *Engineering Management in Production and Services*, 11(2): 70–86, 2019.

9 Butler, W., Davies, S. and Zhao, P. Autonomous vehicle technology report. *IEEE Transactions on Intelligent Transportation Systems*, 21(11): 4567–4575, 2020.

10 Bıyık, C., Sadigh, B., and Sadigh, D. Smart mobility adoption: A review of the literature. In *Proceedings of the International Conference on Future Transportation Technology*, Istanbul, Turkey, pages 44–52, 2021.

11 Algherbal, E. A. and Ratrout, N. T. Smart cities and intelligent transportation systems: Integration of autonomous vehicles and their impact on congestion and safety. *Transportation Research Procedia*, 84: 504–511, 2025.

12 Tyagi, R. and Sreenath, N. Integrating autonomous vehicles with intelligent transportation systems: Opportunities and challenges. In *Proceedings of the International Conference on Emerging Smart Computing and Informatics*, pages 120–126, 2022.

13 Dimitrakopoulos, G., Demestichas, P. and Adamopoulou, E. Connected vehicles and smart infrastructure in future transport. *IEEE Vehicular Technology Magazine*, 15(2): 52–59, 2020.

14 Aldakkhelallah, A. and Simic, M. Design challenges of intelligent connected vehicle systems. In *Proceedings of the IEEE International Conference on Smart Vehicular Technology*, pages 63–70, 2021.

15 Faisal, A., Dixon, M., Hebbani, F., and McDonald, J. Connected and autonomous vehicles: A review of cooperative systems. *IEEE Access*, 7: 99860–99873, 2019.

16 Ahangar, R., Pahlavan, K., and Milstein, L. B. Security and privacy in autonomous vehicle communication: Challenges and solutions. In *Proceedings of the IEEE International Conference on Connected Vehicles and Smart Transportation*, pages 77–84, 2021, 77–84.

17 Zhang, Y., Liu, H., Ma, X. and Li, W. Deep learning for autonomous vehicle decision-making: A survey. In *Proceedings of the IEEE International Conference on Intelligent Transportation Systems*, pages 152–159, 2022.

18 Shi, S., Ling, N., Jiang, Z., Huang, X., He, Y., Zhao, X., Yang, B., Bian, C., Xia, J., Yan, Z., Yeung, R. W., and Xing, G. Soar: Design and deployment of a smart roadside infrastructure system for autonomous driving, In *Proceedings of the ACM International Conference on Mobile Computing and Networking (MobiCom)*, Washington D.C., USA, Pages 1–16, 2024.

19 Sun, Y., Salami Pargoo, N., Jin, P. J., and Ortiz, J. Optimizing autonomous driving for safety: A human-centric approach with LLM-enhanced RLHF. In *Proceedings of the Conference on [To Be Confirmed]*, Woodstock, NY, USA, pages 1–5, 2024.

20 He, X. and Lv, C. Toward intelligent connected e-mobility: Energy-aware cooperative driving with deep multiagent reinforcement learning. *IEEE Vehicular Technology Magazine*, 18(2): 67–75, 2023.

21 Wang, K., Li, Z., Yu, T., and Sakaguchi, K. Smart mobility digital twin for automated driving: Design and proof-of-concept. In *Proceedings of the IEEE International Conference on Connected and Automated Vehicle Systems (CAVS)*, Tokyo, Japan, pages 1–6, 2023.

22 Liu, S., Wang, D., Li, Y., and Li, L. Sensor fusion for autonomous vehicles: A review. *IEEE Transactions on Intelligent Vehicles*, 7(1): 17–35, 2022.

23 Shladover, R. Connected and automated vehicle systems: Introduction and overview. *Journal of Intelligent Transportation Systems*, 22(3): 190–200, 2018.

24 Zhang, Y., Liu, H., Ma, X., and Li, W. Deep learning for autonomous vehicle decision-making: A survey. *IEEE Transactions on Intelligent Transportation Systems*, 23(2): 894–912, 2022.

25 Tyagi, R. and Sreenath, N. Integrating autonomous vehicles with intelligent transportation systems: Opportunities and challenges. In *Proceedings of the International Conference on Emerging Smart Computing and Informatics*, pages 120–126, 2022.

26 Ghosh, A., Misra, J. and Singla, P. Cybersecurity challenges in autonomous vehicle technology. *Computer Science Review*, 39: 100344, 2021.

27 Kuutti, S., Bowden, R., Jin, Y., Barber, P., and Fallah, S. A survey of deep learning applications to autonomous vehicle control. *IEEE Transactions on Intelligent Transportation Systems*, 22(2): 712–733, 2021.

28 Schwarting, M., Alonso-Mora, J., and Rus, D. Planning and decision-making for autonomous vehicles. *Annual Review of Control, Robotics, and Autonomous Systems*, 1: 187–210, 2018.

5

Artificial Intelligence (AI) for Smart and Sustainable Urban Transportation

Ishika Gupta, Hriday Gupta, Siddharth Gupta, and Prerna Ajmani*

School of Information Technology, Vivekananda Institute of Professional Studies-Technical Campus, Delhi, India

*Corresponding author: ishikakumari377@gmail.com

5.1 Introduction

Rapid urbanization, climate change issues, and the growing demand for smart transportation solutions have accelerated the transformation of cities in the world toward sustainable smart cities. Three transformative technologies, namely artificial intelligence (AI), the internet of things (IoT), and 6G Networks, are at the heart of this change. These technologies are redefining urban life as they build connected, flexible, and people oriented system which actually unite transportation, public services, governance, healthcare, and environment monitor systems. Smart cities are based on AI and IoT to automatize infrastructure, enhance common services, and promote the quality of life of residents. Smart decision-making is assisted by AI in analyzing data. Contrarily, IoT guarantees real time information gathering of the interconnected systems such as energy grids, traffic lights, environmental sensor and city utilities. These systems are able to collaborate with each other immediately once the 6Gs fast, reliable, and AI-friendly capacities are implemented, offering responsive, strong, and efficient services. The major component of this transformation is the emergence of electric vehicles (EVs). EVs will not be transport only when coupled with AI and IoT systems. They are mobile nodes in city networks; capable of communicating with infrastructure, utilities and even other smart nodes in real time. This communication is also realized through 6G network, which supports functions such as vehicle to everything (V2X) communication, real time diagnostics, autonomous navigation, and decentralized energy management. In the chapter, the author discusses a collaboration between AI, IoT, and 6G in relation to smart cities and EVs. It examines supporting technologies, fundamental functions, advantages, and obstacles to the implementation of these systems. Moreover, it discusses the ways in which AI and IoT can enhance sustainability, efficiency, and safety in the management of smart cities and transportation, where 6G provides the basis in highspeed and low-latency connectivity.

5.2 Background

5.2.1 Introduction to 6G, Smart Cities, and EVs

High rate of urbanization, combined with the desire to be sustainable, has seen the emergence of smart cities. These cities apply high technologies to enhance the quality of life, resource efficiency, and environmental sustainability. Electric vehicles (EVs) are at the heart of this transformation because they are revolutionizing the way transportation is conducted. In order to achieve the full potential of smart cities and EVs; good communication infrastructure is required. Here lies 6G (sixth generation) wireless communication technology. It is anticipated to be beyond the effective capacities of 5G. 6G delivers ultra high-speed data transmission, reaching up to 1 Tbps, very low latency of less than 1 ms, and AI friendly design that will be an ideal basis of highly connected urban environment. Such skills are critical toward the successful management of smart cities and electric cars [1].

- In smart cities, 6G supports intelligent traffic systems, energy-efficient buildings, real-time environmental monitoring, and responsive public services.
- In EV ecosystems, it facilitates vehicle-to-everything (V2X) communication, real-time diagnostics, optimized route planning, and smart charging infrastructure.

Such a 6G/smart cities and EVs combination is a significant step to building high-tech and clean urban areas that are people-centric.

5.2.2 Evolution from 5G to 6G: Key Features and Capabilities

The transition between 5G and 6G is a significant advance in the technology of communication. This transformation is driven by increasing the demand for higher data rates, extremely low latency, and extensive connectivity. There were significant improvements made with 5G; nevertheless, 6G is looking to go a step higher. The following are some of the main features and capabilities that make 6G superior to 5G:

- Terahertz Communication (THz Band): 6G uses the terahertz spectrum (0.1–10 THz), unlike 5G, which uses sub 6 GHz and millimeter wave frequencies. This can support up to 1 Tbps (terabits per second) of data rates and extremely low latency times of less than a millisecond.
- AI-native Networking: 6G does not add artificial intelligence as an additional feature, but it is a fundamental component of the system. Network optimization, intelligent routing, and dynamic resource management are some of the areas where AI is applicable. The result of this method is a very flexible and self-organized network.
- Holographic Communication: 6G will support real-time holographic communications that will provide immersive experiences in the implementation of virtual and augmented reality.
- Integrated Sensing and Communication (ISAC): 6G networks differ from the past generations, as they integrate sensing and communication functionalities.
- Quantum Communication and Security: 6G is expected to incorporate quantum key distribution (QKD) to enhance data security and provide unbreakable encryption.
- Ultra-massive MIMO: Expanding on Massive MIMO technology, 6G will feature ultra-massive multiple-input and multiple-output systems to support extremely high-capacity data transmission.
- Zero-energy Communication: Through energy harvesting and efficient power management, 6G will enable low-power communication for IoT devices, significantly contributing to sustainability.

5.2.3 Concept of Sustainable Smart Cities, EVs, and Their Importance

5.2.3.1 Sustainable Smart Cities

A smart city is sustainable as it employs technology to improve the quality of life of its citizens and limits the effect it has on the environment. The key concept is to streamline urban infrastructure and services with the help of such technologies as AI, IoT, and 6G networks. Key aspects include:

- Efficient Resource Management: Smart cities aim to reduce energy and water consumption through real-time monitoring and automated control systems.
- Reduced Carbon Emissions: By adopting green technologies and renewable energy sources, sustainable smart cities minimize their carbon footprint.

- Enhanced Quality of Life: Citizens benefit from smart healthcare, efficient public transportation, and seamless connectivity.
- Economic Growth: Smart cities foster innovation and entrepreneurship by providing a digitally advanced ecosystem.
- Social Inclusivity: The aim is to have digital equity and accessibility to all groups.
- Resilience and Adaptation: Smart cities are constructed in a way to cope with environmental challenges. They employ the climate monitoring and disaster management systems.

The concept of smart cities sustainability is significant. The fast urbanization has led to increased pollution, drainage of resources, and inequality in society. The incorporation of sustainability into the smart cities planning will avert these concerns. It also ensures that environmental protection and technological advancement can coexist.

5.2.3.2 Electric Vehicles

EVs represent a huge transition in the transport sector. They no longer use traditional internal combustion engine (ICE) cars, but cleaner and more sustainable ways. EVs are powered by electric motors fed by rechargeable battery packs that, in most cases, are powered by electricity that is generated from renewable sources. The various categories of EVs are battery electric vehicles (BEVs), plug-in hybrid electric vehicles (PHEVs), and fuel cell electric vehicles (FCEVs) [2]. EV is necessary in developing sustainable and smart urban ecosystems as they help in the following:

- Environmental Sustainability: EVs help reduce urban air pollution and greenhouse gas emissions, aligning with climate goals and carbon neutrality initiatives.
- Urban Mobility Transformation: Integrated with AI and IoT, EVs support smart mobility systems that reduce traffic congestion and travel times.
- Energy Optimization: EVs interact with smart grids to manage demand and supply, enabling vehicle-to-grid (V2G) services that stabilize electricity networks.
- Economic Innovation: EV adoption stimulates the green economy through investments in clean energy infrastructure, manufacturing, and services.
- Public Health Improvement: Cleaner air and quieter roads from EVs directly enhance the well-being of urban populations.

As cities adopt 6G technologies, the role of EVs will increase. They offer real time connectivity, autonomous functions, and connection with digital cities.

5.2.4 Role of AI and IoT in Building Smart and Sustainable Urban Environments

IoT and AI constitute the key elements of the creation of sustainable smart cities. Their role includes:

- Data-driven Decision-making: AI-powered analytics enable urban planners to make informed decisions based on real-time data from IoT sensors.
- Autonomous Infrastructure Management: From smart lighting systems to automated waste management, AI algorithms control and optimize city functions, reducing operational costs and resource wastage.
- Predictive Maintenance: AI-based predictive maintenance systems use data from IoT devices to detect potential failures and schedule timely repairs, minimizing downtime.
- Smart Mobility Solutions: AI enhances traffic management, route optimization, and public transportation efficiency, reducing congestion and emissions.
- Citizen Engagement Platforms: AI-driven chatbots and virtual assistants improve interaction between residents and city administrations, ensuring transparent and efficient public service delivery.
- Energy Optimization: AI algorithms predict energy demand and manage supply through IoT-connected smart grids, ensuring optimal usage and reduced energy loss.
- Emergency Response Systems: AI-driven surveillance and monitoring systems detect unusual activities and potential threats, enhancing public safety and emergency response.

In the context of 6G networks, AI and IoT work synergistically to create a hyper-connected ecosystem where devices communicate seamlessly and respond intelligently to dynamic conditions. This intelligent connectivity is important to automate city operations and achieve sustainable development objectives.

5.2.5 Role of AI and IoT in Building Electric Vehicles

The convergence of AI and the IoT is creating a novel era of innovation in the development of EV. The technologies are significant in making the current EVs more intelligent, efficient, and user-friendly.

5.2.5.1 AI in Electric Vehicles

- Battery Management Systems (BMS): AI optimizes battery performance by predicting degradation, managing thermal conditions, and extending battery life.

- Autonomous Driving: Advanced AI models power driver-assist systems and autonomous navigation by processing data from sensors (LiDAR, cameras, radar) in real time.
- Predictive Maintenance: AI algorithms analyze historical and real-time data to detect faults before they occur, reducing downtime and maintenance costs.
- Personalized Driving Experience: AI enables adaptive cruise control, smart route planning, and user-specific settings for comfort and energy efficiency.
- Energy Optimization: AI models forecast energy consumption patterns and optimize charging schedules based on usage trends and grid demand.

5.2.5.2 IoT in Electric Vehicles

IoT links EV to a network comprising of devices, infrastructure, and cloud platforms. It is able to share real-time data and control smartly.

- Remote Diagnostics: IoT sensors monitor critical vehicle parameters like motor temperature, battery health, and tire pressure, alerting users or service centers automatically.
- Vehicle-to-everything (V2X) Communication: EVs use IoT to communicate with traffic lights, pedestrians, other vehicles, and charging stations, enhancing road safety and traffic efficiency.
- Smart Charging Infrastructure: IoT facilitates interaction between EVs and charging networks to manage peak loads, availability, and payment systems.
- Fleet Management: Connected EVs help businesses monitor and improve fleet operations in real time. This boosts logistics and lowers environmental impact.

5.2.5.3 Synergy of AI, IoT, and 6G

In the case of 6G networks, there will be even a more comfortable implementation of AI and IoT in EVs: Very low latency guarantees an increased reaction of vehicles in an emergency situation (e.g., collision avoidance). Massive data bandwidth: It enables continuous data streams for high-resolution mapping, infotainment, and predictive modeling. Edge computing and AI-native networks will enable real-time processing in the vehicle or in the infrastructure in its surroundings without using a cloud data center. To conclude, AI and IoT are at the center of the construction of smarter, safer, and more efficient EVs. Their connectivity not only improves the EV experience, but it also facilitates the larger aims of the connected smart cities.

5.3 Enabling Technologies

The 6G communication systems will use enabling technologies that will transform the wireless networks into efficient, secure, and extensive connectivity networks. Its major innovations are reconfigurable intelligent surfaces (RIS),

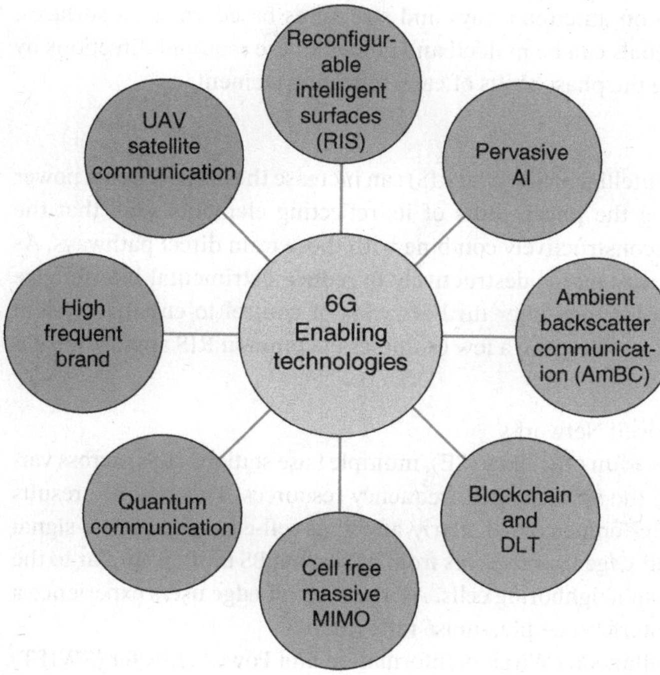

Figure 5.1 Enabling technologies.

which modulate the signal coverage and eliminate interference by regulating the phase differences between reflected signals, and cell-free Massive MIMO, which increases coverage and capacity by splitting the point of access in very large regions. Another technology, which utilizes ambient radio frequency signals to transmit data, is the AmBC, which utilizes less power sources since this technique does not require specialized sources. Quantum communication provides greater protection, distributed learning and decision-making can be achieved through AI pervasiveness, and satellite communication by UAVs will have greater coverage with reduced physical barriers. These technologies have potential, yet they require integration complexity, scalability, and energy efficiency that will have to be taken care of to roll out 6G effortlessly [3].

Figure 5.1 gives a pictorial guide to the enabling technologies.

A. Reconfigurable Intelligent Surfaces (RIS)
A reconfigurable intelligent surface (RIS) is a flat surface that is a composite of multiple passive reflecting elements. The necessary phase shift can be applied to the incoming signal individually by each element. RIS can be divided into two forms based on the materials utilized for the reflecting elements:

structures based on antenna arrays and structures based on meta surfaces. The reflected signals can be molded and guided in the required directions by precisely varying the phase shifts of each reflecting element.

RIS Applications:

A Reconfigurable Intelligent Surface (RIS) can increase the desired signal power by cleverly modifying the phase shifts of its reflecting elements such that the reflected signals can constructively combine with those from direct pathways. As an alternative, they may merge destructively to reduce detrimental interference from several users. RISs thus offer further layers of control to enhance system performance even more. Here are a few examples of common RIS applications in different systems.

i. RIS-aided Multicell Networks
To enhance spectrum efficiency (SE), multiple base stations (BSs) across various cells reuse the same limited frequency resources. This practice results in inter-cell interference, particularly affecting cell-edge users. The signal power that a cell-edge user receives from its serving BS is often similar to the interference from neighboring cells. As a result, cell-edge users experience a low signal-to-interference-plus-noise ratio (SINR).

ii. RIS-aided Simultaneous Wireless Information and Power Transfer (SWIPT) Networks
Simultaneous wireless information and power transfer (SWIPT) networks facilitated by RIS A BS with a steady power supply broadcast wireless signals to information receivers (IRs) and energy receivers (ERs) simultaneously in SWIPT, a promising method of delivering cost-effective electricity to energy-constrained IoT networks. The fact that the ERs and IRs in SWIPT systems function under different power supply needs presents a significant difficulty.

iii. RIS-aided Non-orthogonal Multiple Access (NOMA)
The future of multiple access strategies in wireless networks is bright thanks to NOMA. Compared to conventional orthogonal multiple access (OMA), this approach greatly improves the spectrum efficiency by allowing numerous concurrent users to share each orthogonal resource block. However, NOMA might not be the ideal option in some particular circumstances where the channel vectors of the users are orthogonal to one another.

B. Pervasive AI
Pervasive AI is referred to as the disbursal of artificial intelligence into all manner of devices and environments in order to make decisions within real-time and in context, which is at the edge. It attempts to make AI capabilities ubiquitous through seamless integration within daily objects and systems. This is achieved by the distribution of AI models within a network

of edge devices, where data is processed locally, hence relying less on the centralized cloud infrastructure. This decentralized method helps reduce latency, conserve bandwidth, and also enhance privacy by localizing sensitive data. Moreover, federated learning and other approaches where data is never sent to a central location and collaborative model updates are completed without sending raw data to each other are exploited by pervasive AI. The other extremely significant one is the multi-agent reinforcement learning, where many agents are used to solve real-time complex problems in interaction with one another. These decentralized AI agents base their decisions on their engagements with the surroundings; therefore, they give more dynamic and adaptive systems. The struggles are, however, to evenly distribute the computing power among devices, to control the network latency, as well as to make sure that the scarce resources are well utilized to keep the systems running and in a scalable manner. Pervasive AI can be used to make intelligent decisions in a broad spectrum of situations, such as smart cities and IoT ecosystems, and it is basically achieved by implementing AI as part of the physical world.

C. Ambient Backscatter Communication (AmBC)

AmBC system involves transmission of data using radio frequency signal like radio, television, and mobile telephony, without a power grid or direct power connection. It uses these signals, which are reflected off the ambient to alter them and encode data. Antennaed devices are able to pick up these signals and act upon them. In the AmBC system, backscatter devices communicate using broadcast signals from different RF sources, including cellular base stations, FM towers, and TV towers. The backscatter transmitter modulates the data onto the surrounding ambient signal and reflects it to the backscatter receiver. As a result, AmBC removes the need for a dedicated frequency spectrum for data transmission. This system has several advantages over other communication methods. First, by not needing a dedicated spectrum, AmBC improves spectrum use. Second, it does not require a dedicated RF source, making both the maintenance and deployment cost significantly less. Moreover, AmBC does not require the use of conventional power grids and communication networks. This is why it is a good choice for energy-saving, low-cost, and scalable communication within resource-constrained or remote regions.

D. Blockchain (BC) and DLT

A blockchain is made up of a number of fairly connected or chained blocks of transactions, as the name implies, in order to create a digital record. A "block" is created by combining verified transactions that occurred during a given time frame. Consequently, there are a certain number of transactions in each block. Cryptographic hashes are used to securely link these blocks in the order in which they were created. This indicates that each block's "previous block hash" field contains the hash value of the block before it. There are no antecedents to the first block, sometimes referred to as the "genesis block." As a result, the genesis block's prior block hash field is set to all zeros. The digital ledger is encrypted and contains all previous transactions in sequential order. Each node in a peer-to-peer blockchain network copies the ledger. Because write operations are limited to append mode, the distributed ledger can expand over time.

According to the hype, DLT has a lot of potential to help 6G networks and their uses grow. DLT uses the same protocol to allow for the safe and dispersed storage of a digital ledger at each participating node in the underlying network. Thus, an identical copy of the digital ledger is autonomously maintained by each node. BC is one of the most well-known forms of DLT and has drawn widespread interest. The distributed ledger is created and updated by BC according to a predetermined structure. A hash-based chainCell connects the timestamped, immutable blocks (which include a set of valid transactions) in this cell. To put it another way, in BC, transactions are verified, grouped together, and cryptographically sealed into a block.

Blockchain's immutable and decentralized nature makes it particularly effective in ensuring transparency, security, and accountability within the network. With the ongoing development of 6G networks, the incorporation of Blockchain and DLT will be the key factor in the creation of a safe, scalable, and efficient communication system.

E. Cell Free Massive MIMO

Future wireless networks may use cell-free massive multiple-input multiple-output (CFmMIMO) as a physical layer technology since it can take advantage of all the benefits of network MIMOs and distributed antenna systems (DASs), including link reliability, high channel capacity, and macro-diversity gain. By offering a substantial array of gains and spatial degrees of freedom, CFmMIMO systems aid in mitigating the intrinsic inter-cell interference (ICI) present in cellular networks. Several distributed access points (APs) collaborate with precoding and detection processing in these networks to provide services to multiple users while sharing time-frequency resources. All users, regardless of where they are in the network, can always receive efficient service from any AP thanks to the option of having one or more antennas.

F. Quantum Communication

The 6G technology is going to bring a significant transformation in the telecom sector. Although the 5G networks are still being rolled out throughout the world, scientists and technocrats are already considering the sixth generation of wireless technology, 6G. It is one of the most thought-provoking aspects of 6G as it may be connected to the communication with quantum, which employs the principles of quantum mechanics to ensure the security of data transmission. In this analysis, we will examine the possibility of how the merger of 6G and quantum communication will transform connectivity and enhance security in the coming decade. Quantum communication can be used to provide a very secure transfer of data, which prevents eavesdropping, with the assistance of quantum entanglement and quantum key distribution (QKD). The advantage of privacy in a more interconnected world is that 6G will have virtually flawless encryption and robust security due to the application of quantum concepts. In addition to being able to accelerate data transmission and enhance robust communication systems, quantum networks may have a considerable impact on other segments of the economy such as national security, medical care, and the finance sector. The possibility of self-fulfilling 6G and quantum communication has both culminated in making the paramount global connectivity ever more thrilling and secure.

G. High Frequent Brand

The conventional radar is unable to track objects over so long distances or when the line of sight is disrupted. It is due to the usefulness of high-frequency radar systems. HF over-the-horizon radar offers surveillance over a long distance. It is able to locate and monitor ships, airplanes as well as ocean weather patterns. This radar is able to view the objects much further since it bounces the radio waves off the ionosphere. Also, the HF OTH radar is essential in identifying ballistic missiles because it can identify any turbulence in the ionosphere that is due to the passage of a missile. It is also able to scan the conditions at the ocean by monitoring the Doppler changes in surface wave patterns which provide the speed and direction of the wind.

H. UAV Satellite Communication

The high altitude unmanned air vehicles, particularly new high altitude platform stations, have transformed wireless networks. These self-driving aircrafts are elevated above the normal airliners. Most weather conditions do not affect them, and this provides them with inimitable advantages in terms of covering extensive distances and getting around physical barriers. This introduction examines the development of high-altitude unmanned aerial vehicles and explains how their importance in contemporary communication networks has been on the increase. It discusses how the platforms have revolutionized telecommunications and enhanced the unmanned aerial system technology. This paper provides a strong focus on how critically HA-UAVs will be in the future of international communications networks.

5.4 Components

5.4.1 Smart City

Smart cities improve the quality of life of citizens, their sustainability, and their security through technological and data-driven solutions. Connected transportation infrastructure is an intelligent transportation system, buildings that are energy efficient, waste management, and air quality monitoring in real time are examples of connected infrastructure that make a smart city very essential. These components that are being guided by IoT, AI, and big data allow cities to make informed choices that improve the living conditions in urban areas, reduce the amount of energy used, and ensure the safety of people. By utilizing smart technologies in their lives, cities can enhance the quality of life of their citizens, improve the services they provide, and manage resources in a more efficient way [4].

Figure 5.2 represents the enhancement of the development of smart cities.

Figure 5.3 represents the elements through which we can accomplish the objectives of the smart city.

A. Smart Economy
   The task of a smart economy is to create new economic growth through the use of data-driven decision-making and the use of advanced technologies. It contributes to the growth of companies and startups by creating an environment where creativity, teamwork, and investment are appreciated.

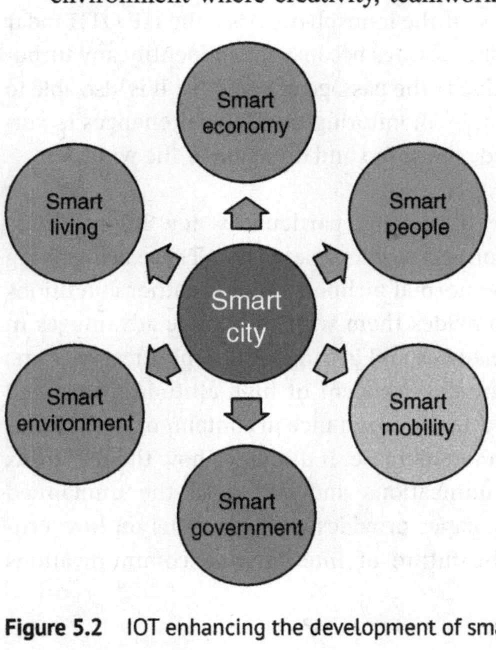

Figure 5.2 IOT enhancing the development of smart cities.

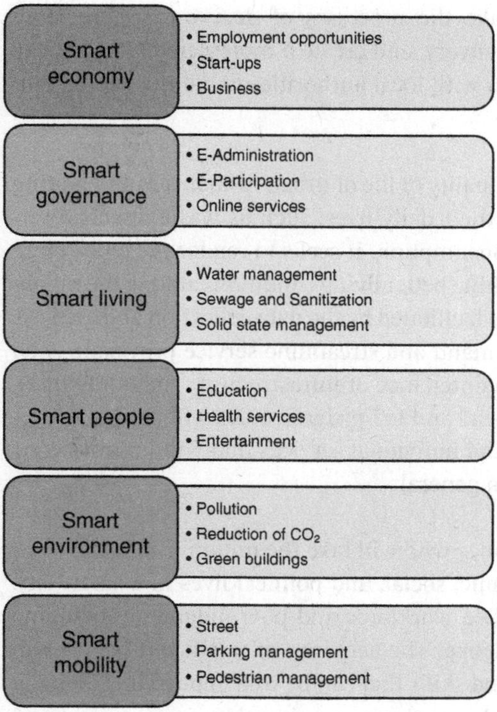

Figure 5.3 Components of smart city.

Nevertheless, competition between the cities in attracting businesses is one of the possible disadvantages, and it could create economic inequality instead of common prosperity. To achieve maximum benefits of the smart economy, the cities must strike a balance between long-term investment and short-term objectives to achieve sustainability and a fair allocation of development. The cities can execute their plans based on SWOT analysis that determines the strengths, weaknesses, opportunities, and threats to formulate their unique plans that can favor better economic development, interrelation with the suppliers, and adaptability of the region.

B. Smart Governance

There is a need, especially in smart governance, for the management of increased urban spaces as the population grows rapidly. An ICT-related standardization policy that encourages transparency of the political processes, coupled with citizen participation in the political processes, also helps to advance the goal because every citizen has direct access to his or her portion of services; inclusiveness and efficiency are therefore nurtured. There is less investment in governance infrastructure. Technology smart governance leveraging can be used to make the processes of decision-making more streamlined, enhance service delivery, and create a more democratic society with active citizen participation with local authorities to aid in shaping policies and regulations.

C. Smart Living

Smart living also promotes the quality of life of urban residents by integrating technology in most aspects of their daily lives, such as waste, medical services, sanitation, and energy consumption. It seeks to render the cities more sustainable in regard to establishing better living conditions among the inhabitants. Smart living solutions are facilitated by the data collection and analysis in order to anticipate future demand and streamline service provision, such as healthcare, transport, and maintenance of infrastructure. Including intelligent technologies such as sensors and IoT gadgets, cities will be able to provide more attentive, effective, and individual services that will improve civic security and living conditions in general.

D. Smart People

The smart citizens will be the ones who will take the initiative in using technology to enhance their economic, social, and political lives in a smart city. Nevertheless, the lack of a skilled workforce and poor judgment, including ethnicality can be a great challenge. The activities of intelligent people will involve imparting knowledge and skills that can be well applied in a digitally dominated society. Smart people can leverage education, innovations, and creativity to promote economic growth, social growth, and cultural enrichment. Through IoT and wireless technologies, smart services in healthcare, education, and social welfare can be provided by the cities. This will create a setting that will encourage life-long learning and self-development.

E. Smart Environment

Living and sustainable cities cannot be developed without a smart environment. It lays emphasis on environmental quality optimization by applying technology, such as cloud computing, data analytics, and IoT devices. A smart environment aims at minimizing pollution, safeguarding resources, encouraging the use of renewable energy, and promoting urban agriculture. Cities can also combat climate change by utilizing the available natural resources and developing green areas that are beneficial to people and animals. The use of technology will enable the monitoring of the environmental state in real-time and swiftly react to such problems as air pollution or water scarcity. This assists in having a more sustainable and healthy urban ecosystem.

F. Smart Mobility

Smart movement enhances the efficiency of transport, decreases traffic, and promotes the adoption of environmentally friendly traveling options. It is a major component of smart cities in sustainability. Smart mobility systems apply IoT, sensors, and data analytics to enhance real-time traffic, optimizing the routes of public transport as well as developing smarter parking. Parking management is intelligent and saves time hunting for parking spaces. Smart streets are fitted with sensor-connected devices with the role of monitoring and controlling traffic flow in real time. Besides providing real time information about the traffic movement and the timetable of the public transport, smart transportation systems also improve the overall mobility experience to the citizens and visitors of the city [5].

Figure 5.4 represents the elements of smart mobility, which are the integral elements for sustainable smart cities.

i. Smart Street
The foundations of smart mobility are smart streets. These streets are fitted with sensors, cameras, and wireless networks that spy on the traffic,

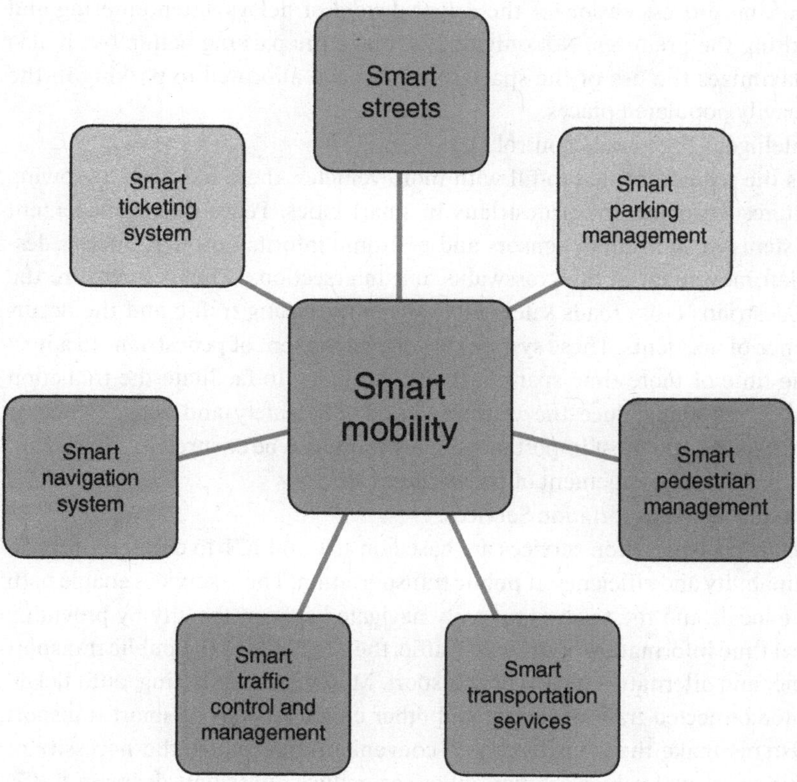

Figure 5.4 Elements of smart mobility.

road conditions, and the environment. They run traffic through real-time information optimization, better safety, and energy savings. Renewable energy is also integrated in smart streets such as solar panels to operate streetlights and other infrastructures, and also minimize the carbon footprints in the city. Moreover, smart streets also help to minimize energy consumption, security, and mobility through automatic changes in relation to the traffic lights, environmental conditions, and provision of important information to city planners [6].

ii. Smart Parking Management
Smart parking management is aimed at making the process of finding and making a ready parking spot easier in urban areas. It would also notify the driver of a parking area through the assistance of sensors and real-time information to minimize congestion and save time. Smart parking also implies automated billing, when the time spent in the parking lot is counted and charged, and cashless payment is offered to make it even more convenient. Lastly, automated entry and exit systems make the entire parking process easier for the driver devoid of delays when entering and exiting the premises. Not only does it make the parking better, but it also maximizes the use of the space that has been allocated to parking in the heavily populated places.

iii. Intelligent Pedestrian Control.
As the roads continue to fill with more vehicles, there has been a growing interest in managing pedestrians in smart cities. Pedestrian management systems include smart sensors and real-time information that track pedestrian movement at the crosswalks and intersections. This is to ensure the pedestrians cross roads safely and efficiently, easing traffic and the occurrence of accidents. These systems use the movement of pedestrians to adjust the time of more time spent at the traffic lights to facilitate the transition processes and reduce the waiting period. The safety and convenience of road users, and most importantly pedestrians, can be ensured in smart cities through the management of pedestrians [7].

iv. Intelligent Transportation Services.
Smart transportation services are based on IoT and ICT to enhance the sustainability and efficiency of public transportation. These services enable both the locals and the tourists to easily navigate between the city by providing real time information on the city traffic, the timetable of the public transport line, and alternative means of transport. Moreover, car sharing, auto ticket, interconnected train or metro, and other characteristics of smart transport systems make the transport more convenient and reduce the necessity of using personal vehicles. Smart cities can reduce emissions, decrease traffic jams, and make transportation efficient and sustainable [8].

v. Control and Management of Traffic Intelligence.
Smart traffic control systems are highly advanced technologies that are applied to enhance traffic flow and reduce congestion in cities. These systems will comprise smart light traffic systems, real-time monitoring, and managing incidents, which are aimed at making traffic movement smooth and reducing delays to a minimum. The incident management systems rapidly address street closings and accidents, and the smart traffic management systems modify the traffic lights' timing according to the current circumstances. Connected vehicles and smart sensors offer valuable information to enhance the efficiency and safety of traffic by tracking the conditions of roads. These measures will be able to mitigate traffic problems and enhance the general travelling experience of the people of the city [9].

vi. Smart Navigation System
The smart navigation systems operate based on real-time information flows and IoT to offer travelers the optimal routes to take without experiencing roadblocks and delays. These systems offer facilities such as crowd information, alternative routes, and real-time updates on incidents, which assist drivers and pedestrians in making informed decisions. Smart navigation systems give priority to the routes in case of an emergency, such as ambulance services, when there is a need to have a quick response to a situation. When communication networks are integrated with GPS, these systems provide effective and current information to enhance the effectiveness and safety of transportation within the city.

vii. Smart Ticketing System
The smart ticketing system enables easy purchase of tickets on the means of transport. It renders a cashless procedure to be comfortable. Digital platforms and mobile apps have ensured that travelers can purchase tickets, examine times, and obtain details regarding their respective trips. The smart ticketing system has an automatic seat reservation and crowd analysis feature, making sure that public transport is accessible to everyone. The smart ticketing systems, consequently, eliminate physical tickets and provide an easier and more convenient way of paying the fare, therefore, enhancing the user-experience and increasing the overall efficiency of the public transportation systems in smart cities.

5.4.2 Electric Vehicles

Figure 5.5 represents the components of EVs.

- The battery management system (BMS) is necessary to ensure the efficiency and well-being of an electric vehicle's battery. The AIs complement the BMS

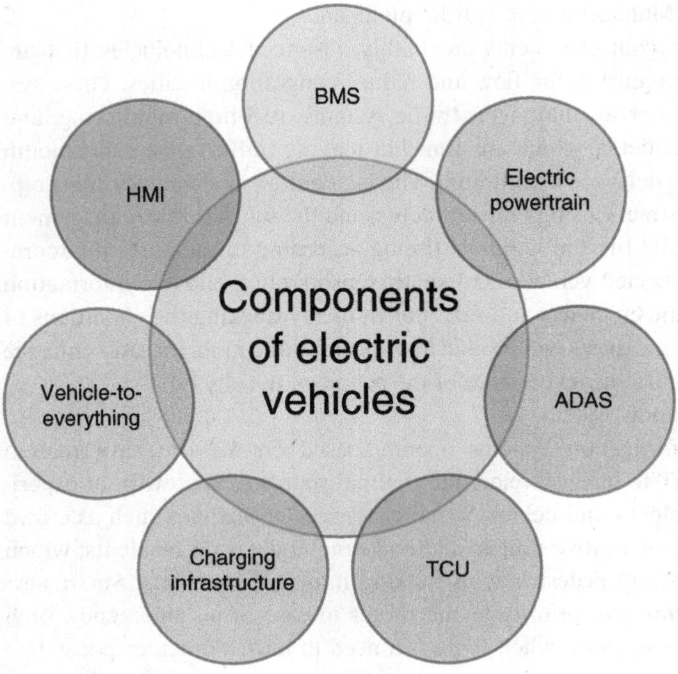

Figure 5.5 Components of electric vehicles.

by forecasting battery degradation, controlling the trends of energy consumption, and allowing smart charge strategies depending on the usage patterns. BMS units, when paired with IoT, have the opportunity to communicate with other systems like charging stations and energy grids in real time, which would make it possible to conduct power-efficient operations and maintain the equipment proactively.

- The 6G networks are set to be the fastest networks with low latency, which will enable them to exchange data continuously with high frequency, which will help in making energy distribution smarter and result in smooth integration with renewable energy sources, making EVs more environmentally friendly and responsive to the smart infrastructure of the city.
- ENE The electric powertrain contains the motor, controller, inverter, and the drive train. EVs are run by the mechanical system. AI enhances its functionality based on real-time information and machine learning. These technologies vary the torque and the speed as per the circumstances and patterns of the road.
- The IoT sensors monitor the components and environmental conditions and make sure that they perform well and address faults at an early stage.

- The 6G will help coordinate vehicles because of the high data transfer rates and high response time. It also involves coordinated driving like platooning, particularly on smart highways. This will assist in decongesting and energy saving.
- Advanced driver assistance systems (ADAS) relate to AI technologies such as computer vision, deep learning, and decision-making algorithms that are used to automatize such functions as emergency braking, adaptive cruise control, and lane departure warnings. These systems heavily rely on the IoT to communicate between the V2X. This enables the electric cars (EVs) to communicate with the surrounding cars, road signs, and pedestrians.
- ADAS will be enhanced in a 6G environment because the environment will have very low latency and the capacity to link numerous devices. It will facilitate real-time driving, and this collaborative driving will make the roads safer, and fully autonomous electric mobility that fits smoothly with the evolving city systems.
- The telematics control unit (TCU) is what makes EVs gather and transmit data regarding vehicle performance, position, and diagnostics. This is the data that AI processes to give predictive information, like the optimal routes based on traffic patterns, recommendations on saving energy, and alerts on remote repairs.
- IoT provides an interface to cloud services, which makes it possible to update it over-the-air (OTA) and integrate it with fleet management systems.
- With the introduction of 6G, the world will be connected smoothly, have real-time telematics, and have enjoyable driver experiences. This will enhance efficiency and satisfaction with operations in smart urban settings.
- EV charging infrastructure is becoming intelligent. The algorithms of AI will be able to simulate demand on power, propose the optimal time to charge, and even regulate the price to avoid overloading the grids.
- IoT is used to link EVs with smart charging stations and energy grids, which enables efficient scheduling and monitoring of energy consumption in real-time. The advent of 6G will introduce such functions as high-speed wireless charging, decentralized energy exchanges, and highly responsive charging networks. It will make more individuals embrace EVs and contribute to the sustainability of related urban settings.
- Vehicle-to-Everything (V2X) communication is one of the most important systems where EVs communicate with one another (V2V), with the infrastructure (V2I), with pedestrians, and with energy systems (V2G). AI interprets such a huge amount of data in order to enhance the decision-making process, traffic optimization, and cooperative driving. With the IoT, this data communication will be integrated into a broader ecosystem where EVs are used to stabilize the grid and the efficiency of mobility in the city [10].

- In 6G, such interactions are real-time, highly reliable, and scalable, and can support millions of simultaneous connections and enable real-time and decentralized intelligence in congested spaces.
- Last but not least is the Human-Machine Interface (HMI) that is essential in enhancing human interactions with their cars. It operates touchscreens, voice commands, as well as heads up displays. The assistance of AI is made through the capability to process natural language and include personalization features based on user preferences and behavior. The IoT helps the HMI with additional external data sources, e.g., weather, traffic, and calendar events, to give useful suggestions. 6G will transform HMI by adding immersive technologies, e.g., augmented and virtual reality, real-time translation languages, and extremely responsive voice assistants. This will not only make electric cars efficient but also smart and interactive [11].

5.5 AI-driven Sustainable Solutions

The concept of sustainable smart cities relies on the application of technology to enhance the utilization of resources, reduce the amount of carbon and greenhouse gases, and increase the standards of living in cities. AI plays a critical role in such an environment to foster sustainable solutions. It aids in intelligent decision-making, forecasting, and controlling city activities. Combined with the high speed and the high connectivity of 6G networks, AI will transform cities into effective, low-carbon ecosystems [12]. The following are some of the main AI-based sustainable solutions that may affect the future of smart cities:

Figure 5.6 represents the solutions driven by AI.

A. Energy-efficient 6G Networks through AI Optimization
Energy consumption is one of the largest issues facing next-generation networks [13]. The following ways can be used in addressing these challenges using AI-driven energy optimization:
  - Smart Energy Management: AI algorithms dynamically allocate network resources based on real-time demand, minimizing unnecessary power consumption. This ensures that energy-intensive components are activated only when required.
  - Self-learning and Adaptive Techniques: AI-enabled self-learning systems continuously analyze network traffic and adapt power usage patterns, leading to significant energy savings.
  - Sleep Mode Algorithms: AI algorithms can intelligently switch off idle network nodes or base stations, conserving energy without compromising connectivity.

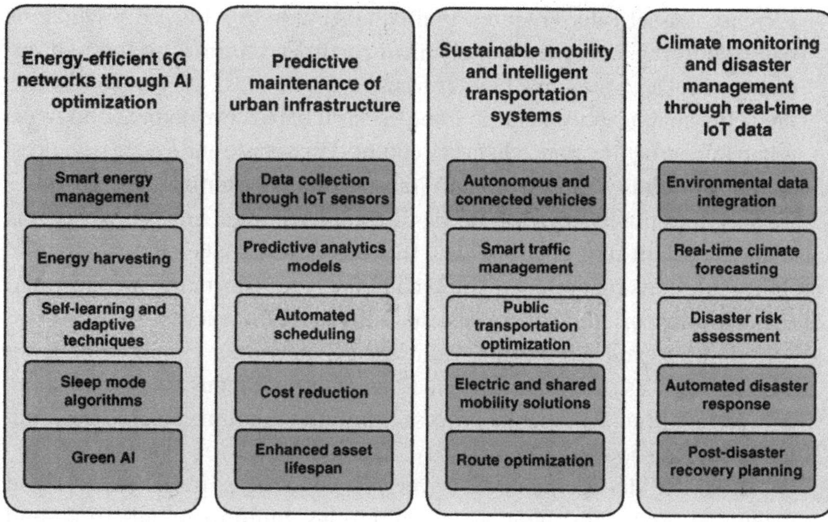

Figure 5.6 AI-driven sustainable solutions.

- Green AI: Implementing AI models that are energy-efficient themselves, optimizing both network operations and computational overhead.

  With the implementation of AI-based solutions in the 6G structure, network operators will be able to minimize carbon footprint at an extraordinary rate without losing performance and connectivity.

B. Predictive Maintenance of Urban Infrastructure

Urban infrastructure such as roads, bridges, water supply, and government houses require round the clock checks and maintenance. The conventional ways of maintenance tend to result in a reactive response, which creates disturbances and increased expenses. Predictive maintenance is an AI-driven forecasting system that offers a proactive solution to the issues that may arise ahead of their development [14].

- IoT Sensor Data Collection: Live collection and analysis of data collected by embedded sensors of parts of the infrastructure are used to identify the early indicators of wear and tear.
- Predictive Analytics Models: AI models are used to handle previous data to anticipate probable failure and service. These models have the ability to detect trends of degradation, corrosion, or mechanical load.
- Automated Scheduling: Maintenance will be automatically planned in terms of priority to reduce the impact on community services and increase safety.
- Cost Reduction: Cities will be able to spend resources more efficiently by preventing expensive emergency recovery through the prediction of failures.

- Increased Asset Life: Frequent maintenance due to AI-based predictions will extend the life of the major infrastructures, minimizing the requirement to replace them on a regular basis.

  Predictive maintenance helps to improve sustainable development and make urban infrastructure more reliable by cutting down waste and saving resources.

C. Sustainable Mobility and Intelligent Transportation Systems

Transport does not just contribute significantly to carbon emissions, but it is also an important part of urban life. It is possible that 6G connectivity and mobility solutions that offer AI could completely transform the management of the city transport and ensure sustainability and efficiency:

- Autonomous and Connected Vehicles: 6G networks enable vehicle-to-everything (V2X) communication, allowing autonomous vehicles to share data in real time. AI systems process this data to optimize traffic flow and reduce congestion.
- Smart Traffic Management: AI-driven systems analyze traffic patterns and make real-time adjustments to signal timings, minimizing idle times and cutting down emissions.
- Electric and Shared Mobility Solutions: AI-enabled fleet management systems will make the transportation of cities use their electric and shared vehicles efficiently and minimize their carbon footprint.
- Route Optimization: The time spent on routes and fuel use is minimized by allowing real-time route planning on the basis of dynamic data analytics.

  Through fostered mobility solutions, cities will reduce congestion, enhance air quality, and assist in their transition to low-carbon transportation systems.

D. Climate Monitoring and Disaster Management through Real-time IoT Data

With the rising rate of extreme weather, cities need to be ready to track the weather and take corrective actions during disasters. From AI-based climate change monitoring and disaster management systems:

- Environmental Data Integration: AI integrates data from diverse IoT sensors monitoring air quality, temperature, humidity, and other environmental parameters.
- Real-time Climate Forecasting: Advanced machine learning models predict weather patterns, enabling cities to prepare for adverse conditions proactively.
- Disaster Risk Assessment: AI-driven simulations and predictive models assess the potential impact of natural disasters, guiding preparedness measures.
- Automated Disaster Response: AI-powered decision support systems automate emergency response plans, optimizing evacuation routes and resource allocation.
- Post-disaster Recovery Planning: AI systems analyze damage data to prioritize restoration activities, ensuring a swift return to normalcy.

Through real-time data and AI analysis, cities have an opportunity to be more effective in their mitigation and adaptation efforts to climatic effects, human life protection, and reduced destruction of infrastructure. To sum it up, the construction of smart cities of the future is based on AI-driven sustainable solutions, in particular, 6G networks. These solutions make cities more resilient, energy-saving, and dynamic in challenging situations. It is possible to monitor the climate, maximize energy consumption, introduce predictive maintenance, and improve transportation in the cities to minimize environmental impact and increase the quality of life. With an increased convergence of AI, IoT, and 6G, sustainable smart cities will transition beyond imagination to real applications, which will create a more sustainable and green future.

5.6 Security and Privacy in AI and IoT for Smart Cities and Electric Vehicles

5.6.1 Smart Cities

With the evolution of smart cities under AI, IoT, and 6G networks, security and privacy have become significant issues. This high interconnectedness of devices, systems, and applications exposes urban infrastructure to multiple cyber risks, data breaches, and unauthorized access. Despite the fact that 6G networks are much faster and more connected than ever before, they also introduce new security threats that should be addressed to ensure the safety and privacy of citizens and urban properties. The section examines the threats and weaknesses that may be associated with smart cities with 6G and discusses AI-based security, with references to how blockchain can be used to achieve data transaction security [15].

A. Potential Threats and Vulnerabilities in 6G-Enabled Smart Cities
Smart cities expose numerous weaknesses that can be exploited by evil factions due to the use of 6G networks. The following are some of the principal dangers associated with 6G-powered smart cities:

a. Data Breaches and Privacy Violations

- Unauthorized Access: Weak authentication mechanisms can lead to unauthorized access to smart city systems, compromising data integrity and confidentiality.
- Data Aggregation Risks: The integration of diverse data from multiple sources heightens the risk of data leakage during aggregation and processing.

b. Cyber Attacks and Network Vulnerabilities
- o Distributed Denial of Service (DDoS) Attacks: DDoS attacks may be capable of disabling the smart city services significantly through overloading network resources.
- o Man-in-the-Middle (MitM) Attacks: In the MitM attacks, the cybercriminals intercept communications between machines and networks. In the process, they can corrupt or steal important information.
- o DDoS Using Rogue IoT Devices: Rogue IoT devices may be used as access points to cyber attacks. They are capable of spreading malware and breaking the security of the network.
- o Advanced Persistent Threats (APTs): Attackers can adopt sophisticated techniques to stay inside the networks of cities over a long period and gather information and cause long-term harm.

c. AI-related Security Risks
- o Adversarial Attacks: With the AI models provided with manipulated data, bad actors can use the vulnerabilities inherent in the models to generate inaccurate outcomes or conclusions.
- o AI Model Poisoning: Attackers can contaminate the AI model and cause the model to make incorrect predictions or security breaches by manipulating training data intentionally.
- o Data Privacy Concerns: Data analytics and surveillance powered by AI raise moral concerns concerning the invasion of privacy and unauthorized collection of data.

B. AI-driven Security Solutions: Anomaly Detection and Threat Prediction
Threat Prediction and Anomaly Detection. Smart cities can greatly benefit the safety of artificial Intelligence due to its real-time threat detection, anomaly analysis, and proactive response plans. This is the way AI-powered solutions can reduce security threats:

a. Real-time Detection of an anomaly.
- o Behavioral Analysis: The AI models can detect anomalies that can be indicative of security violations by acquiring the normal patterns of network traffic and device interactions.
- o Dynamic Anomaly Detection Models: AI is capable of identifying unusual activity, such as unauthorized access or data leakage, by reading sensor and network logs continuously.

b. Early Warning and Threat Prediction Systems.
- o Threat Intelligence Integration: With the integration of threat intelligence feeds and AI systems, prediction is more precise, and new and emerging threats can be discovered.
- o Proactive Mitigation strategies: AI can propose preventive measures, such as the improvement of encryption codes or security measures, in case a potential threat is detected. c. Adaptive Security Policies.

C. Blockchain Integration for Secure Data Transactions

The blockchain technology is proving to be a viable solution to privacy and security concerns in the smart cities. IoT, blockchain, and AI atop 6G networks provide a decentralized, impermeable system of identity management and transactions of data.

a. Decentralized Data Management

- Immutable Ledgers: Blockchain guarantees the integrity of data and eliminates any desire to alter data, since it is stored in an immutable ledger.
- Transparent Transactions: All data transactions are recorded transparently, enabling traceability and accountability while maintaining user privacy.
- Secure Data Sharing: Blockchain-based platforms can facilitate secure data exchange between city departments and service providers without relying on centralized control.

b. Secure Identity and Access Management

- Digital Identities: Blockchain enables secure and verifiable digital identities for citizens and devices, reducing identity theft and impersonation risks.
- Owen Control Mechanisms: The smart contracts of the blockchain provide very strict controls to access, ensuring that only users or devices that are authorized can access specific data or services.

c. Enhancing Data Privacy with Zero-knowledge Proofs

- Zero-knowledge Proofs (ZKPs): This kind of blockchain system enables one side of the transaction to convince the other that a statement is correct without telling the other side what data was used to validate the statement. This saves privacy and does not compromise data authenticity.
- Privacy-preserving Analytics: A combination of blockchain privacy-preserving properties and artificial intelligence can help cities to safely analyze data without any release of personal information.

D. Real-world Implementations and Case Studies

- Dubai Smart City: Minimizes the threats of data manipulation through blockchain technology to ensure safe identity management and data transactions.
- Singapore Smart Nation: Applies both blockchain and AI-based anomaly detection to the public safety systems to allow the government organizations to exchange data safely.
- Estonia e-Government: It applies blockchain technology to protect official records and protect the data of citizens against unauthorized access.

E. Challenges and Future Directions

In spite of the security advantages of using blockchain and AI in combination, there remain problems, including:

- Scalability Problems: Scalability can be a problem when large amounts of data are processed by blockchain networks.

- o High Computational Costs: It takes a lot of processing power to run blockchain nodes and AI-driven security models.
- o Interoperability Issues: It is still technically difficult to guarantee compatibility across various AI, blockchain, and IoT systems.
- o Regulations Regarding Data Privacy: It is a wise practice that requires planning and compliance to realize smart city solutions without violating the laws of data protection, such as GDPR.

Finally, in the era of smart cities, where networked systems, real-time data transfers, and similar concepts become the norm, security and privacy are essential. Through the detection of anomalies and predicting threats, especially using AI, cities will be able to prevent cyber threats proactively. In addition to providing a high level of protection through the provision of safe identity checks and decentralization of information, the integration of blockchain technology provides an additional layer of protection. It is important to have an overall security architecture to secure smart city ecosystems because the 6G networks will bring unheard-of connectivity and performance. To facilitate safe and resilient smart city infrastructures, the proposed research needs to focus on enhancing the robustness of AI, enhancing the performance of blockchain, and creating standardized protocols.

5.6.2 Security and Privacy in AI and IoT for Electric Vehicles

Given that EVs are becoming more interconnected with the help of AI and IoT technologies, the question of security and privacy is growing. These huge amounts of real-time information transfer among vehicles, infrastructure, cloud services, and user devices pose considerable vulnerabilities. There is a need to ensure robust cybersecurity practices and privacy to keep the users' trust, intercept malicious practices, and ensure that these vehicles operate safely [16].

5.6.3 Key Security Challenges in Connected Electric Vehicles

- EVs have software-controlled control units, which make them vulnerable to cyberattacks on Vehicle Systems. These systems are subject to cyberattacks that may compromise the safety of the drivers [17].
- Privacy Invasions and Data Leaks: EVs can gather sensitive user information, such as their trip history, driving habits, and preferences. This information can be accessed by unauthorized parties due to weak encryption of data or a lack of authentication.

- Malicious Firmware Updates: Over-the-air (OTA) updates powered by IoT are convenient but can be compromised if not properly secured.
- IoT Device Vulnerabilities: IoT sensors in EVs and charging stations may become entry points for malware or ransomware attacks.
- Cyberattack on Vehicular Systems: EVs use software-controlled control units in critical tasks such as braking, steering, and battery control. These systems have vulnerabilities that can be exploited by cyberattacks, and this may endanger the safety of the drivers.
- Malicious Firmware Updates: OTA updates are computed by the IoT often, and are convenient, but can be exploited in the event they are not properly secured.
- IoT Device Vulnerabilities: IoT sensors in EVs and charging stations may serve as entry points for malware or ransomware attacks.

5.6.3.1 Role of AI in Security for Electric Vehicles

- Anomaly Detection: AI systems monitor the behavior of cars in real time and identify any deviations from conventional functioning that could indicate a failure or an intrusion. Threat prediction machine learning algorithm uses past data to identify and forestall any potential threat.
- Context-aware Security Policies: AI devices can modify security measures according to the driving circumstances (e.g., location, car speed, user profile) [18].

5.6.3.2 Role of Blockchain and DLT for Privacy and Trust

- Secure Identity Management: By enabling decentralized identity verification for users and vehicles, blockchain guards against spoofing and illegal access.
- Immutable Data Logging: Events such as maintenance history, software updates, and charging transactions can be securely recorded on a blockchain ledger.
- Privacy-preserving Mechanisms: The sensitive data can be checked with the help of technologies such as Zero-knowledge Proofs (ZKPs), and actual content is never disclosed.

5.6.3.3 6G and the Future of EV Cybersecurity

As 6G networks are rolled out, EV security frameworks will be required to change:

- Quantum-resistant Cryptography: Securing data in the future in case of threats of quantum computing.
- EVs using AI and IoT should be deployed safely based on security and privacy. The safety of user data and system integrity will be critical to EVs as they become essential elements of sustainable smart cities, along with improving mobility and energy efficiency [19].

5.7 Case Studies and Real-world Implementations

5.7.1 Smart Cities

An example of this is seen in the realization of the concept of sustainable smart cities in various innovative cities across the world with the assistance of AI, IoT, and 6G networks. Pilot projects and research efforts are being generated in large numbers to exhaust the possibilities of 6G technology as the cities abandon their legacy systems and 5G functionality in favor of the thrilling prospects of 6G. This section highlights research projects and pilot projects aimed at sustainable urban solutions, provides case studies from current smart cities making the transition to 6G, and talks about the difficulties and lessons discovered by earlier generations [20].

A. Existing Smart Cities Transitioning Toward 6G

a. Seoul, South Korea: The Ultra-connected Smart City

The first to be mentioned in terms of smart city development is Seoul, which is known to have one of the most advanced technological infrastructures. The city has been proactively preparing the transition between 5G and 6G by focusing on ultra-connectivity and decision-making that is based on data [21].

- 6G Readiness Initiatives:
  - A strategic plan has been implemented by the South Korean government to create 6G technology by 2028, and Seoul is one of the test areas.
  - The city is in the process of enhancing its current 5G communication to support terahertz communication and ultra-low latency capability of 6G.
- Smart Mobility and Transportation:
  - Seoul has installed smart traffic management systems based on real-time sensors of the IoT and AI-driven autonomous buses.
  - Vehicle-to-everything (V2X) seamless connectivity is under test in an attempt to enhance pedestrian safety and reduce traffic congestion.
- Public Safety and Healthcare:
  - Remote surgery trials and immediate monitoring of the health condition of patients through wearable IoT devices are the 6G-enabled healthcare projects of the city.
  - AI-driven public safety systems are being implemented to detect emergencies and respond with automated protocols.
- Lessons Learned:
  - The city discovered that maintaining robust cybersecurity is critical as connectivity increases.
  - Integrating legacy systems with new technologies posed challenges, highlighting the need for backward compatibility.

b. Dubai, UAE: Toward a Sustainable and Digital Future
Dubai is one of the first of its kind, having actively adopted smart city projects, and focusing on incorporating the latest technologies into enhancing sustainability and life in the city.

- Smart Infrastructure and Connectivity:
  - Dubai is also upgrading its 5G networks to 6G, which will allow continuous connectivity of IoT devices to the infrastructures used by people.
  - The city will be concentrated on intelligent buildings whose energy management and maintenance are automated and predictive.
- AI-driven Urban Services:
  - The use of chatbots and digital assistants that operate on AI is utilized to guide residents in getting public services and real-time information.
  - The identity management based on Blockchain provides secure access to the government services.
- Sustainable Mobility:
  - Roads and Transport Authority (RTA) of Dubai is running autonomous taxis in 6G that will be used to navigate the roads and provide safety to passengers.
  - Smart parking solutions maximize traffic space optimization and time-saving in traffic search.
- Challenges and Lessons Learned:
  - The implementation of energy-efficient solutions required significant investment and policy alignment.
  - Data interoperability between different smart systems remains a challenge due to diverse technology providers.

c. Singapore: A Smart Nation Blueprint
The Smart Nation initiative by Singapore is an integrated strategy of using technology in both sustainable and smart urban growth. This nation is also becoming a pioneer in the adoption of 6G to improve its smart city structure.

- 6G Research and Development: The Singapore government has spent a lot of money on R&D to investigate the use of 6G with heavy emphasis on ultra-reliable, low-latency communication.
- The collaboration between telecom companies and universities is also targeting the creation of AI-native networks that can optimize themselves.
- Smart Public Services: Real-time monitoring of air quality and energy-saving systems powered by artificial intelligence are being introduced in the city. The IoT-enabled waste bins that notify of the need to gather the waste optimize waste management.

- o Healthcare Innovation: Telemedicine and remote diagnostics 6G-enabled are under pilot testing so that hospitals can minimize the load on them.
- o The disease prediction models based on AI are the models that use information on the public health databases to detect new health-related risks. Even with the difficulties and the lessons learned, the project they participated in was successful.<|human|>Difficulties and Lessons Learned:
- o The problem of Data privacy has emerged due to a widespread level of surveillance and data gathering, and it needs robust data governance regulations.
- o There was a challenge in integrating new technologies with the current systems in terms of technical and financial aspects.

B. Sustainable Urban Solutions Pilot Projects and Research.

Since the 6G technology is a fresh concept, many pilot projects and research projects are being initiated in different parts of the world to determine the feasibility of the technology in ensuring sustainable urban development. The following are some of the prominent ones:

a. 6G Smart Campus in Finland.

- o Project Scope:
  Aalto University in Finland is undertaking pilot projects in order to find out the use of 6G in the establishment of a smart campus setting.
- o Technological Focus:
  - Digital twins of campus infrastructure powered by AI for energy management and predictive maintenance.
  - IoT sensors connected to 6G testbeds for real-time environmental monitoring.
- o Research Outcomes:
  - Improved energy efficiency and predictive fault detection through real-time data analysis.
  - Enhanced connectivity between devices, reducing latency in data communication.

b. Smart Mobility Pilot in Tokyo, Japan

- o Project Objective:
  - To test autonomous vehicles and intelligent traffic systems using 6G communication.
- o Key Innovations:
  - Real-time V2X communication to coordinate vehicle movement and optimize traffic flow.
  - AI-powered traffic prediction models to dynamically adjust traffic signals.

- o Expected Impact:
  - Reduced travel time and decreased carbon emissions from idling vehicles.
  - Enhanced public safety through collision detection and automatic braking.

C. Challenges and Lessons Learned from Previous Generations

Some of the challenges of previous generations offer valuable insights as smart cities move to 6G networks based on 4G and 5G frameworks:

a. Data Privacy and Data Security.

Smart cities can now be threatened with data security threats never before heard of, including breaches and illegal data access, with the help of 5G networks. Some of the lessons to be learned include the use of a strong encryption scheme and multilayered security mechanisms.

b. Infrastructure Compatibility

The transition from 4G to 5G required major investments in infrastructure upgrades. Cities realized the importance of establishing versatile and modular systems that would be able to accommodate future upgrades with minimum disquiet.

c. Interoperability and Integration.

Introduction of new technologies, especially in such aspects as transport and government services, has been a major challenge when it comes to the integration of the old systems.

d. Acceptance by the public and the effects on society.

Public acceptance cannot be denied even in the light of technological development. The importance of the issue of privacy and its education to the population was underlined by the previous smart city projects.

To sum up, real applications and pilot projects prove that the transition to 6G-powered smart cities is a complicated process that requires community involvement, policy coordination, and technological advancement. Cities such as Seoul, Dubai, and Singapore are leading by integrating 6G with AI and IoT in order to optimize the living and sustainability of urban life. In order to ensure the successful deployment, however, the problems related to the issues of data privacy, systems integration, and social acceptance have to be addressed. The world's cities can fully achieve the potential of sustainable, resilient, and digitally inclusive urban areas by employing the latest research and experience gained over time.

5.7.2 Electric Vehicles

The transition to sustainable mobility at the global level is already going on, and many cities and organizations are adopting the strategies of EVs supported by AI, IoT, and new technologies of 6G [22].

A. Singapore—Smart Nation Electric Mobility
Project Overview:
The Smart Nation initiative in Singapore includes a strategic motivation to go massively integrated with EVs and infrastructure.
Key Technologies Used:
- AI-driven predictive traffic and energy analytics.
- IoT-based smart charging stations with real-time availability tracking.
- EVs connected via 5G and early-stage 6G pilot networks for V2X communication.

Outcomes:
- Reduced traffic congestion and emissions through smart route planning.
- Enhanced public transport integration with shared EV fleets.
- Smart charging platforms reduce peak load stress on the city's grid.

B. Oslo, Norway—Europe's EV Capital
Project Overview: Oslo is the European leader in EV adoption in order to become a zero-emission city.
Key Technologies Used:
- Demand forecasting of the electricity grid using AI.
- Secure IoT-based data sharing by the government.

Outcomes:
- Over 80% of new vehicle registrations are electric.
- Dynamic pricing for EV charging based on renewable energy availability.
- Use of EVs in public services (e.g., buses, garbage trucks).

C. Tesla—Over-the-Air AI Updates and Real-Time Diagnostics
Project Overview: Tesla applies AI and IoT in its fleet to facilitate smart operations of the fleet and improve user experience in real time [23].
Key Technologies Used:
- AI-powered autopilot system with continuous updates.
- IoT sensors collecting vehicle data for diagnostics and performance tracking.
- Blockchain research for future secure charging and transaction models.

Outcomes:
- OTA updates reduce the need for service center visits.
- Predictive maintenance lowers downtime.
- Customer data privacy protocols are improving through edge AI deployment.

D. Amsterdam, Netherlands—V2G (Vehicle-to-Grid) Integration
Project Overview: Amsterdam has also been the first to introduce V2G to make EVs a part of its renewable energy system.
Key Technologies Used:
- IoT-enabled bidirectional charging stations.
- AI algorithms to balance energy demand and EV charging patterns.
- Blockchain for secure energy transaction records.

Outcomes:

- EVs act as temporary energy storage during peak solar/wind output.
- Real-time optimization of energy flows across neighborhoods.
- Enhanced grid resilience and reduced reliance on fossil fuels.

E. Delhi, India—EV Integration in Public Transport

Project Overview: The Delhi government initiated massive plans to have buses and three-wheelers electrified to address the pollution problem.

Key Technologies Used:

- AI-based fleet optimization software for electric buses.
- IoT for battery monitoring and route tracking.
- Pilot testing 6G-ready communication nodes at charging depots.

Outcomes:

- Improved public transportation efficiency and passenger safety.
- Better vehicle uptime due to predictive diagnostics.
- Increasing public trust in EV viability.

These case studies point to the way that cities and organizations are adopting EVs based on the strategic deployment of AI, IoT, and next-gen communication networks. The three crucial success factors are factual planning, safe infrastructure, and combined public and private endeavors.

5.8 Challenges for 6G

To achieve successful deployment, several issues need to be addressed because 6G networks embrace AI and IoT, which will make smart cities sustainable. One of the challenges is to maintain the network ecosystem open and customizable, which is essential to innovation and proprietary technology may hinder it. The other significant issue is achieving 100% coverage, especially in remote and underdeveloped locations where there are still geographical constraints and high costs of infrastructure. Moreover, the use of terahertz communication, which offers vast bandwidth, presents technical hurdles that need to be overcome. With more reliance on AI and IoT devices, ensuring network security becomes a major issue and needs the advanced solution of post-quantum cryptography and quantum key distribution to maintain data integrity and privacy when the era of 6G is reached.

Figure 5.7 presents the pictorial representation of the challenges for 6G.

A. Openness And Customization

From one angle, openness and sharing are the fundamental principles of the Internet, as we are all aware, and they facilitate its growth. From a different angle, the ecosystem of mobile communication networks restricts their development to a certain degree by using unique technologies.

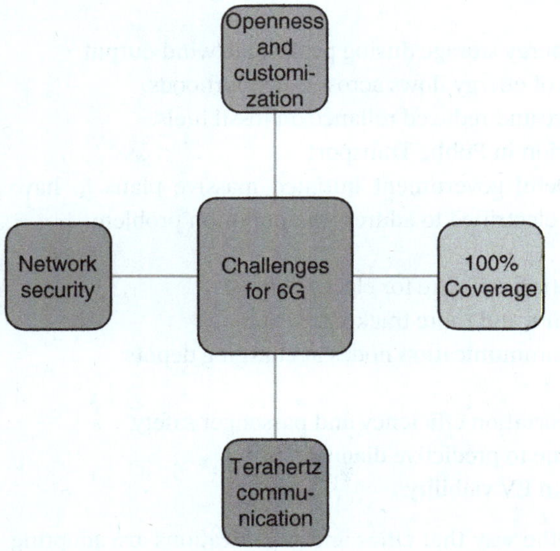

Figure 5.7 Challenges for 6G.

B. 100% Coverage
Even while a mobile phone can now be used to live a basic and appropriate life, more than three billion people worldwide still lack access to the Internet. The main causes of the dearth of network infrastructure in remote locations are difficult geographic circumstances and the high expense of installing base stations and optical fiber cables.

C. Terahertz Communication
The 6G era is anticipated to make use of the terahertz frequency band, which spans from 100 GHz to 10 THz. This frequency range has never been used before due to its large bandwidth, suggesting that there are no restrictions on its use.

D. Network Security
Network security is a significant element of the digital economy's growth. The 5G components consist of low latency, unprecedented reliability, gigabit capacity, and, most importantly, network security. With the advent of 6G, the technology of QKD (Quantum Key Distribution) and PQC (Post-quantum Cryptography) will be introduced into networks to ensure the total safety of the network.

5.9 Future Directions

The future research will focus on eliminating the existing problems and exploring new opportunities for sustainable development of smart cities with the development of 6G networks. The combination of AI, the IoT, and 6G will open unprecedented opportunities to enhance digital inclusion, environmental sustainability, and urban living. There are, however, several areas of research that need to be explored to reap the full benefits of the same. This section aims to focus on the future research directions in 6G and sustainable smart cities by placing emphasis on the aspects of technical, environmental, social, and governance [24].

A. Advanced 6G Communication Technologies

The next generation of wireless communication will have to overcome formidable technical constraints to enable ultra-high data rates, low latency, and massive connectivity. Key areas of interest in this area of study will include:

a. Terahertz (THz) Communication

- High Frequency Spectrum Investigation: To achieve data rates of up to 1 Tbps, consider the use of THz frequencies (0.110 THz).
- Channels Modelling and Signal Propagation: Design accurate models to understand the transmission of THz signals in urban environments.
- Adaptive Beamforming and Massive MIMO: Improve beamforming and multimedia antenna methodology to prevail over path loss and attenuation of the signal at high frequencies.

b. Intelligent Reflecting Surfaces (IRS)

- Reconfigurable Surfaces to Signal Enhancement: Research IRS to reconfigure the reflection of electromagnetic waves, enhancing signal coverage in dense urban spots.
- Energy Optimization: Minimization of power consumption, signal strength, and reliability maintenance.

c. AI-native Network Architecture

- Self-organizing Networks: Develop AI-controlled network systems capable of adaptive control to changing device densities and traffic patterns in real time.
- Robotic Network Maintenance: To identify potential problems in advance; apply predictive analytics.

B. Sustainable and Resilient Smart City Infrastructure

The construction of sustainable smart cities must be completed with the incorporation of energy-efficient technologies and sturdy infrastructures. Further studies in this field will be aimed at adhering to the following fields:

a. Energy Harvesting and Wireless Power Transfer

- Self Powered IoT Devices: Design systems that use the ambient power sources, like solar, kinetic, or radio frequency to power sensors and devices.
- Efficient Power Management Protocols: Design protocols to be used to reduce energy use and guarantee continuous operation.

b. Green Communication Technologies

- o Low Power 6G Transceivers: Develop energy-efficient equipment to minimize the carbon footprint of communication systems.
- o ENERGY Aware Routing Protocols: Introduce routing algorithms that trade-off between transmission speed of data and energy usage.

c. Climate-Resilient Infrastructure

- o Disaster Ready Communication Networks: Have networks that can carry on functioning even when there is a natural calamity, ensuring that the people are safe.

C. Data Privacy and Cybersecurity in Smart Cities

The issue of data security and privacy will be essential, with the smart cities becoming more networked and data-driven. Future studies must deal with the following issues:

a. Privacy-preserving Data Sharing

- o Homomorphic Encryption: An encryption algorithm that allows the calculation of encrypted information without decryption.

b. Quantum-resistant Security Protocols

- o Make cryptographic algorithms; which resist quantum computing attacks to secure data in the long term.
- o Quantum Key Distribution (QKD): Study the safe communication methods that can ensure the safety of data through the application of quantum properties.

c. Automated Threat Detection and Mitigation

- o AI-based Intrusion Detection Systems (IDS): The systems are machine learning-based to detect any anomaly patterns that may indicate cyberattacks.
- o Blockchain-enabled Identity Management: Minimize risk of identity theft by enhancing security using decentralized identity solutions.

D. Human Centric and Inclusive Smart City Solutions

Inclusive and human-centered design is a future research priority that needs to be applied to ensure that smart cities innovations are beneficial to all their inhabitants.

a. Civil Society and Digital Literacy.

- o Participatory Platforms: Produce applications that can allow citizens to be actively engaged in the process of urban planning and governance.
- o Digital Literacy Programs; Educate the public awareness and knowledge of emerging technologies, in order to encourage inclusive adoption.

b. Social and Ethical Factors.

- o Ethical Frameworks to Surveillance: Design principles; to make sure that surveillance technologies do not violate the privacy of citizens; and independence.

E. Intelligent Transportation and Mobility Management
Sustainable and efficient city mobility is one of the key aspects of smart cities. Further research should focus on:

a. AVNs.
- Vehicle to Everything (V2X) Communication: To support autonomous driving, enhance real time information exchange among automobiles and infrastructure.
- Traffic Prediction Algorithms: Use AI to anticipate traffic congestions/ and suggest optimal paths.
- Traffic Prediction Algorithms: Make use of AI to predict traffic jams/ and recommend the best routes.

b. Integrated Multimodal Transportation Systems
- Mobility as a Service (MaaS): Develop platforms that integrate public transport, ridesharing, and micromobility solutions.
- Dynamic Traffic Signal Control: Implement real-time adaptive signal systems to reduce congestion and emissions [25].

F. AI and IoT for Urban Environmental Monitoring
Sustainability goals can only be attained through both environmental monitoring and data-driven management.

a. Monitoring Air and Water quality. Smart Sensor Networks: Develop low-cost, energy-efficient sensors/ to be used to continuously monitor the environment. RealTime Data Analytics: Study the levels of pollutions and predict the trend based on AI models.

b. Smart Waste Management Automated Waste Sorting: Train robots to sort the waste and promote recycling.
Waste Collection: Reduce emissions and fuel consumption: Optimize waste collection/routes using real time data to be an IoT-driven waste collection.

G. Governance and Policy Frameworks 6G Smart Cities.
Sustainable smart cities cannot be implemented and maintained without proper governance. This research should focus on the following issues:

a. Data Governance Policy Development. ELDA Laws: Data Sovereignty Deal with cross-border data flows and data ownership.

b. Regulatory Sandboxes: These allow controlled experimentation of new technology and assessing the potential risks.

c. Public–Private Partnerships (PPPs)
- Collaborative Innovation Models: Encourage partnerships between governments, academia, and private enterprises to drive smart city projects.
- Funding and Investment Strategies: Explore sustainable financing models to support long-term smart city initiatives.

H. Advanced AI Models for Predictive Urban Planning
Predictive models can significantly enhance urban planning by predetermining issues and using resources as effectively as possible.

a. Digital Twin Cities
- o Simulation and Scenario Analysis: Use digital twins to model urban scenarios and evaluate the impact of new policies.
- o Predictive Maintenance: Apply AI to detect maintenance needs for public infrastructure before failures occur.

b. Sustainable Urban Design
- o Energy Consumption Prediction: Leverage AI to forecast energy demand and develop conservation strategies.
- o Climate Adaptation Planning: Model the impact of climate change on urban areas and propose mitigation strategies.

In conclusion, the future research on 6G and smart cities should be multidisciplinary and combine new technology with social, ethical, and environmental aspects of the issue. By focusing on data privacy, smart transportation, green infrastructure, new communication technologies, and participatory governance, the researchers are able to design technologically advanced, resilient, equitable, and environmentally concerned cities. Investing in these research areas will provide the foundation of the next-generation smart cities that would meet the evolving needs of city residents and improve the sustainability of the city in the long term.

5.10 Conclusion

In other words, the intersection of AI, IoT, and 6G networks is one of the key steps toward the complete potential of sustainable ecosystems of smart cities and EVs. These technologies allow cities to be more robust, smarter, and green. 6G has high-speed and low-latency to coordinate the coordination of multiple processes. IoT ensures the acquisition of real-time data and interoperability. AI provides the ability to automate and make decisions. Such technological synergy is quite advantageous to the EVs, which become the critical components of the smart cities and help to promote energy-efficient mobility, smart charging, and harmonious interaction with the infrastructure of the city. In this same manner, the smart cities collectively benefit due to enhanced safety of citizens, optimal resource management, enhanced service provision as well as cleaner infrastructure created by such digital technologies.

This vision, however, requires robust security, privacy, governance, and interoperability structures alongside technological progress in order to realize the vision. The continuous study, the collaboration across sectors, and creative policymaking would be critical in overcoming emerging challenges and ensuring equitable access to such benefits as cities experience a more rapid digital transformation.

The use of AI and IoT in the 6G framework could ensure clean, clever, and sustainable urban living, which would be offered to future generations.

References

1. Mathew, E. Intelligent transport systems and its challenges. In *International Conference on Advanced Intelligent Systems and Informatics*. Springer International Publishing, Cham, 2019.
2. Suske, David, Touko Tcheumadjeu, L. C., Sohr, A., and Xiaoxu, B. Traffic information systems for smart mobility as part of smart cities. In *International Symposium on Smart Mobility for Smarter Cities (SmartCity 2016)*. 2016.
3. Paiva, S., Ahad, M. A., Tripathi, G., Feroz, N., and Casalino, G. Enabling technologies for urban smart mobility: Recent trends, opportunities and challenges. *Sensors*, 21(6): 2143, 2021.
4. Manfreda, A., Ljubi, K., and Groznik, A. Autonomous vehicles in the smart city era: An empirical study of adoption factors important for millennials. *International Journal of Information Management*, 58: 102050, 2021.
5. Alonso, C. M., Aletà, N. B., and Ruiz, R. M. A. Smart mobility in smart cities. In *CIT2016 – XII Congreso de Ingeniería del Transporte València*, pages 1209–1219. Universitat Politècnica de València, 2016.
6. Rath, M. Smart traffic management system for traffic control using automated mechanical and electronic devices. *IOP Conference Series: Materials Science and Engineering*, 377(1): 012201, 2018.
7. Pau, G., Campisi, T., Canale, A., Severino, A., Collotta, M., and Tesoriere, G. Smart pedestrian crossing management at traffic light junctions through a fuzzybased approach. *Future Internet*, 10(2): 15, 2018.
8. Pulugurtha, S. S., Puvvala, R. K., Pinnamaneni, R. C., Duddu, V. R., and Najaf, P. Buses as probe vehicles for travel time data collection on urban arterials. In *T&DI Congress 2014: Planes, Trains, and Automobiles*. 2014.
9. Qadri, S. S. S. M., Gökçe, M. A., and Öner, E. State-of-art review of traffic signal control methods: Challenges and opportunities. *European Transport Research Review*, 12: 1–23, 2020.
10. Anthopoulos, L. G. and Tzimos, D. N. Carpooling platforms as smart city projects: A bibliometric analysis and systematic literature review. *Sustainability*, 13(19): 10680, 2021.
11. Akca, T., Sahingoz, O. K., Kocyigit, E., and Tozal, M. Intelligent ambulance management system in smart cities. In *2020 International Conference on Electrical Engineering (ICEE)*. IEEE, 2020.
12. Moran, O., Gilmore, R., Ordóñez-Hurtado, R., and Shorten, R. Hybrid urban navigation for smart cities. In *2017 IEEE 20th International Conference on Intelligent Transportation Systems (ITSC)*. IEEE, 2017.

13 Carli, R., Dotoli, M., and Epicoco, N. Monitoring traffic congestion in urban areas through probe vehicles: A case study analysis. *Internet Technology Letters*, 1(4): e5, 2018.

14 Carter, E., Adam, P., Tsakis, D., Shaw, S., Watson, R., and Ryan, P. Enhancing pedestrian mobility in smart cities using big data. *Journal of Management Analytics* 7(2): 173–188, 2020.

15 Savithramma, R. M. and Sumathi, R. Road traffic signal control and management system: A survey. In *2020 3rd International Conference on Intelligent Sustainable Systems (ICISS)*. IEEE, 2020.

16 Bazzi, A., Masini, B. M., Pasolini, G. and Andrisano, O. Smart navigation in intelligent transportation systems: Service performance and impact on wireless networks. *International Journal on Advances in Telecommunications* 6: 57–70, 2013.

17 Srinivasan, R., Sharmili, A., Saravanan, S., and Jayaprakash, D. Smart vehicles with everything. In *2016 2nd International Conference on Contemporary Computing and Informatics (IC3I)*. IEEE, 2016.

18 Al-Mayouf, Y. R. B., Mahdi, O. A., Taha, N. A., Abdullah, N. F., Khan, S., and Alam, M. Accident management system based on vehicular network for an intelligent transportation system in urban environments. *Journal of Advanced Transportation*, 2018(1): 6168981, 2018.

19 Ajmani, P., Singh, N., and Verma, P. Internet of vehicles taxonomy and evaluation: Architectures, protocols, and issues. In *2022 5th International Conference on Contemporary Computing and Informatics (IC3I)*. IEEE, 2022.

20 Al Mahairzi, Z. S. A. A. and Reddy, N. S. Smart road technology for traffic management and ITS infrastructure assessment: A case study of muscat express highway. *International Journal of Advanced Engineering, Management and Science*, 3(5): 239857, 2017.

21 Kazi, S., Bagasrawala, M., Shaikh, F., and Sayyed, A. Smart e-ticketing system for public transport bus. In *2018 International Conference on Smart City and Emerging Technology (ICSCET)*. IEEE, 2018.

22 Ismail, M. M. M., Jusoh, M., Sabapathy, T., Osman, M. N., Rahim, H. A., Yasin, N., and Fazilah, A. F. M. (2019). IoT based smart parking system. *Journal of Physics: Conference Series*. 1424: 012021. https://doi.org/10.1088/1742-6596/1424/1/012021.

23 Luque-Vega, L. F., Michel-Torres, D. A., Lopez-Neri, E., Carlos-Mancilla, M. A., and González-Jiménez, L. E. Iot smart parking system based on the visual-aided smart vehicle presence sensor: SPIN-V. *Sensors*, 20(5): 1476, 2020.

24 Trubia, S., Severino, A., Curto, S., Arena, F., and Pau, G. Smart roads: An overview of what future mobility will look like. *Infrastructures*, 5(12): 107, 2020.

25 Bamwesigye, D. and Hlavackova, P. Analysis of sustainable transport for smart cities. *Sustainability*, 11(7): 2140, 2019.

6

Smart Mobility: Integrating AI for Sustainable Urban Transportation Solutions

A.Jothi Kumar

Thinkcove Technologies, Chennai, Tamil Nadu, India

Corresponding author: jothi1977@gmail.com

6.1 Introduction

The United Nations estimates that by 2050, nearly 70% of the global population will reside in urban areas, which will be an unprecedented level of urbanization during the twenty-first century. Coupled with this rapid urban growth are significant infrastructure development challenges, particularly in the domains of urban mobility and transport. There is a growing demand for efficient, sustainable, and equitable transportation systems as cities grow in size and number [1]. Urban transit systems that are run ragged are typically characterized by congested roads, inadequate traffic flow, longer travel times, noise and air pollution, and higher rates of accidents. These systems often do not have the responsiveness and flexibility to accommodate real-time traffic pattern changes or combine multimodal transportation options. Moreover, most urban areas have experienced decreasing quality of life, energy loss, and inefficiencies due to reactive planning instead of proactive planning. The redefinition of the design, monitoring, and management of urban transportation networks is enabled by the advent of artificial intelligence (AI). To make intelligent, real-time decisions, AI offers the ability to take advantage of a wide variety of disparate data sources, such as those from mobile apps, GPS devices, Internet of Things (IoT) devices, and connected vehicles [2]. With the use of sophisticated machine learning, computer vision, deep learning, and reinforcement learning algorithms, AI systems are able to assess the state of affairs now, forecast future situations, and carry out the best possible actions with little assistance from humans.

AI-powered traffic management systems are able to improve traffic flow, identify and react to problems, redirect vehicles, and dynamically regulate traffic signals. These technologies impact environmental sustainability, as well as traffic reduction, because they allow for reduced carbon emission, reduced idling, and better integration of shared and electric mobility service [3]. AI also facilitates more general smart city goals such as enhancing public safety, supporting the transition to autonomous vehicles, and ensuring equal access to transport. By providing a comprehensive analysis of recent applications of AI in traffic management, mobility optimization, and sustainable city planning, this research investigates how AI can assist with urban mobility challenges.

The ethical, legal, and social concerns involved in introducing intelligent traffic systems are also analyzed, and there is a critical review of effective implementations in leading smart cities. The research contributes to the growing body of literature seeking to shape the future of urban transport by introducing AI as a strategic enabler of policy and societal change as well as a technological solution.

6.2 AI in Traffic Management Systems

To convert urban traffic management systems from static, reactive systems to dynamic, adaptive, and data-driven environments, AI has become a facilitator of choice. By leveraging algorithms that can learn, predict, and optimize in real time, AI introduces intelligence to infrastructure that was previously constrained by deterministic programming and human control. This section addresses the main features and applications of AI-based traffic control systems.

6.2.1 Real-time Traffic and Incident Detection

Conventional traffic monitoring systems are usually slow to react, since they rely primarily on human observation and predefined criteria to identify congestion or problems. However, AI-driven systems analyze visual information from traffic cameras, aerial cameras, and CCTV using computer vision, deep learning, and sensor fusion [4]. For instance, convolutional neural networks (CNNs) are greatly employed in counting traffic flows, identifying types of vehicles, and detecting anomalies, such as illegal parking and accidents (Figure 6.1).

More recently, edge computing has been paired with AI to allow for localized processing of sensor data and video feeds, reducing latency and bandwidth consumption. In addition to their ability to automate the detection of traffic violations and malfunctions, these AI-based surveillance systems also augment traffic operators' situational awareness so that they can make better and faster decisions.

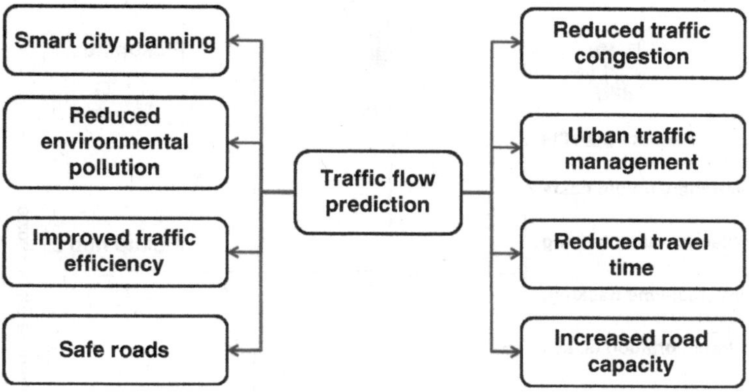

Figure 6.1 Traffic flow prediction.

6.2.2 Modeling Predictive Traffic Flow

One of AI's most significant contributions to traffic management is its predictive power. Through the analysis of historical traffic patterns and real-time data, artificial intelligence algorithms are able to predict congestion hotspots, trip durations, and demand surges. Time-series models such as long short-term memory (LSTM) networks and hybrid deep learning techniques are frequently used for short- and medium-term traffic forecasts [5].

Using these predictive data, traffic management centers can take proactive measures such as rerouting traffic, sending mobile units to high-risk areas, or anticipatorily modifying signal timings. Also, these forecast models come in very useful when conventional traffic controls are insufficient, such as during emergencies, construction works, or festivals.

6.2.3 Dynamic Traffic Signal Optimization

Fixed-time schedule-based static signal control often results in unnecessary delays and increased emissions. Adaptive traffic signal control is made possible by AI, in which reinforcement learning (RL) algorithms, such as Q-learning or deep Q-networks (DQNs), learn the best timing and signal phasing rules from actual traffic situations. These systems adaptively modify traffic signals by maximizing cumulative traffic flow and minimizing mean vehicle waiting times.

An example from the real world is the Surtrac system in Pittsburgh, USA, where decentralized AI signal controllers collaborate to reduce emissions by 20% and travel times by over 25%. To support more general mobility and safety goals, such systems also give priority to emergency vehicles and public transport.

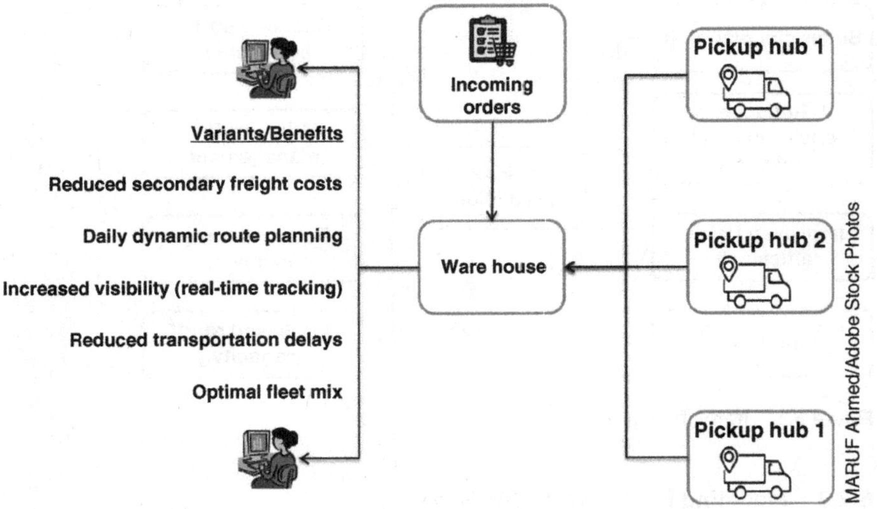

Figure 6.2 Route optimization and traffic demand management.

6.2.4 Route Optimization and Traffic Demand Management

Demand-responsive traffic management relies heavily on AI to control and guide travel behavior through intelligent pricing, scheduling, and routing. Waze and Google Maps are two examples of mobile apps and navigation devices that use AI to suggest alternate routes based on traffic, accidents, or road closures (Figure 6.2) [6].

To ensure optimal route planning, these technologies employ ensemble models and real-time data fusion across connected devices.

Certain cities are integrating AI with digital twins, which are virtual models of transportation networks that allow scenario testing and optimization under different infrastructure constraints and traffic volumes [7]. AI enhances the planning and operational decision-making process by helping to assess the impact of interventions such as toll pricing, lane closures, or traffic diversions.

6.2.5 Integration with Connected and Autonomous Vehicles

AI traffic management solutions need to develop to seamlessly coordinate with vehicle-to-infrastructure (V2I) and vehicle-to-vehicle (V2V) technologies as cities prepare to have connected and autonomous vehicles. AI automates real-time traffic coordination decision-making, making autonomous fleets travel more accurately around intersections, deal with merges, and avoid collisions more effectively than manually operated automobiles (Figure 6.3) [8].

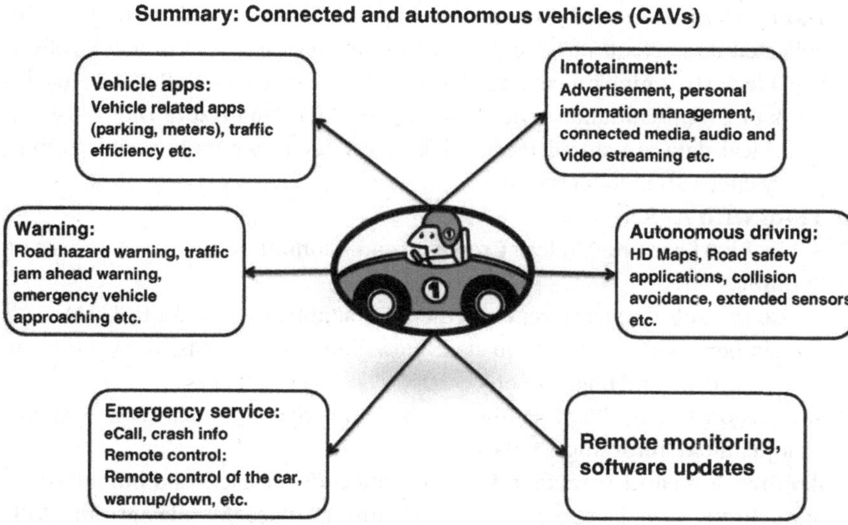

Figure 6.3 Integration with connected and autonomous vehicles.

To achieve maximum mobility across the city in a decentralized but coordinated fashion, future traffic management systems will be expected to operate as multi-agent systems where AI agents at infrastructure and automobiles interact through common protocols. AI for traffic management from detection and prediction to control and optimization involves a complete ecosystem of smart components interacting on multiple layers [9].

6.3 Virtual Architecture for AI-based Traffic Management Systems

A traffic management system based on AI normally uses a virtual structure on several layer suits made from the following components:

1. **Perception Layer**: This lower level includes information collection equipment for vehicles such as GPS devices, drones, IoT sensors, traffic cameras, and devices on board (OBIU). These units capture real-time data on traffic, speed, road conditions, weather, and vehicle categories.
2. **Networking and Communication Layers**: This team ensures that data are sent well using technologies such as 5G, dedicated short-distance communication (DSRC) and edge/cloud-based systems. The real-time section of the real-time situation between connected vehicles and infrastructure provides great convenience through the vehicle-to-some (V2X) communication protocol.

3. **Data Integration and Processing Layers**: To collect and normalize the collected data, cloud or edge data platforms are used. Real-time analysis is capable of streaming processing platforms such as Apache Kafka and Apache Flink or Cloud-Outland services such as AWS IOT and Azure Digital Twins. Historical data are stored in data lakes and good owners to enable offline policy check and model training.
4. **Team with AI Services**:
   - Machine Learning Models: Crowds, impressions of events, and forecasts of traffic.
   - Use controls that learn reinforcement for adaptive signal control.
   - This behavior benefits from data view lines for analysis, recognition of license plate and object of object.
   - Decision Engine: They synthesize data and trigger actions such as signal adjustment, rerouting, or alarm.
5. **Applications and Interface Layers**: Data ends for users and human operators through the dashboard, Control Centre Interface, Mobile app, and API. Traffic flow visually
6. **Simulation and Digital Twin Teams (Optional)**: To guess how traffic will behave in different scenarios, for example, during high hours or during special events, virtual urban models are developed. Before the AI model is used in the real world, the AI model is trained and evaluated in these virtual surroundings [1]. Generally used devices include unit-based simulation platforms, such as Matsim and Sumo (simulation of urban mobility).

6.3.1 Cloud-edge Cooperation

The distribution of the sky-edge hybrid system is scalable, but low-oppression traffic control is possible. While cloud server model training, historical data analytics, and large-scale coordination, edge knots, such as roadside units or local server processing time-sensitive activities such as object recognition or signal control administration [10].

It provides quick, local decision-making facilities, which are important for emergency preparedness and autonomous vehicle coordination, and it also reduces the number of networks and guarantees continuity in the loss of connection.

6.3.2 Security and Computer Management

Cyber security framework is part of a strong virtual architecture, which guarantees the privacy and integrity of traffic data. Strategies are used to protect against molesting, seizures or rejection of service, computer encryption, secure API, blockchain-based computer information, and strategies including an AI-based infiltration system detection system [11].

In addition, rules for managing data on the management of data on use of data, similar techniques, and access checks in accordance with laws such as GDPR and regional urban mobility ordinances.

Scalable, intelligent, and flexible urban transport is made possible in large parts of virtual infrastructure that supports AI-based traffic control systems. It provides an adaptable framework for continuous innovation, facilitates the communication between heterogeneous devices, and lets distributed artificial intelligence processing take place [6]. As smart cities grow, virtual architectures will become indispensable for the integration of autonomous cars, new mobility services, and environmental transportation goals.

6.4 AI Applications in Sustainable Urban Mobility

6.4.1 Predictive Maintenance of Public Transport Infrastructure

The sensor data (e.g., tracks, brakes, and motors) from buses, trains, trams, and infrastructure that comes with them can be tracked and analyzed by AI algorithms [11]. Predictive maintenance models can detect wear or failure by using machine learning techniques, such as deviations and the use of time series. This strategy: (1) Reduces reactive repair emissions and resource waste; (2) The asset increases the service life; (3) The service reduces obstacles; (4) And the dependence on public transport and users improves confidence (Figure 6.4) [2].

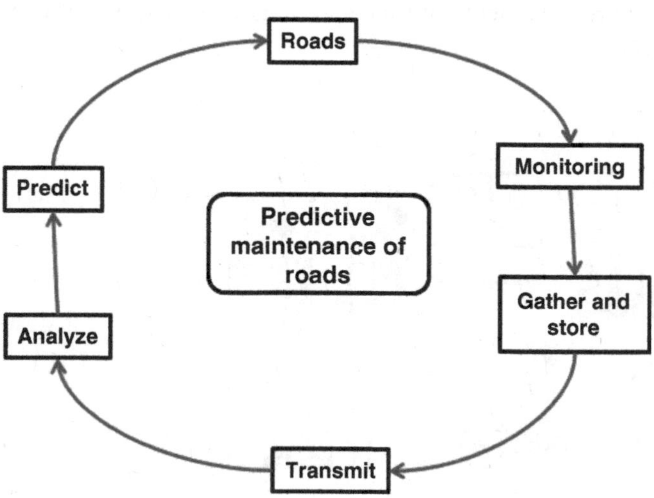

Figure 6.4 Predictive maintenance of public transport infrastructure.

6.4.2 Dynamic Riding Sharing and Micro Mobility Integration

Maximizing the requests from the trip, optimizing pickups and delivery sites, and dynamically adding together to demand shared mobility services (for example, shared taxis, e-scooters, and bicycles) by connecting customers with cars using AI [9]. AI can

- Reduce emissions per passenger kilometers,
- Reduce the number of trips and traffic on the roads, and
- Increase the accommodation rate for the vehicle using geographical data in real time and combinatory adaptation.

AI promotes a polymorphic, permanent urban dynamic ecosystem by improving the gap between various micro-movement platforms and traditional public transport networks.

6.4.3 Electric Vehicles (EV) Charging Optimization and Fleet Handling

AI is used to increase fleet operation and EV charging infrastructure as cities move toward electrical mobility (Figure 6.5) [5]. AI can:

- Predict demanding charging demand based on weather and travel patterns;
- Plan fleet charging to reduce grid loads and expenses;
- Suggest the best places for charging stations;
- Optimize battery health and energy use.

The carbon further purpose of neutrality and encourages the inclusion of EVS in urban transport networks

Figure 6.5 Electric vehicles (EV) charging optimization and fleet handling.

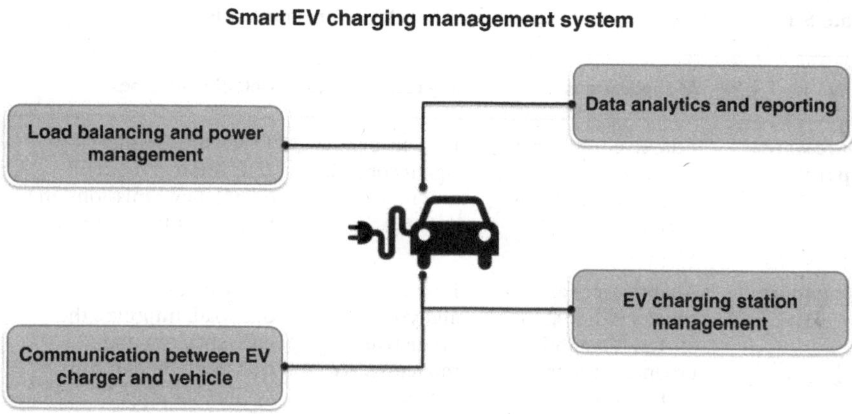

Figure 6.6 Data-driven mobility solution.

6.5 Data-driven Mobility Solution

AI in transport is based on its ability to use heavy datasets from different sources:

- IoT Sensors: Movement, traffic volume, and surroundings were placed on roads, intersections, and cars to collect information.
- GPS and Mobile Devices: Provide geolocation information for behavioral analysis, navigation, and vehicle tracking [10].
- Connected and Autonomous Vehicles (CAVS): Product data with high loyalty, real-time dynamics that can be used to adapt to the system (Figure 6.6).

This data is processed by big data analytics technologies, which produce the changed urban traffic conditions and real-time response to the action-rich insight.

6.6 Case Studies of AI Implementation in Smart Cities

Table 6.1 presents a comparative summary of AI implementation in smart cities.

6.7 Moral and Political Views

Strong moral frameworks and inclusive rules should correct the use of AI in transport:

- Data security and Privacy: Concerns for abuse and monitoring are raised by the collection of dynamic data. Techniques should be used for anonymization and encryption.

Table 6.1 Comparative summary of AI implementation in smart cities.

City	AI applications	Key technologies	Notable outcomes
Barcelona, Spain	Adaptive traffic lights, prioritization of emergency vehicles, urban mobility scheme	IoT sensors, AI signal control, mobility analytics	Travel time reduced by 21%, improvement of emergency, emissions and noise pollution reduced
Singapore	Adaptive signal control (Glide), off-test, demand prognose, timer customization	Real-time analytics, predictive modeling, AV systems	Reduced top hour overload, improves the efficiency of public transport, model shifts against stability
Los Angeles, USA	ATSAC system for real-time traffic optimization, AI Dashboard for operators	ML algorithms, edge computing, V2X infrastructure	12–16% inactive time reduction, smooth traffic flow, support for equity in the mobility scheme
Hangzhou, China	Video analysis, Explore of the Event, the city of Alibaba for smart coordination	Deep learning, cloud computing, real-time video feeds	Reduction of emergency time by 49%, 15% increase in traffic velocity, scalable AI infrastructure
Amsterdam, Netherlands	Cycle traffic optimization, delivery of the last meal, public mobility modeling	ML models, smart logistics, open data platforms	Dynamic cyclist signal control, low-supply discharge, data-informed policy

- Inclusion: The AI system should be designed to fulfil the needs of all urban population, including deprived groups, the elderly, and the disabled [12].
- Openness and Accountability: In order to guarantee trust and dependence in automated systems, the AI-declaration model must be sensible and auditable.
- Politics and Governance: In order to promote innovation and protect public interest, decision-makers must ensure that the rules are in line with the AI progress [13].

6.8 Future Trends and Innovations

The following are some new developments in AI for transport:

- Use Edge AI for Real-time Decisions: AI models running on the edge unit reduce local reaction and signal processing delay.

- Federated Learning: Cross-jurisdictional cooperative AI enables training without the exchange of personal information.
- Integration with Digital Twins: Urban Transport Systems allows imitation of virtual model interventions and adaptation of policy options.
- AI with 5G Synergy: Vehicle-to-infrastructure (V2I) interaction is improved for real-time coordination through high-speed, low-delegation connection.

6.9 Conclusion

Artificial intelligence changes city transport by facilitating wise, data-driven decisions in a durable mobility system and traffic management. AI improves productivity, security, and environmental stability in different types of applications, including optimization of electric vehicles, future indication, and dynamic signal control. The AI system can respond in an adaptable way to questions about urban mobility using IoT tools, connected cars, and real-time data from public infrastructure. These technologies reduce overload, low carbon emissions, and improve the general quality of life in cities. In addition, AI is encouraged by incorporating AI into a multimodal transport network. The practical benefits of AI-based solutions are depicted through real applications in smart cities. However, in order to guarantee responsible implementation, morality, legal, and data issues must be solved.

References

1 Chen, L., Jin, J., and He, Z. AI-powered traffic control in smart cities: A Case Study of Hangzhou City Brain. Sustainable Cities and Society, 63: 102476, 2020. https://doi.org/10.1016/j.scs.2020.102476.

2 Mohammadi, M., Al-Fuqaha, A., Sorour, S., and Guizani, M. Deep learning for IoT big data and streaming analytics: A Survey. *IEEE Communications Surveys & Tutorials*, 20(4): 2923–2960, 2018. https://doi.org/10.1109/COMST.2018.2844341.

3 Shaheen, S., Cohen, A., and Chan, N. Mobility on demand (MOD) and urban sustainability. *Current Sustainable/Renewable Energy Reports*, 6: 29–36, 2019. https://doi.org/10.1007/s40518-019-00125-y.

4 Ajuntament de Barcelona. Barcelona smart city strategy, 2020. Retrieved from https://ajuntament.barcelona.cat/digital/en/smart-city.

5 Zhang, Y., Yang, M., and Liu, S. Smart urban traffic management: A review of AI applications in city traffic systems. *IEEE Access*, 9: 103740–103758, 2021. https://doi.org/10.1109/ACCESS.2021.3099022.

6 Kumar, N., Mallick, P. K., and Zomaya, A. Y. (2020). Intelligent transportation systems: A review of machine learning and deep learning applications. *IEEE Transactions on Intelligent Transportation Systems*, 22(7): 3874–3885, 2020. https://doi.org/10.1109/TITS.2020.2995520.

7 European Commission. Artificial intelligence for the urban environment: Use cases and policy implications, 2020. Retrieved from https://ec.europa.eu.

8 Alibaba Cloud. City brain: Empowering smart cities with artificial intelligence, 2019. Retrieved from https://www.alibabacloud.com/solutions/city-brain.

9 Li, Y., Zheng, Y., Zhang, H., and Chen, L. (2018). Traffic prediction in a bike-sharing system. In *Proceedings of the 24th ACM SIGKDD International Conference on Knowledge Discovery & Data Mining*, pages 318–327. https://doi.org/10.1145/3219819.3219885.

10 Wang, X., Wang, F. Y., Zhang, Y., and Wang, L. (2021). Reinforcement learning for traffic control: Models and applications. *Journal of Advanced Transportation*, 8812543, 2021. https://doi.org/10.1155/2021/8812543.

11 Land Transport Authority (LTA). Smart mobility 2030: Intelligent transport systems strategic plan. Government of Singapore, 2022. Retrieved from https://www.lta.gov.sg.

12 Gemeente Amsterdam. Amsterdam smart mobility programme. Amsterdam Smart City, 2021. Retrieved from https://amsterdamsmartcity.com.

13 City of Los Angeles Department of Transportation. ATSAC system overview and performance metrics, 2021. Retrieved from https://ladot.lacity.org.

7

Reinforcement Learning for Energy-efficient Urban Freight Transportation

Nancy Jasmine Goldena and R. Rashia Subashree

Department of Computer Applications and Research Centre, Sarah Tucker College (Autonomous), Affiliated to Manonmaniam Sundaranar University, Tirunelveli, Tamil Nadu, India

Corresponding authors: nancy_lordwin@rediffmail.com; rasiyakrishnan@gmail.com

7.1 Introduction

Urban goods transport is fundamental to the function of today's cities, which causes a continuous flow of goods in complex and overloaded urban landscapes. As the global change to e-commerce increases and urbanization increases, the amount and frequency of freight movements have increased [1]. However, this growth is accompanied by significant challenges: increasing fuel consumption, increasing traffic stops, and increasing environmental footprints due to carbon emissions. These concerns require more intelligent and durable logistics solutions that can suit real-time variations in demand, infrastructure, and traffic conditions.

The traditional freight management system is mainly based on static routing and planning frameworks. Although these methods can offer an estimated production under stable conditions, they have a lack of necessary flexibility and adaptability in today's dynamic urban environment [2]. Static systems are often less adaptable to energy consumption, react to unexpected traffic disorder or adapt to change the customer's needs. When urban emissions are reduced and green areas are used, the boundaries of traditional logistics strategies have become clear quickly.

A subdomain of AI, reinforcement learning (RL), has proven to be a powerful alternative. RL enables agents to learn optimal decisions by interacting with the environment, getting responses and continuously improving performance over time [3]. In the case of urban freight transport, RL offers a data-driven, adaptive approach that can increase the decision in fleet management, distribution planning and route planning. By learning from real-time data and long-term results, the RL system has the ability to reduce operating costs, increase distribution efficiency and significantly reduce energy [4].

The purpose of this chapter is to find RL applications in energy-efficient transport of urban products. It focuses on three major RL Techniques: Marl, PPO, and DQN and discusses their practical implications, integration with smart logistics techniques and real-world implementation challenges. Through this discussion, the paper wants to provide a basis for understanding how RL can make urban systems more durable, intelligent and responsible components in future smart cities.

7.2 Fundamentals of RL

The relevance of urban shipping replacement learning is a type of machine learning where an instrument learns to decide by taking action in an environment and receiving a reaction, as shown in Figure 7.1. Unlike monitored learning, where the model learns from a label data set, RL depends on exploration and utilization [2]. The agent examines different strategies and utilizes those who receive the highest cumulative price over time.

The primary components of RL contain:

- Agent: To make a decision and interact with the environment.
- Environment: The external system with which the agent interacts.
- Condition: Represent the current environment.
- Action: A set of all sorts of tricks that can be created by the agent.
- Reward: An environment based on action.
- Policy: Agent's method of choosing action based on states.
- Price function: An estimate of future prices that helped the agent determine whether conditions or tasks are desired.

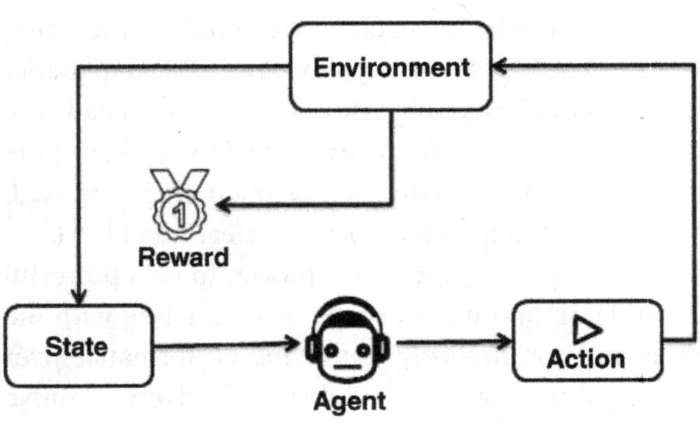

Figure 7.1 Fundamentals of reinforcement learning.

When it comes to urban freight logistics, the environment represents the city's infrastructure, traffic patterns, distribution points and the availability of the fleet. The agent can be a delivery vehicle, a route system or a central transfer check. Activities may include selection of delivery paths, adjusting speeds, or assigning vehicles [5]. The reward structure can be designed to reflect important performance indicators such as fuel efficiency, distribution time, customer satisfaction and carbon emissions.

The urban freight systems are naturally dynamic and complex. Time decisions should be taken over unexpected factors such as unexpected factors such as traffic consequences, road closure, and delivery priorities such as unexpected factors RL's adaptive learning capabilities make it a good fit for these kinds of environments. For example, it allows logistics suppliers to adapt to vehicle routes dynamically because new information is available, which improves distribution performance by reducing energy consumption [1]. Moreover, RL can be integrated with other technologies such as IoT sensors, GPS tracking, and data analytics platforms to form smart logistics networks. These networks can provide rich, real-time data that enhances the learning process and leads to more accurate and efficient decision-making.

In summary, RL provides a robust and flexible framework for tackling the challenges of modern urban freight transportation [6, 7]. By learning from continuous feedback and adapting to changing conditions, RL-based systems can drive significant improvements in energy efficiency, cost-effectiveness, and sustainability in logistics operations.

7.3 RL Applications in Energy-efficient Urban Freight Transportation

Reinforcement Learning has been increasingly adopted to solve a range of complex problems in urban freight logistics, as shown in Figure 7.2. Its applications are particularly promising in three core areas: vehicle routing, delivery scheduling, and fleet management. Each of these areas is critical to the efficiency and environmental impact of freight systems and stands to benefit significantly from adaptive, data-driven decision-making.

7.3.1 Dynamic Timetable (OR) Route Customization

Strengthening learning vehicles (RL) enabled to choose the optimal routes that are beneficial in reducing energy consumption and travel time as shown in Figure 7.3. By incorporating real-time traffic data, road conditions and vehicle-specific energy profiles, RL agents such as DQN or PPO learn to avoid overloading or high-breakup overload [8]. Over time, these agents improve the decision by using the response

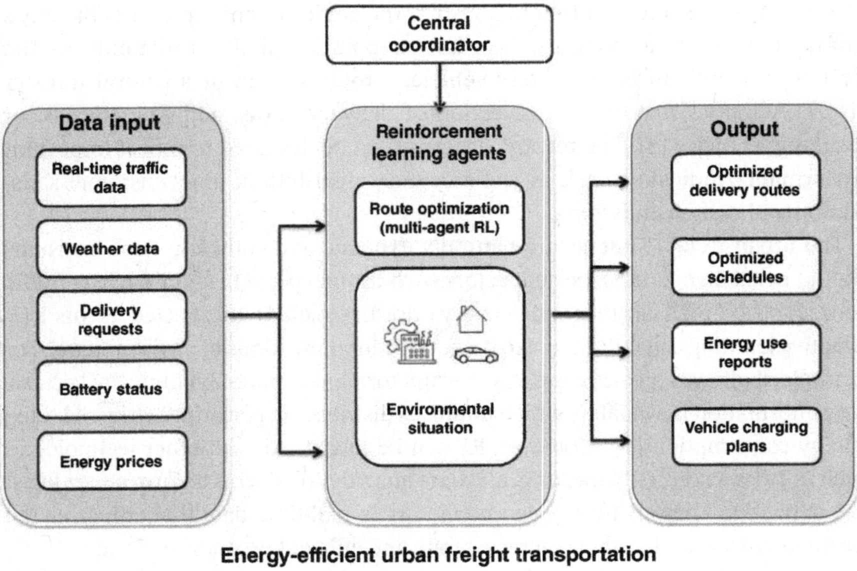

Figure 7.2 Energy-efficient urban freight transportation.

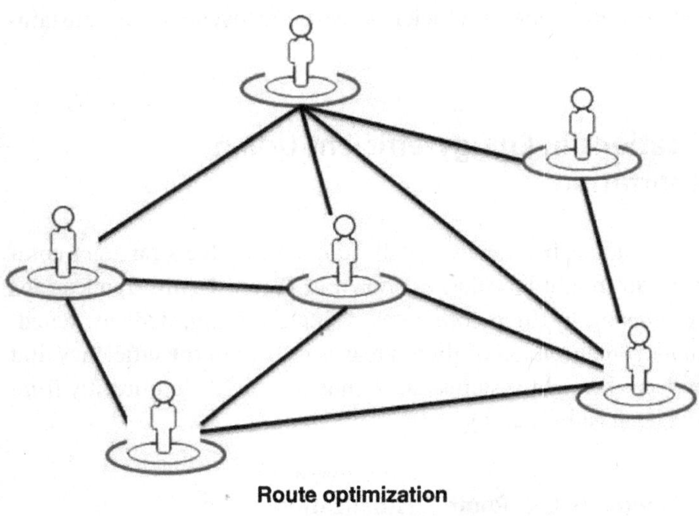

Figure 7.3 Dynamic timetable (OR) route customization.

from the environment, resulting in fuel-capable and timely delivery. Simulation platforms such as Sumo (simulation of urban mobility) are often used for training and testing these funds under realistic traffic conditions.

Marl is especially useful in landscapes related to many vehicles [4]. In such cases, individual RL agents (each vehicle) are in a shared environment, not only a teacher from their own actions, but also from the behavior of other vehicles. This approach helps to handle the amount better and adapt the routes when multiple vehicles are involved in a delivery system.

7.3.2 Delivery Scheduling

RL can adapt it to not only how the vehicles go, but also when delivery should be made. By accounting for traffic flows, customer accessibility and vehicle load time, RL algorithms help develop adaptive programs that balance timely limitation of time with energy efficiency [7]. In time-sensitive logistics scenarios, such adaptation can improve the reliability and use of resources to a large extent. Here, a role plays in optimizing the schedule in a fleet of marl vehicles, coordinating their routes and time to ensure that the distribution process is personal and efficient as a collective system. For example, PPO or DQN can be used in a multi-agent setup, where each vehicle adopts its plan based on local conditions with a view to the functions of other vehicles [9].

7.3.3 Fleet Management

Urban products that carry a fleet contain a variety of vehicles that need to be assigned and maintained efficiently. RL facilitates the allocation of an intelligent fleet by predicting demand and assigning vehicles, accordingly, as shown in Figure 7.4 [3]. It also helps to make decisions in real time for vehicle transmission, imbalance and maintenance planning, and eventually leads to downtime and energy waste.

In fleet management, the Marl algorithm can be used to handle interactions between different vehicles or even vehicles and centralized transmission systems [10]. For example, multi-agent deep deterministic policy gradient (MADDPG) can be used to improve coordination in the allocation of vehicles and rebalancing, ensuring that each vehicle is used based on its current location, load and other factors.

7.3.4 Vehicle Platooning and Coordinated Driving

In platooning, a group of freight vehicles will be allowed to run at the same time with synchronized speeds to reduce the use of aerodynamic drag and fuel.

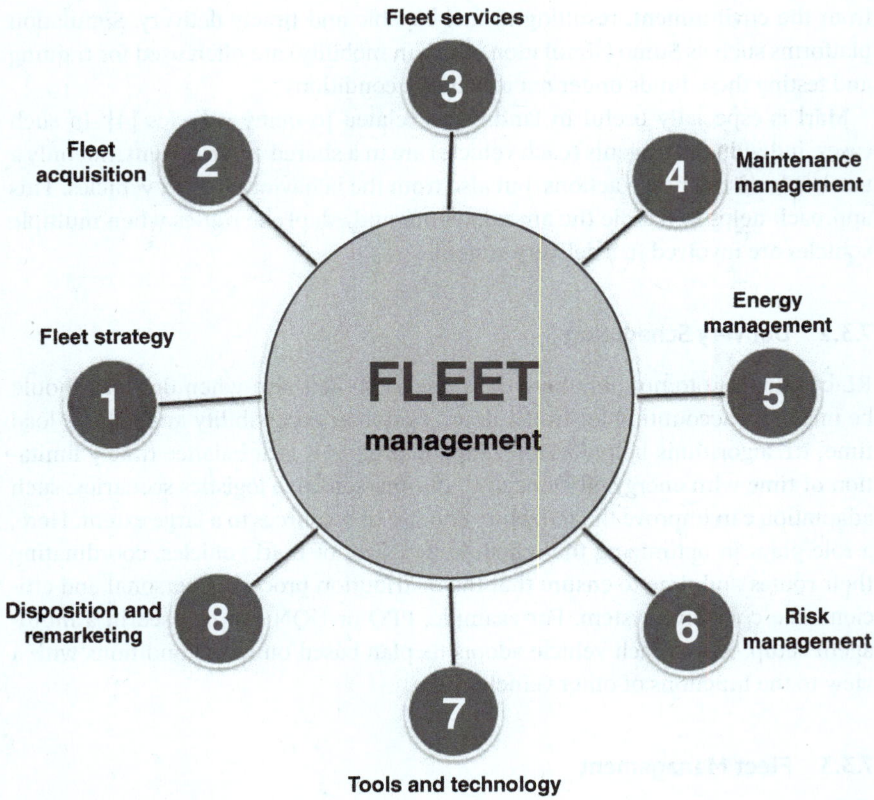

Figure 7.4 Fleet management.

Cooperative reinforcement learning (CO-RL) is used to handle the distance between vehicles, acceleration patterns and orbit change [11]. Algorithms such as MADDPG are able to work independently by maintaining coordination with others, as shown in Figure 7.5. This technique is especially useful in motorway scenarios, where frequent driving speeds and formation checks can have sufficient energy savings in the fleet.

In Platooning, Marl ensures that many vehicles can learn to cooperate without a centralized control, each vehicle adjusts its tasks based on other people's positions and functions, increasing efficiency and reducing energy consumption [12].

7.3.5 Examples and Use of Cases

Logistics involves projects from companies such as Amazon and DHL in logistics, which use RL-based systems to optimize final meal distribution. These systems

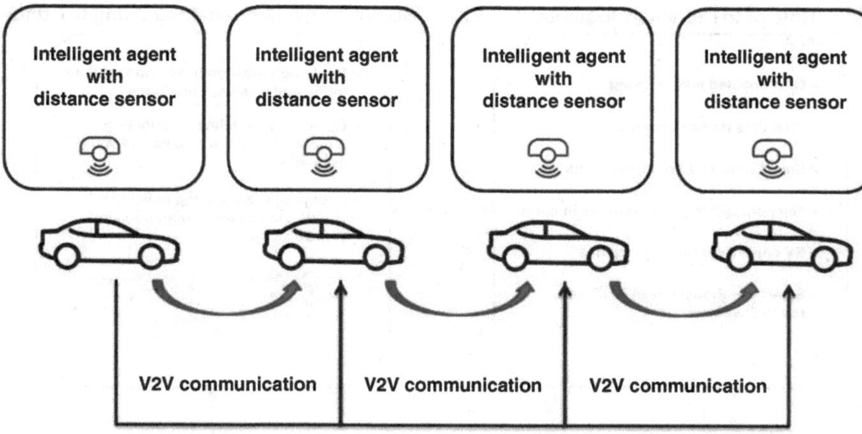

Figure 7.5 Vehicle platooning and coordinated driving.

often appoint PPOS or DQN to adapt to the passage of individual vehicles [5]. Urban pilot programs have also discovered to use Marl for coordinating several autonomous distribution robots, which aim to reduce overlapping routes and energy consumption. These cases highlight RL in specific benefits and logistics sectors [3].

7.4 Integration of RL Applications with Smart Logistics and IoT

Figure 7.6 illustrates the integration of RL applications with smart logistics and IoT.

7.4.1 IOT's Role in Smart Logistics

IoT makes sensory and connection paths of smart logistics. This allows the system to see, interpret and react to dynamic real conditions.

Key IoT Component:

- **GPS Capable Floor Training**: Enables real-time monitoring and passage optimization.
- **Traffic and Weather Sensors**: Provide environmental references that affect routing and safety [10].
- **Telemetry System for Smart Vehicles**: Monitor vehicle health matrix, driving behavior and fuel/battery consumption.

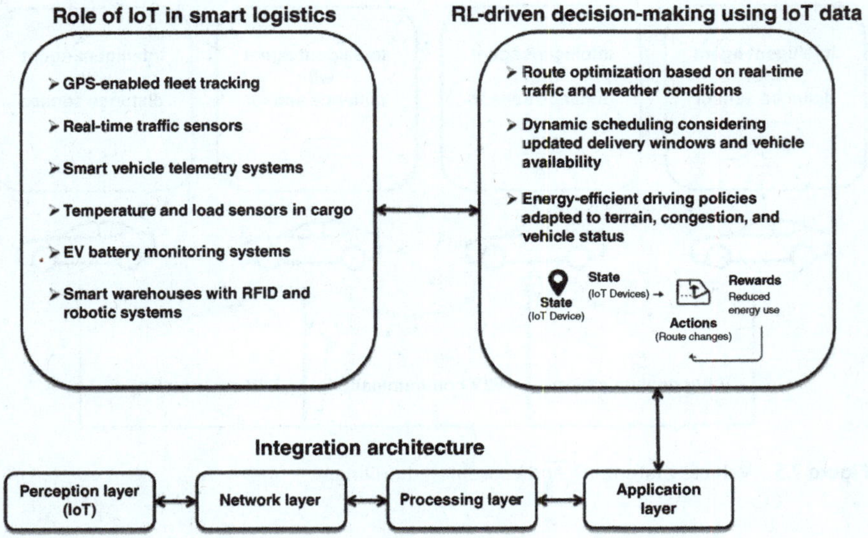

Figure 7.6 Integration of RL applications with smart logistics and IoT.

- **Last Status Sensor**: Tracing temperature, moisture, shock and load safety in sensitive shipments.
- **EV Battery Monitoring System**: Real-time charge status, battery health and range estimate supplies [13].
- **Smart Warehouse**: UTNEV RFID system, computer vision and autonomous robotics to manage inventory and fulfil orders.

This wide sensor network generates large versions of high-resolution, time-stamped data, which acts as observation (condition) for RL models.

7.4.2 Takes RL-powered Decisions Using IoT Data

Reinforcement learning (RL) is well-suited for logistics, where the system must make decisions in a constantly uncertain and dynamic environment [9].

How It Works:

- **State**: IoT data (e.g., vehicle, battery level, and traffic conditions).
- **Action**: Rooting options, planning updates, maintenance decisions, etc.
- **Reward**: Quantity based on CPI as fuel savings, distribution time ban, energy use, or customer satisfaction.

Applications of RL with IoT Input:

- **Route Optimization**: RL learns talented trails using live GPs and traffic data to reduce the time of driving and energy. DQN or PPO can be used to make routing decisions based on real-time traffic data [7].
- **Dynamic Planning**: Load priority, road status and driver's real-time updates for delivery paths based on the availability of the driver [13]. Marl can be used to coordinate multiple marriages, which ensures dynamic adjustment based on the functions of several agents in the system.
- **ECO Drawing Strategies**: Learn energy-capable acceleration, braking and path-transformation behavior in different areas. PPO here is an appropriate alternative for continuous adaptation of driving policy.
- **Resource Allocation**: Shipping weights or adaptive allocation of vehicles based on demand forecast and real-time availability. Marl can be used to allocate resources in many vehicles, enabling real-time coordination in fleet management [6].

RL funds continuously improve over time through new patterns in IoT data, testing and testing and reaction.

7.4.3 Integration Architecture

A layered architecture enables spontaneous interaction between IoT sensor and RL algorithm.

Layer:

- **Perception Layer (IoT Unit)**: Physical logistics gathers unprocessed environmental data.
- **Networking**: Uses protocols such as 5G, LoRaWAN, NB-OT, or edge computing for little distances.
- **Treat Team (RL Engine)**: Policy performs updates, action choices, and feedback integration.
- **Application Layer**: Interface with fleet management systems (FMS), transport management systems (TMS), warehouse management systems (WMS), and robotic controller.

Key Architectural Features:

- **Edge AI Payment**: RL-INTRANCE running at the edge (e.g., vehicle units) for low-option decisions.

- **Cloud-edge Collaboration**: Cloud-based, edge-projected, scalable, and effective training.
- **Closed Loop-reaction**: Transportation for the RL model to continually learn from actions, feedback to it.

7.4.4 Use Cases in Urban Goods

Smart Dispatch and Routing

- RL agents consider dynamic vehicles by assessing factors such as battery level, load, driver hours, and traffic forecasts.
- The Marl model can be used to ensure that many vehicles are coordinated, and multi-agent settings adapt to real-time environmental changes, improving the efficiency of the system [14].
- For example, Amazon and DHL have begun using AI-based routing for adjusting final meals, which may include PPO or Dix.

Predictive Maintenance

- IoT-based health data (engine temperature, vibration, and fluid level) triggers RL-based guidelines only for time maintenance [12].
- Marl can optimize maintenance planning in many vehicles, reduce undeserved shutdown and adapt to fleet-wide maintenance strategies.

Stock Automation

- Robot equipped with IoT and RL systems learn to adapt to shelf socks, pick squares, and order sequentially [15].
- PPO or DQN can be used here for training robot systems to maximize the flow and reduce operating costs in the storage.

EV Recharging Adaptation

- RL agents Learn Guidelines for optimizing charging plans by assessing real-time network load, dynamic pricing, and vehicle usage patterns.
- DQN can be especially useful for learning an optimal charging policy based on past experiences and predictions on the demand for battery [4].

The Cold Chain Management

- RL models adjust the cooling system based on temperature sensors, travel time, and package sensitivity [16].
- Marl can be used to coordinate the functions of several cooled vehicles, ensuring optimal environmental conditions during transport.

The Most Important Insight on RL-algorithm Integration with IoT:

1. **Data Flow**: Real-time sensor from IoT units feed the data directly into the RL model, which makes decisions and determines.
2. **Taken Adaptive Decisions**: Integration means that RL can be beneficial to make decisions based on the upcoming data (e.g., change the routes due to new traffic conditions).
3. **Multi-agent Learning (Marl)**: Make sure many vehicles or robots can coordinate their functions, to learn from both their own behavior and others to maximize the general system efficiency of the fleet.

7.4.5 Benefits of RL–IoT Integration

Table 7.1 summarizes the benefits of RL–IoT integration.

The integration of RL, IoT and Marl in smart logistics systems unlocks significant capacity to increase energy efficiency, reduce operating costs, and increase stability [3]. The use of PPO, DQN, and Marl ensures that logistics operations are continuously adapted in real-time data, improvement, routing, planning, fleet management, and maintenance processes. This dynamic, adaptive system changes logistics in urban goods, and drives both operating improvement and environmental benefits.

Table 7.1 Benefits of RL–IoT integration.

Benefit	Description
Energy Efficiency	Through optimal routing, eco-driving, and smart charging.
Operational Cost Reduction	By reducing fuel, idle time, maintenance, and resource wastage.
Improved Reliability	Real-time adaptation to disruptions ensures consistent delivery performance.
Scalability	**Marl** models scale easily across large logistics fleets and warehouse networks.
Automation and Autonomy	Enables autonomous decision-making in warehousing, delivery, and fleet management.
Sustainability	Lower emissions via intelligent energy use and resource allocation.
Data-driven Continuous Learning	**RL agents** improve over time, learning from every new delivery, failure, and success.

7.5 Challenges and Limitations

Here are the challenges and limitations of integrating RL with smart logistics and IoT, which are broken into individual points in energy efficient urban transport:

7.5.1 Data Quality and Availability

Challenge: RL models depend on high quality data from the IoT sensors. Problems such as GPS units, sensor operation or lack of data can adversely affect the model's learning and decision-making quality.

Specific Limit: IoT networks, especially in urban areas, may experience coverage intervals, especially in a dense environment where there may be insufficient data for signal intervention or inadequate infrastructure. In addition, the sensors can degrade over time, affecting their accuracy and stability.

Example: Traffic sensors may not always reflect the status of the actual road and create the wrong route plan.

7.5.2 Calculation Complexity

Challenge: The RL model in large-scale logistics systems require important calculation resources due to the complexity of the training decisions and the amount of data from sensors. The model must handle real-time data processing, which is calculated intensively.

Specific Limit: In edge computing environment (e.g., vehicle treatment), limited calculation power can make it difficult to use RL models that require intensive processing. This can lead to a slow decision-making or lack of adaptability in real-time logistics operations.

Example: Real decision in traffic routing may not be possible with limited calculation resources, which can lead to delays or suboptimal performance.

7.5.3 Integration with Legacy Systems

Challenge: Many logistics operations depend on inheritance systems (e.g., traditional fleet management software, inventory management systems) that were not created to integrate with modern IoT units or RL-based systems. Integration into an integrated system can be expensive and time-consuming.

Specific Limit: Problems with compatibility between new RL algorithms and existing systems often require extensive upgrading or system overhaul, which can increase both deployment time and cost of distribution. In addition, data silos between inheritance systems and new IoT infrastructure can prevent seamless data flow.

Example: A logistics company's old fleet management software cannot support real-time data input from the IoT sensors such as the vehicle's telemetry or traffic updates.

7.5.4 Security and Reliability Problem

Challenge: The autonomous RL agents in urban shipping transport, especially for the delivery of final mile, are susceptible to unprotected or unexpected decisions whose conditions in the real world are not strictly trained and tested.

Specific Limit: Even with well-designed RL systems, real-time decisions (e.g., urban traffic, pedestrian interaction) can introduce risks in the very dynamic environment. RL agents can take measures that are unsafe, especially in scenarios where training data is limited or where unexpected events occur (e.g., closed roads, accidents).

Example: The delivery drone using RL can make wrong decisions about obstacles or pedestrians due to unexpected factors, leading to potential accidents.

7.5.5 Moral and Regulatory Questions

Challenge: There is a lot to use RL in urban goods that introduce moral concerns for data privacy (e.g., vehicle placement), monitoring (e.g., continuous monitoring of driver behavior), and algorithm bias (e.g., decisions to determine in favor of any areas or demographics).

Specific Limit: RL regulatory structures in logistics and autonomous systems are still developing. Privacy laws (e.g., GDPR for data protection) and vehicle safety standards (e.g., autonomous vehicle regulations) are a developed challenge. Failure to follow these rules can delay or stop the distribution of RL solutions.

Example: The use of RL-based systems in autonomous distribution cars may require strict government approval and site tracking can lead to challenges related to privacy problems.

7.6 Future Directions

- Integration of RL and IoT in urban goods logistics is ready to go significantly through several major directions in the future. An important development area is the development of scalable Marl systems, which enables dynamic coordination between large fleets of asymmetrical distribution agents such as trucks, drones and autonomous robots. These systems should support making decentralized decisions with minimal communication overhead to act effectively in real-time urban environment.

- The progress of learning routine and lifetime will allow RL agents to withdraw from scratches without being adapted to new cities, regulatory environment and traffic patterns, which will improve long-term distribution efficiency. In order to ensure the safety and compliance of regulations, there is an increasing requirement for secure and clear RL techniques that can provide transparent decisions to respect the real-life barriers of the world.
- In addition, hybrid frameworks that combine RL with traditional adaptation techniques and forecast models are expected to be more widespread. These better routes plan, energy consumption forecasts, and will be able to demand prediction, both benefit from the power-driven learning force (PPO and DQN) and classic planning strategies such as algorithms. The distribution of the real world will be expanded beyond the increase of edge data and mild RL models, able to estimate units of vehicles or warehouses, which reduces deletion and dependence on cloud infrastructure.
- Integration with smart network infrastructure and renewable energy sources will also be important, as RL can be used to sync vehicle charge and operation with demand for the grid and energy availability. Finally, the implementation of Federated and Privacy Conservation RL will allow logistics suppliers to allow trained models in a disguised fleet without compromising on sensitive operating data, which paves the way for a safer and intelligent logistics system.

7.7 Conclusion

RL and IoT convey urban goods logistics by enabling convergence, intelligent, adaptive and energy-efficient operation of technologies. Through applications such as dynamic root optimization, delivery planning, fleet handling and smart stock automation, RL agents, especially the algorithms such as algorithms such as proximal policy-rating (PPO), deep Q-network (Dix) and Ful-Research. The real-time data collected through the IoT sensor acts as a rich base for these models, providing the possibility of continuous learning and system-wide adaptation. In addition, the integration of RL paves with age calculation, smart online coordination and future maintenance the way for more durable and autonomous logistics ecosystems. As this area develops, future research should solve challenges such as scalability, security, clarity and privacy, and ensure that the

RL-powered logistics systems are not only effective, but also strong, transparent and broadly distributed. Finally, the RL IoT paradigm represents a transformative approach to creating smart, green areas and more flexible urban goods law.

References

1 Nazari, M., Oroojlooy, A., Snyder, L., and Takác, M. Reinforcement learning for solving the vehicle routing problem, *arXiv preprint arXiv:1802.04240*, 2018. https://arxiv.org/abs/1802.04240.

2 Ulmer, M. W., and Thomas, B. W. Same-day delivery with time windows and uncertain customer requests. Transportation Science, 52(2): 392–414, 2018. https://doi.org/10.1287/trsc.2016.0726.

3 Yaqoob, I., Hashem, I. A. T., Ahmed, A., Gani, A., Imran, M., and Guizani, M. Internet of Things architecture: Recent advances, taxonomy, requirements, and open challenges. *IEEE Wireless Communications*, 24(3): 10–16, 2017. https://doi.org/10.1109/MWC.2017.1600421.

4 Yu, Y., Chen, X., Zhu, J., and Jiang, Y. Federated reinforcement learning: Techniques, applications, and open challenges. *Information Fusion*, 90: 1–20, 2023. https://doi.org/10.1016/j.inffus.2022.10.009.

5 Dong, Y., Qi, Y., Yang, B., and Jin, J. Multi-agent deep reinforcement learning for urban delivery dispatch optimization. Expert Systems with Applications, 200, 117020, 2022. https://doi.org/10.1016/j.eswa.2022.117020.

6 Mnih, V., Kavukcuoglu, K., Silver, D., Rusu, A. A., Veness, J., Bellemare, M. G., Graves, A., Riedmiller, M., Fidjeland, A. K., Ostrovski, G., Petersen, S., Beattie, C., Sadik, A., Antonoglou, J., King, J., Kumaran, D., Wierstra, D., Legg, S., and Hassabis, D. Human-level control through deep reinforcement learning. *Nature*, 518(7540): 529–533, 2015. https://doi.org/10.1038/nature14236.

7 Mohsen, B. M. AI-driven optimization of urban logistics in smart cities: Integrating autonomous vehicles and IoT for efficient delivery systems. *Sustainability*, 16(24): 11265, 2024. https://doi.org/10.3390/su162411265.

8 Tassiulas, L. and Ephremides, A. Stability properties of constrained queueing systems and scheduling policies for maximum throughput in multihop radio networks. *IEEE Transactions on Automatic Control*, 37(12): 1936–1948, 1992. https://doi.org/10.1109/9.182479.

9 Schulman, J., Wolski, F., Dhariwal, P., Radford, A., and Klimov, O. Proximal policy optimization algorithms. *arXiv preprint arXiv:1707.06347*, 2017. https://arxiv.org/abs/1707.06347.

10 Mallik, S., Qin, H., Nandi, S., Alam, M. Y., Roy, A., and Sardar, T. H. *Applications of Computational Learning and IoT in Smart Road Transportation*

System. Springer Tracts on Transportation and Traffic, Springer Cham, 2025. https://doi.org/10.1007/978-3-031-87627-1.

11 Zhang, K., Yang, Z., and Basar, T. Multi-agent reinforcement learning: A selective overview of theories and algorithms. *Handbook of Reinforcement Learning and Control*, 321–384, 2021. https://doi.org/10.1007/978-3-030-72109-5_9.

12 Niu, T., Ma, T., and Zeng, Z. Reinforcement learning in multi-agent delivery systems: A review and outlook. *Applied Sciences*, 10(24): 8857, 2020. https://doi.org/10.3390/app10248857.

13 Lowe, R., Wu, Y., Tamar, A., Harb, J., Abbeel, P., and Mordatch, I. Multi-agent actor-critic for mixed cooperative-competitive environments. *arXiv preprint arXiv:1706.02275*, 2017, https://arxiv.org/abs/1706.02275.

14 Chen, J., Liu, H., and Zhang, W. *A survey on reinforcement learning for transportation systems. IEEE Transactions on Intelligent Transportation Systems*, 23(5): 3930–3948, 2022. https://doi.org/10.1109/TITS.2021.3065341.

15 Torres, M. F. M., Aguilar, J. J. R., Domínguez, I. J. T., and Del Mar Cerbán Jiménez, M. A cost-optimized model for dynamic port hinterland delineation: An application to the Spanish port system. *Applied Sciences*, 15(1): 96, 2024. https://doi.org/10.3390/app15010096.

16 Zhao, Z., Chen, W., Wu, X., Chen, P., and Zhang, J. Deep reinforcement learning for vehicle routing problem with stochastic demand. *Transportation Research Part C: Emerging Technologies*, 97: 657–681, 2019. https://doi.org/10.1016/j.trc.2018.11.022.

8

Advancements and Challenges in Autonomous Vehicles and Smart Mobility: The Role of AI in Transforming Transportation

A. Jane[1,], K. Merriliance[1], and*
Mary Immaculate Sheela Lourdusamy[2]

[1] *Department of Computer Applications and Research Centre, Sarah Tucker College (Autonomous), Affiliated to Manonmaniam Sundaranar University, Tirunelveli, Tamil Nadu, India*
[2] *Department of Computing and Engineering, School of Engineering Technology and Applied Science (SETAS), Heritage Christian University College (HCUC), Accra, Ghana*

*Corresponding author: jane77johnson@gmail.com

8.1 Introduction

The rapid advancement of artificial intelligence (AI), sensor technology, and networking has fueled the advent of self-driving cars and smart mobility, fundamentally altering how we transport people and commodities. These autonomous vehicles navigate roads with little to no assistance from humans by utilizing a combination of sensors, cameras, radar, and AI-powered algorithms. Several companies and researchers worldwide are fighting for the first completely autonomous cars as the race to build these vehicles heats up. Self-driving cars have many advantages. Since human error is a leading source of accidents, they have the potential to greatly increase road safety.

People who are unable to drive, such as the elderly and disabled, get more freedom and mobility. Due to their ability to reduce or even eliminate human error, autonomous cars have the potential to significantly improve vehicle safety and revolutionize the automotive industry. These vehicles will be equipped with advanced technology like sensor fusion, computer vision, path planning, control systems, and problem detection. While these technologies were implemented using standard algorithms, deep learning has emerged as a potent substitute in the past ten years, surpassing conventional techniques in a range of applications and

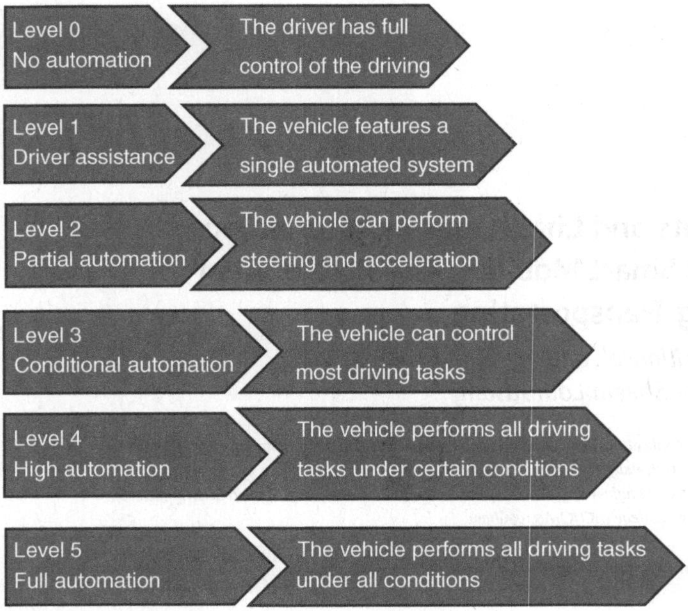

Figure 8.1 Levels of vehicle autonomy.

fields of study. It has had a striking effect on how autonomous cars are designed and function.

Autonomous vehicle technology is based on deep learning, reinforcement learning, and neural networks, which enable cars to carry out tasks including sensing, planning, and control. Convolutional neural networks, in particular, are excellent at processing sensor input, including camera images, for object detection and scene comprehension. By concentrating on maximizing rewards for good actions, reinforcement learning—more especially, deep reinforcement learning—allows cars to learn intelligent decision-making techniques in dynamic environments. The Society of automotive engineers (SAE) categorizes these cars into six levels (0 to 5) of driving automation, which have been approved by the United States Department of Transportation (Figure 8.1).

Level 0 (No Driving Automation): The majority of cars in use today are in this category, with manual controls for every function. Even with the help of technology, such as an emergency braking system, the driver is still responsible for the "dynamic driving task." It doesn't operate the car, hence it doesn't qualify as automation.

Level 1 (Driver Assistance): Here, one of the automatic features is either for steering or acceleration, moreover like cruise control. Other driving tasks, such as braking and steering, still fall under the responsibility of the human behind the wheel. For instance, adaptive cruise control that helps to maintain a safe distance from the car ahead fits into this Level 1 category.

Level 2 (Partial Driving Automation): The automobile can manage steering and acceleration/deceleration, but it falls short of achieving full self-driving capabilities because a human is always in the driver's seat and can take over at any time.

Level 3 (Conditional Driving Automation): This level includes cars that can detect their surroundings and make decisions on their own. For instance, speeding up to pass a stationary object can be automatically decided by the car. However, a human's presence in any place of necessity is required, rendering the driver to take over an unhandled situation by the car.

Level 4 (High Driving Automation): In Level 4, the vehicles focus on autonomous driving as reality. It comes with a limitation that they can only operate in particular areas, mostly urban environments, where speeds hover around 30 mph. This limitation is termed as geofencing. Because of this catch, most Level 4 cars are highly focused on ride-sharing services at the moment.

Level 5 (Full Driving Automation): The "dynamic driving task" is no longer an issue with Level 5 automobiles. These automobiles do not have braking or acceleration pedals. They're free from the limitation of geofencing, and they can handle any driving task that a skilled human driver can manage. Although there are several locations around the globe that test fully autonomous vehicles, none are available for public use just yet.

Smart mobility is all about a wide aspect of transportation that blends together digital technologies, shared mobility, and sustainable solutions to build urban mobility systems that are efficient, safe, and eco-friendly. This concept includes everything from connected vehicles and ride-sharing services to electric vehicles (EVs) and smart traffic management, along with mobility-as-a-service (MaaS) platforms that provide smooth transportation options. As a new idea, smart mobility is highly in tune with the Sustainable Development Goals set by the United Nations for 2030. Our thoughts about mobility, especially in relation to city planning and the logistics of moving goods and people, are set to change drastically in the years ahead. Moving forward, the essential components of smart mobility will cover the key aspects like sustainability, active transport, the use of green fuels, and engaging with citizens. The impact of these changes will be wide-ranging, touching on sustainability, the economy, and quality of life, all of which will directly influence citizens and government entities, similarly (Figure 8.2).

Two major concepts that contribute in shaping the future of smart mobility in the years ahead are the transportation of goods and the shift in how we think about moving people, which is leaning toward a mobility-as-a-service model. While observing the arena of transporting goods, we can already find various robot prototypes around the globe delivering essential items right at people's homes. This

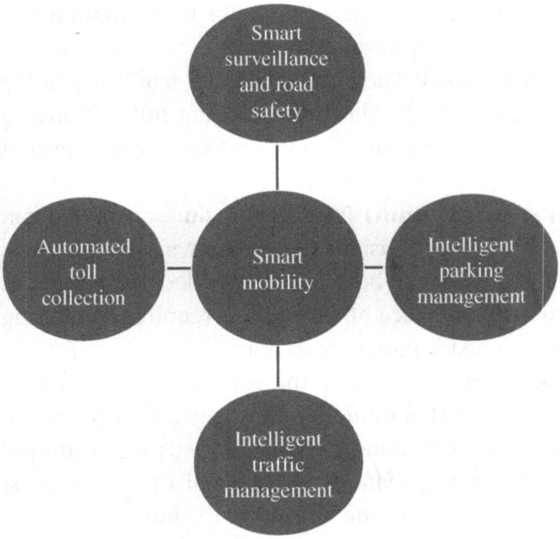

Figure 8.2 Benefits of smart mobility.

shift will require businesses like supermarkets and delivery services to adapt their operations. Technologies like the internet of things (IoT), big data, and artificial intelligence (AI) will be vital in developing these new solutions. It's important to note that the changes in the upcoming years, which will reshape the jobs of the future in an inevitable manner. In terms of human transportation, the trend toward mobility-as-a-service will significantly reduce the number of private vehicles on the road, paving the way for lighter, more compact, shared, electric, and frequently autonomous vehicles.

8.2 Advancements in Autonomous Vehicles and Smart Mobility

Advances in smart mobility and driverless automobiles are causing a massive transformation in the transportation industry. Through the use of cutting-edge technology like improved sensors, AI, and networked infrastructure, these advancements seek to not only make travel safer and more effective but also to offer ease. They are also laying the foundation for more accessible and dependable travel choices for anyone.

The automobile industry has been impacted by AI, especially in the area of autonomous vehicles, or self-driving cars. These vehicles' sophisticated AI systems let them to navigate and operate without the assistance of a human.

AI is essential to these cars' ability to sense their environment, make decisions in real time, and safely drive on highways. AI uses a range of technologies, including computer vision, sensor fusion, machine learning algorithms, and improved connectivity, to make sure the automobile operates effectively.

When combined, these technologies allow autonomous cars to interpret data from multiple sensors, analyze that data, and make decisions that are on par with or even better than those made by human drivers. Developing cars that are not only self-driving but also enhance passenger enjoyment, safety, and traffic control is the ultimate objective. AI technologies enable automobiles to detect their surroundings using a range of sensors, such as LIDAR, cameras, RADAR, and ultrasonic sensors. Extensive training is used by machine learning algorithms to assess incoming data, identify objects, anticipate movement patterns, and make snap decisions.

A perception system that decodes sensory data, strong data processing driven by machine learning, and advanced communication systems (such as V2V, V2I, and V2X) that enable vehicles to communicate with their environment and with each other are all necessary for AI in self-driving automobiles.

8.3 Artificial Intelligence in Autonomous Vehicles

8.3.1 Understanding AI in the Context of Self-driving Cars

AI is crucial for self-driving cars because it integrates technologies such as sensor fusion, computer vision, and machine learning. By combining these technologies, cars may comprehend their environment, form opinions, and drive themselves without a driver. This viewpoint highlights how AI is necessary to guarantee the safe and independent operation of self-driving cars.

8.3.2 Machine Learning in Autonomous Vehicles

AI and machine learning are necessary for cars to analyze sensor data, learn from the past, and make informed judgements. AI algorithms control information processing, which is necessary for the safe and adaptable operation of autonomous cars, while machine learning promotes continuous improvement. Autonomous cars rely on machine learning algorithms to learn and adjust based on the data they have gathered. By analyzing and comprehending vast amounts of sensor and prior experience data, these algorithms continuously enhance their decision-making capabilities. Since it enhances their capacity to navigate, anticipate, and react to a variety of road conditions, this feature is essential for autonomous cars and highlights their crucial role in developing self-driving technology.

In the field of autonomous vehicles, machine learning has already gained traction and is predicted to become even more significant in upcoming innovations.

By examining some of its applications, a deeper comprehension of machine learning in autonomous vehicles can be attained. Driving monitoring allows one to keep an eye on the driver and their behavior via neural networks. This stops unauthorized personal usage and aids in the detection of car theft.

Furthermore, occupancy detection techniques can improve the overall experience for all passengers in the vehicle.

Various vehicles and their surroundings can be classified by training neural networks using machine learning algorithms, through image processing. These networks come in handy to recognize different objects and shapes of specific items. Machine learning aids in analyzing an object's proximity and speed direction, hence in calculating clearance and in making lane change decisions. Accuracy in the identification of objects is a key challenge encountered by autonomous vehicles of this age, as even a minor change in the incoming data can lead to a catastrophic decision. This is where a more improved and generalized approach to machine learning can be put to use, enhancing object perception and identification. Every sensor has unique advantages and disadvantages. Independent operations, owing to the production of varying outputs, complicate decision-making, whereas with the interference of machine learning, these outputs can be integrated, such that they complement each other, giving an upper hand in decision-making. In the instance of a camera, a passive sensor, and a radar, which is active, under brutal weather, the images captured by the camera can become distorted, complicating data analysis, whereas the radar retains accuracy even in such situations. Massive scale need of data base to identify and classify objects, and time, alongside effort to minimize data distortion, is another setback. Contamination of data can endanger both the passengers and the pedestrians, thus it demands some sort of understanding pertaining to the responses of humankind under the context of these situations.

On the roads, as humans, we take into account the hand gestures and eye contact to predict the decisions of fellow passengers in changing lanes, whereas this simple act is complex for the machine learning principles to process, demanding trends under the field. The general population fears the system's reliability, and exhibits a constant lack of trust and skepticism about autonomous vehicles and self-driving cars, though scientists are optimistic about it. A complete assurance of the safety of the user cannot be guaranteed by the autonomous vehicles, as the technological advancements of the present day are not yet well equipped to combat and surpass human intelligence, given their inaccuracy in detecting and classifying objects on the course of the vehicle.

8.3.3 Deep Learning in Autonomous Vehicles

Deep learning is a special area of machine learning that uses the human brain as its model, drawing inspiration from the foundations of artificial neural networks

(ANNs). The algorithms of deep learning consist of layers of ANNs, that filters out contents of high priority and importance from raw and massive datasets. This peculiar ability to sift key pointers makes them the most sought-after technology in the development of a wide variety of applications, such as natural language processing, image and speech recognition, and even autonomous driving. The development of AlexNet, a deep convolutional neural network, which marked its milestone in the field of deep learning, with its solutions provided to the ImageNet visual recognition challenges, further catalyzed the development of self-driving cars. Advancements in powerful computing resources and the availability of large, high-quality training datasets and other progressions in this field have resulted in its widespread utilization. You can find a summary of some existing deep learning structures in Table 8.1.

Table 8.1 Deep learning structures.

Deep learning type	Description
Autoencoder	The system consists of an encoder and a decoder, and it's specifically designed to learn a compressed representation of the input data, allowing for the original data to be reconstructed [1]. To help autonomous cars determine the proper steering angle during navigation, autoencoders are combined with end-to-end deep learning algorithms [2].
Convolutional neural networks (CNNs)	The CNN, or Convolutional Neural Network, employs convolution operations to pull out and learn important features from data. This approach is critical for identifying data patterns that would be difficult to detect with typical algorithms. Its structure is layered, with top layers addressing more complex qualities and lower sections concentrating on learning simpler ones.
Deep Belief Networks (DBNs)	The deep belief network is made up of several layers of RBMs (Restricted Boltzmann Machines). To create a more complex deep belief network, these RBMs—which may be shallow or have two layers—are stacked [3]. In addition to being taught through unsupervised learning, these are also essential in a variety of applications, such as natural language processing, computer vision, and speech recognition, particularly in autonomous navigation, picture detection, and classification [4].
Recurrent neural networks (RNNs)	Sequential data processing is a strength of Recurrent neural networks. They have proven effective in a range of autonomous vehicle applications, such as time series forecasting, voice recognition, and natural language processing, due to their unique ability to capture temporal dependencies and patterns [1]. However, it's important to keep in mind that RNNs can be quite sensitive to the order of the input data.

8.4 Perception and Fusion of Sensors for AI-powered Automobiles

Through a perception system, the AI adapts to its environment, which utilizes a variety of sensors and technologies, such as computer vision, to recognize objects, pedestrians, road signs, and obstacles in real time. The skill of accurately perceiving the surroundings is essential for autonomous vehicles, facilitating them to make smart decisions and navigate safely without human assistance. Since an autonomous vehicle is critically a self-driving car, it must be able to sense its environment to confirm whether it can proceed, apply brakes, accelerate, steer, and change lanes, among other actions. Addressing these challenges can be done by employing sensors, which perform like the vehicle's sense organs. They detect barriers in the path and relay feedback to the controller, allowing it to take appropriate actions.

Global positioning system (GPS), also known as Global Navigation Satellite Systems, allows vehicles to specifically understand their surroundings with accuracy up to 1 cm in three-dimensional space (Figure 8.3). It does this by transmitting data through cellular networks and satellites in geostationary orbit. Alternatively, passive sensors obtain information from the environment

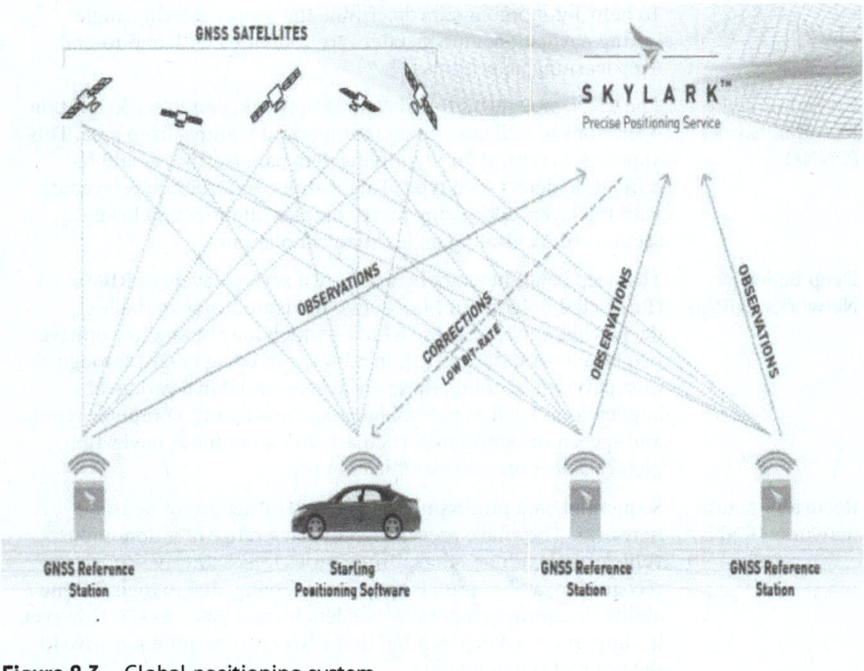

Figure 8.3 Global positioning system.

without sending out any waves. Cameras are an ideal example of this type of sensor. These cameras are used to create a completely 360-degree view around the vehicle, which is absolutely helpful in detecting and recognizing objects. AI algorithms can further process the gathered data for different categories. The Inertial Navigation System (INS) combines data from the Global Navigation Satellite System (GNSS) and the Inertial Measurement Unit (IMU) using sensor fusion software. In difficult situations where GNSS signals could be erratic or nonexistent, like urban canyons, bridges, tunnels, mountains, parking garages, and dense woodlands, the system works especially well. The INS assists the vehicle in tracking the motion of objects in three dimensions by analyzing the data received from the IMU.

Additionally, IMU and GNSS collaborate to improve the accuracy of contemporary navigation systems. While INS can function independently without wanting for the support of GNSS, GNSS is always present to step in and correct any errors that the IMU might make. One of the newest mapping technologies is called LIDAR, or Light Detection and Ranging. The main components that guarantee LIDAR's efficacy are the laser, receiver, scanner, and GPS. By viewing LIDAR as the vehicle's set of eyes, we can observe that it gives the vehicle a full 360-degree picture of its environment. With LIDAR, the host vehicle is positioned at the center and measures distances to construct a 3D model of the surroundings. It can rapidly process countless laser pulses every second. The LIDAR helps the vehicle react by identifying and predicting the behavior of objects as soon as it discovers them. LIDAR is incorporated into the development of self-driving cars in order to make this concept a reality while assuring safety. The laser pulses that are used by the LIDAR system consist of lasers, invisible to human eyes, which make 3D maps and visualize its environment.

RADAR and LIDAR work in similar ways. They are used for detection, but RADAR solely relies on radio waves instead of light waves. We estimate the distance, speed, and direction of an approaching vehicle by determining how long it takes for the radio waves to bounce back. In the domain of autonomous vehicles, RADAR operates at three different frequencies, which makes it classified into short-range radars (SRR), medium-range radars (MRR), and long-range radars (LRR). This technology delivers vehicles with millimeter-level precision and remains highly effective even in poor visibility situations such as clouds, snow, rain, and fog.

- Pulse RADAR: This kind of radar generates a single pulse from the instrument and keeps the frequency constant while it operates.
- Frequency-modulated continuous wave (FMCW) RADAR: In this case, pulses are emitted continuously, with the frequency changing over the duration of the transmission.

We can break down how the camera, which is pivotal for the vehicle to sense its surroundings quickly, works into three main categories:

- The Classifier: This is a traditional method that's already in use. Machine learning plays a crucial role here, as it helps identify and categorize objects like pedestrians and other vehicles.
- Optical Flow: In this stage, the camera transitions to using optical flow. It analyzes three-dimensional structures by tracking corresponding points in images and their motion sequences.
- Artificial Intelligence: This phase employs neural networks and schematic segmentation to assign different categories to various segments of an image.

By integrating all three methods, we can certainly reduce detection errors. Infrared sensors are brought into action to identify objects that cannot be detected by LIDAR, RADAR, or cameras. Far Infrared sensors offer the essential sensing capabilities for level-5 autonomy. They can compile an extra layer of information that other sensors might overlook by tapping into a different part of the electromagnetic spectrum. As different objects emit heat at varied levels, these sensors, which pick signals by detecting the emitted heat, can distinguish between living and nonliving things. A thermal camera is used to convert the FIR signals collected passively into VGA format, followed by the application of learning algorithms to sense and analyze the environment.

8.5 Advantages of AI in Autonomous Vehicles

Autonomous vehicles provide numerous advantages in terms of safety and efficiency. They could drastically cut down on traffic accidents, injuries, and fatalities. Furthermore, autonomous vehicles improve transportation alternatives for marginalized populations, such as the elderly and those with disabilities, giving them more freedom and exposure to opportunities. Aside from that, these cars improve traffic flow by lowering congestion through more effective driving and improved connections with infrastructure and other cars. Environmentally speaking, driverless cars may choose the best routes based on traffic conditions, cut out pointless trips, and minimize stop-and-go driving, all of which improve fuel economy. Finally, by increasing productivity (shorter commutes) and reducing the need for extensive parking facilities, autonomous vehicles may boost the local economy.

8.5.1 Safety Improvements and Accident Reduction

The use of AI in self-driving cars has had a significant impact on society, especially in terms of safety. Modern algorithms enable these vehicles to identify dangers instantly and make well-informed decisions at any time. Artificial intelligence (AI) has the potential to transform road safety by reducing accidents caused by human mistakes by shifting control away from human hands. All passengers and pedestrians travel in greater safety because of these vehicles' advanced sensors and AI algorithms, which respond instantly.

8.5.2 Enhanced Traffic Efficiency and Reduced Congestion

Self-driving automobiles are thought to be revolutionized by AI, which also helps to improve traffic flow and reduce congestion. These AI-powered smart cars are more adept at following traffic patterns than cars operated by people. Through the processing of real-time data and prompt decision-making, AI helps to minimize traffic congestion by identifying optimal routes and reducing road crowding. Traffic flow in our cities might be greatly enhanced by this technology.

8.5.3 Environmental Impact (Reduced Emission)

Self-driving cars are becoming more ecologically friendly thanks in large part to AI, which optimizes driving habits and boosts overall productivity. With the help of AI-powered algorithms, these vehicles can now choose the best routes, cutting down on pointless stops and speeding up travel, which lowers emissions and fuel usage. Since autonomous cars are already more common on our roads, their potential to enhance driving habits and performance could eventually lessen transportation's negative environmental effects.

8.6 Challenges in Autonomous Vehicles and Smart Mobility

Years of perseverance and creativity have made autonomous cars a reality, and they are no longer just a pipe dream. Building and implementing these intricate systems nevertheless presents challenges, notwithstanding the advancements. As autonomous vehicles become more autonomous, the barriers grow even more formidable. Adapting the technology to navigate the complexities of the real world is considered to be one of the major difficulties faced. In addition, components beyond technology, like consumer confidence and actions, are critical in determining the future success of Autonomous vehicles.

Roads and traffic are subjective to the location, making them unpredictable. Unclear marking of the lanes and directional signs at the tunnels, potholes on the routes, winding mountain roads are some of the many instances, in addition to unforeseen weather turbulences. This makes it inevitable to examine the impacts of weather, of any kind and degree on the performance of the vehicle, in order to assure the reliability of the autonomous vehicles, as every single flaw and botch puts the life of the user at stake.

- Traffic patterns are constantly shifting. As autonomous vehicles begin to integrate into everyday transportation, they will face a variety of traffic situations, including interactions with both human drivers and pedestrians. Some environments may be orderly and self-managed, while others may become erratic, particularly when individuals disregard traffic laws or drive irresponsibly. Unexpected developments can arise at any time, and in crowded spaces, even minimal movements can have a considerable impact on the overall flow.
- With a horde of autonomous vehicles making up the traffic, it becomes complicated for the autonomous vehicles to distinguish their own signals from those of the bombarding vehicles in their surroundings, thereby posing Radar interferences as a major challenge to be combated by the self-driving cars. Despite the availability of various radio frequencies for the radar, it is still insufficient to encompass all the cars on the road.
- Big data analytics comes with significant challenges, particularly when data is not managed efficiently, which can hinder technological advancement. To make the most of the massive volumes of data produced by autonomous vehicles, it's essential to implement robust training frameworks and decision-making systems.
- Considering the potential for hacking and disruptions, data privacy and cybersecurity become problematic. For instance, if the automotive sector creates technology that enables cars to send signals to apply the brakes or other information, drivers need to have total faith that the communication is accurate. This degree of assurance is necessary for efficient vehicle-to-vehicle (V2V) communication, but safeguarding these systems against possible security risks is a significant obstacle.
- The current law keeps human drivers accountable for any accidents and lacks clear standards for completely autonomous systems. Self-driving cars allow their drivers to rest because the decisions are made by software. This could put children in danger and lead to a lack of attention when their help is needed.

8.7 Future Directions and Conclusion

Reducing traffic, lowering carbon emissions, improving accessibility for all, and making our roads safer are the main goals of the rise of driverless cars and smart mobility solutions. However, we need to establish regulations, address cybersecurity issues, build the required infrastructure, and get public support before we can embrace new technology completely. Smart mobility and driverless cars have the potential to revolutionize transportation as technology develops, bringing about a safer, more effective, and ecologically sustainable future.

The ability of these self-driving cars to function without human assistance is made possible by AI. AI uses technology like machine learning and computer vision to help these cars comprehend their environment and make judgments quickly while driving. Going forward, we need to advance AI, establish precise regulations, inform the public, and focus on incorporating these vehicles with smart city initiatives in order to establish a secure and effective transportation network. While being seamlessly connected to their surroundings, this technology has the potential to make our cars safer and more self-sufficient.

Autonomous vehicle technology is about to undergo a number of changes that will impact everything from transportation and urban planning to road safety. Building robust AI and machine learning systems that can assist in a range of driving situations and locations is essential. In our ideal future, self-driving automobiles will enhance sustainability and transportation efficiency while simultaneously ensuring safety.

This study concludes by emphasizing the vital importance that deep learning, machine learning, and AI play in the development of self-driving automobiles. Because these technologies enable self-driving cars to sense their environment, make smart decisions, and control their motions, they are truly revolutionizing transportation. Nevertheless, in order for the cars to become a seamless part of our lives, we still have to deal with problems like managing complex traffic situations and guaranteeing safety in inclement weather. Autonomous vehicles are also expected to significantly affect metropolitan infrastructure, road safety, and general mobility. In order to achieve a future of intelligent mobility, this study highlights the necessity of developing AI and machine learning as well as the potential of self-driving cars to enhance the sustainability, safety, and comfort of our transportation systems.

References

1 Golroudbari, A. A. and Sabour, M. H. Recent advancements in deep learning applications and methods for autonomous navigation – a comprehensive review. *arXiv preprint arXiv:2302.11089*, 2023.

2 Pak, A., Manjunatha, H., Filev, D., Tsiotras, P. Carnet: A dynamic autoencoder for learning latent dynamics in autonomous driving tasks. *arXiv preprint arXiv:2205.08712*, 2022.

3 Huang, Y., Panahi, A., Krim, H., Yu, Y., and Smith, S. L. Deep adversarial belief networks. *arXiv preprint arXiv:1909.06134*, 2019.
4 Ren, J., Green, M., and Huang, X. From traditional to deep learning: Fault diagnosis for autonomous vehicles. In *Learning Control*, pages 205–2019. Elsevier: Amsterdam, 2021.

Further Reading

Arkin, R. C., Ulam, P. D., and Duncan, B. An ethical governor for constraining lethal action in an autonomous system. *Georgia Institute of Technology, Technical Report*, 2009.

Bachute, M. R. and Subhedar, J. M. Autonomous driving architectures: Insights of machine learning and deep learning algorithms. *Machine Learning with Application*, 6: 100164, 2021.

Bainbridge, L. Ironies of automation. *Automatica*, 19(6): 775–779, 1983.

Becker, F. and Axhausen, K. W. Literature review on surveys investigating the acceptance of automated vehicles. *Transportation*, 44(6): 1293–1306, 2017.

Bendiab, G., Hameurlaine, A., Germanos, G., Kolokotronis, N., and Shiaeles, S. Autonomous vehicles security: Challenges and solutions using blockchain and artificial intelligence. *IEEE Transactions on Intelligent Transportation Systems*, 24(4): 3614–3637. 2023. https://ieeexplore.ieee.org/abstract/document/10023964 (accessed 7 June 2024).

Berreby, F., Bourgne, G., and Ganascia, J.-G. Modelling moral reasoning and ethical responsibility with logic programming. In *Logic for Programming, Artificial Intelligence, and Reasoning*, pages 532–548. Springer, 2015.

Bonnemains, V., Saurel, C., and Tessier, C. Embedded ethics: Some technical and ethical challenges. *Ethics and Information Technology*, 20(1): 41–58, 2018.

Chen, D., Zhang, J., Wang, J., and Wang, F. Y. Freeway traffic stream modeling based on principal curves. In *Proceedings of the 2003 IEEE International Conference on Intelligent Transportation Systems* Vol. 1 pages 368–372. IEEE, 2003.

Chen, L., Lin, S., Lu, X., Cao, D., Wu, H., Guo, C., Liu, C., and Wang, F.-Y. Deep neural network-based vehicle and pedestrian detection for autonomous driving: A survey. *IEEE Transactions on Intelligent Transportation Systems*, 22: 3234–3246, 2021.

de Moura, N., Chatila, R., Evans, K., Chauvier, S., and Dogan, E. Ethical decision making for autonomous vehicles. In *2020 IEEE Intelligent Vehicles Symposium (IV)*, Las Vegas, NV, USA, 2020.

Endsley, M. R. From here to autonomy: Lessons learned from human–automation research. *Human Factors*, 59(1): 5–27, 2017. https://doi.org/10.1177/0018720816681350.

Estivill-Castro, V. Game theory formulation for ethical decision making, In Inverardi, P. and Ranise, S., editors, *Robotics and Well-Being*. pages 25–38. Springer, 2019.

Gkartzonikas, C. and Gkritza, K. What have we learned? A review of stated preference and choice studies on autonomous vehicles. *Transportation Research Part C: Emerging Technologies*, 98: 323–337, 2019.

Goerick, C., Noll, D., and Werner, M. Artificial neural networks in real-time car detection and tracking applications. *Pattern Recognition Letters*, 17(4): 335–343, 1996.

Gouveia, J. P., Seixas, J., and Giannakidis, G. Smart city energy planning. In *Proceedings of the 25th International Conference Companion on World Wide Web - WWW '16 Companion*, pages 345–350. Montréal, QC, Canada, 11–16 April 2016.

Göhring, D., Latotzky, D., Wang, M., and Rojas, R. Semiautonomous car control using brain computer interfaces. In *Intelligent Autonomous Systems 12*, pages 393–408. Springer: Berlin, Heidelberg, 2013.

Haboucha, C. J., Ishaq, R., and Shiftan, Y. User preferences regarding autonomous vehicles. *Transportation Research Part C: Emerging Technologies*, 78: 37–49, 2017.

Heaton, J. Ian Goodfellow, Yoshua Bengio, and Aaron Courville: Deep learning. *Genetic Programmimng Evolvable Machines*, 19: 305–307, 2018.

Hofmann, M., Neukart, F., and Bäck, T. Artificial Intelligence and Data Science in the automotive industry. *arXiv preprint arXiv:1709.01989*, 2017.

Howard, D. and Dai, D. Public perceptions of self-driving cars: The case of Berkeley, California. In *Transportation research board 93rd annual meeting*. Washington, DC, 12–16 January 2014.

Jebamikyous, H.-H. and Kashef, R. Autonomous vehicles perception (AVP) using deep learning: Modeling, assessment, and challenges. *IEEE Access*, 10: 10523–10535, 2022.

Karnati, A., Mehta, D., and Manu K. S. Artificial intelligence in self driving cars: Applications, implications and challenges. *Ushus - Journal of Business Management*, 21(4): 1–28, 2022. https://www.researchgate.net/publication/374998350_Artificial_Intelligence_in_Self_Driving_Cars_Applications_Implications_and_Challenges (accessed 7 June 2024).

Kaur, K. and Rampersad, G. Trust in driverless cars: Investigating key factors influencing the adoption of driverless cars. *Journal of Engineering and Technology Management*, 48: 87–96, 2018.

Krizhevsky, A., Sutskever, I., and Hinton, G. E. ImageNet classification with deep convolutional neural networks. *Communications of the ACM*, 60: 84–90, 2017.

Kyriakidis, M., Happee, R., and de Winter, J. C. Public opinion on automated driving: Results of an international questionnaire among 5000 respondents. *Transportation Research Part F: Traffic Psychology and Behaviour*, 32: 127–140, 2015.

Marko, K. A., James, J. V., Feldkamp, T. M., Puskorius, G. V., and Feldkamp, L. A. Signal processing by neural networks to create "virtual" sensors and model based diagnostics. In *International Conference on Artificial Neural Networks*, pages 191–196. Springer: Berlin, Heidelberg, 1996.

McLaren, B. M. Computational models of ethical reasoning: Challenges, initial steps, and future directions. *IEEE Intelligent Systems*, 21(4), 29–37, 2006.

Naveen, K., Lokesh, M., Varma, K. N. J. L., and Teja, P. S. V. A review on autonomous vehicles and its components. *Journal of Pharmaceutical Negative Results*, 13(7): 6918, 2022.

Panday, V. The role of artificial intelligence in autonomous vehicles. *International Journal of Science and Research (IJSR)*, 12(12): 1538–1541, 2022.

Parekh, D., Poddar, N., Rajpurkar, A., Chahal, M., Kumar, N., Joshi, G. P., and Cho, W. A review on autonomous vehicles: Progress, methods and challenges. *Electronics*, 11(14): 2162, 2022. https://doi.org/10.3390/electronics11142162.

Payre, W., Cestac, J., and Delhomme, P. Intention to use a fully automated car: Attitudes and a priori acceptability. *Transportation Research Part F: Traffic Psychology and Behaviour*, 27: 252–263, 2014.

Pickering, J. E., Podsiadly, M., and Burnham, K. J. A model-to decision approach for the autonomous vehicle (AV) ethical dilemma: AV collision with a barrier/pedestrian (s). *IFAC-Papers On Line*, 52(8): 257–264, 2019.

Rai, S., Kavana H. K., Eklavya, Krishna Vg, A., Akanksha, P., Jain, D. A study on autonomous vehicles - advancements, benefits & challenges, *International Journal for Multidisciplinary Research (IJFMR)*, 6: 2582–2160, 2024.

Rao, Q. and Frtunikj, J. Deep learning for self-driving cars: Chances and challenges. In *Proceedings of the 1st International Workshop on Software Engineering for AI in Autonomous Systems*, pages 35–38. Gothenburg, Sweden, 28 May 2018.

Ren, J., Gaber, H., and Al Jabar, S. S. Applying deep learning to autonomous vehicles: A survey. In *Proceedings of the 2021 4th International Conference on Artificial Intelligence and Big Data (ICAIBD)*, pages 247–252. Chengdu, China, 28–31 May 2021.

Sanghvi, N., Sanghvi, N., Sanghvi, N., Porwal, A., and Rawalh, N. Artificial intelligence in autonomous vehicles. *International Journal of Engineering Research & Technology (IJERT)*, 13(06): 1–6, 2024.

Schoettle, B. and Sivak, M. A survey of public opinion about autonomous and self-driving vehicles in the US, the UK, and Australia. 2014. https://deepblue.lib.umich.edu/handle/2027.42/108384 (accessed 1 January 2019).

Smaldone, A. M., Kyro, G. W., and Batista, V. S. Quantum convolutional neural networks for multi-channel supervised learning. *arXiv preprint arXiv:2305.18961*, 2023.

Sun, Z., Bebis, G., and Miller, R. Onroad vehicle detection using optical sensors: A review. In *The 7th International IEEE Conference on Intelligent Transportation Systems*, pages 585–590. IEEE, 2004.

Syed, F. U., Filev, D., and Ying, H. Fuzzy rule-based driver advisory system for fuel economy improvement in a hybrid electric vehicle. In *NAFIPS 2007- 2007 Annual Meeting of the North American Fuzzy Information Processing Society*, pages 178–183. IEEE, 2007.

Thornton, S. M., Pan, S., Erlien, S. M., and Gerdes, J. C. Incorporating ethical considerations into automated vehicle control, *IEEE Transactions on Intelligent Transportation Systems*, 18(6): 1429–1439, 2016.

United Nations. Transforming our World: The 2030 agenda for sustainable development. 2020. https://sustainabledevelopment.un.org/post2015/transformingourworld (accessed 21 January 2021).

United Nations. World population prospects. 2020. https://population.un.org/wpp/Graphs/Probabilistic/POP/TOT/900 (accessed 2 March 2021).

Xu, W. From automation to autonomy and autonomous vehicles: Challenges and opportunities for human-computer interaction. *Interactions*, 28(1): 48–53, 2020. https://dl.acm.org/doi/abs/10.1145/3434580 (accessed June 7, 2024).

9

Enhancing Urban Traffic Management with Multi-scale Hierarchical GANs

Ashik Shah Jahangeer and P. Shanmugavadivu*

Department of Computer Science and Applications, The Gandhigram Rural Institute, Dindigul, Tamil Nadu, India

*Corresponding author: ashik.jahangeer@gmail.com

9.1 A System Stuck in Time

The methods that were meant to manage urban traffic have been significantly outstripped by the volume of traffic in urban areas. An issue that was initially a straightforward coordination of signals and junctions has evolved into a task that is characterized by complexity, unpredictability, and size. The concept of traffic control that relies on fixed-time signals and predetermined routing logic becomes insufficient and outmoded as metropolitan areas grow more densely populated and dynamic.

In many cities, the traffic system operates in a manner similar to that of a symphony that does not have a conductor. Intersections are obstructed by vehicles, traffic signals work on predetermined schedules that fail to take into account the current conditions, and sensors covertly capture a large amount of data, the majority of which does not transform into insights that can be put into action. Commuters are willing to put up with the turmoil, while transit authorities are struggling with obsolete tools that were designed for a more straightforward era.

There is a fundamental problem with the structure of the majority of traffic management systems, which is the root cause of this stalemate. These systems have been around for decades, and they are responsible for gathering data from a number of different endpoints, sending it to a central control unit, and relying on either human operators or strict algorithms to determine replies. It is important to note that the paradigm is essentially reactive, which means that actions are only made after the occurrence of congestion or other events has taken place. In spite of the fact that it is automated, it frequently relies on rule-based systems that are drawn from historical patterns. These systems do not take into account the wide variety of events that take place on modern metropolitan roads.

Despite this, the city continues to persist without stopping. Whether it be a truck that breaks down on a critical roadway, a sudden thunderstorm during peak traffic, or a political demonstration that causes several vehicles to be rerouted, each of these scenarios has the potential to cause unexpected disruptions within the system. As time has progressed, real-time adaptability has transformed from a luxury into an essential requirement. Knowledge is necessary for adaptation, and this knowledge starts with data. More precisely, data that is instantaneous, contextual, and tightly tied to the behaviors of dynamic agents is the kind of data that is required.

By this point, the paradox has reached its peak. Internet of Things (IoT) devices, which include smart cameras, car trackers, pollution monitors, weather stations, and integrated infrastructure, are used to network modern cities. These devices contribute to the integration of infrastructure. Messages are constantly being propagated around the city. This is despite the fact that the traffic infrastructure frequently fails to react. Despite the fact that the input arrives more quickly than it can be processed, the reaction mechanisms, which consist of fixed-timing signals and pre-coded control logic, are excessively slow, fragmented, and imprecise.

On top of that, conventional traffic systems only acknowledge certain aspects of the entire. A signal controller is able to view the lengths of the queues at its intersection, but it is not aware of the congestion that is occurring upstream. There is a possibility that a city dashboard will display the average travel speeds, but it will not take into consideration the actions of vehicles who are moving slowly along a congested route. The system is able to identify either the forest or the trees, but it most often does not recognize both at the same time. One of the obstacles that prevents effective management is the absence of a unified model that is capable of comprehending traffic on both the micro and macro levels.

It is impossible to fathom the magnitude of the stakes at this point. The duration of the ambulance response was extended. fuel that was thrown away. There are not enough school buses. Productivity reductions have occurred. Emissions that are increasing. When multiplied across millions of daily journeys, even very slight inefficiencies can result in huge costs for society. The concept of traffic goes beyond the simple passage of vehicles in this day and age of climate change and smart city projects; it encompasses the maintenance of life, the guarantee of access to emergency services, and the reduction of our carbon footprint.

Despite the fact that we have access to more data, improved computing power, and better communication than any previous generation, many cities continue to handle traffic like mechanical toys. This is an obvious example of irony. It is predetermined, it is predictable, and it is unable to see or acquire knowledge. When it is important to have real-time, contextualized, and adaptable intelligence, the majority of systems provide preestablished operational methods. Rather than attempting to forecast future processes, they automate those that have already occurred.

It is necessary for us to have a system that not only responds but also anticipates in order to make progress. This system must be able to learn and react simultaneously. A computer program that is capable of generalization, reasoning, and adaptability without the need for explicit commands for each individual instance. The movement of a single vehicle must be synchronized with the rhythm of an entire city for a system to be considered successful. It is necessary for it to recognize patterns not only in the temporal and spatial dimensions, but also in the behavior and context of the situation.

It is necessary to make a change, not only in terms of technology but also in terms of mentality. The development of cognition as opposed to control. From scripts to intelligence, we are transitioning. Bringing together disparate pieces of information into a coherent whole. This vision ultimately results in the development of cutting-edge AI-driven methodologies, such as multi-scale hierarchical GANs (MSH-GANs). They want to rethink urban planning and the management of movement, but not by changing the mechanisms that are already in place; rather, they want to perceive it as a process that is dynamic and always evolving, rather than one that is static.

Beginning with the question, "How can we design systems that reproduce the dynamics of urban movement?" is the first step in the pursuit of that vision.

9.1.1 Literature Review

Generative adversarial networks, often known as GANs, provide a big step forward in the modeling of intricate and ever-changing urban traffic conditions. Early implementations, as described by Goodfellow et al. [1], demonstrated the fundamental architecture of GANs and their ability to generate realistic data distributions. This made it possible for applications to be developed in a variety of domains, including urban traffic management.

Recent developments have investigated hierarchical GAN designs as a means of addressing the multi-scale complexity that is prevalent in urban traffic data. KI-GAN, as defined by Xu et al. [2], incorporates previous knowledge into the GAN architecture in order to successfully tackle the hierarchical and interaction-intensive nature of urban traffic. This results in a considerable improvement in the realism and predictive accuracy of the data that is generated.

It was explained by Kumar et al. [3] that SpectraGAN utilized enhanced spectrum normalization techniques, which resulted in an improvement in the stability of GAN training and permitted more robust modeling of high-dimensional, real-world datasets that are usually encountered in urban transportation situations.

Research, such as that carried out by Zhang et al. [4] and Wang et al. [5], has advanced this field by integrating GANs with graph convolutional networks (GCNs). This has enabled the effective capture of spatial correlations as well as the underlying structural complexities of urban traffic networks. Through the utilization of the inherent spatial and temporal interdependence of traffic data, these models demonstrated an improvement in their predicted accuracy.

Furthermore, research conducted by Liu et al. [6] highlighted the incorporation of external contextual parameters, such as meteorological variables, into GAN models. This integration considerably improved the accuracy and adaptability of traffic predictions across a variety of environmental circumstances.

Taking into consideration these findings, it is clear that GAN-based models play an essential part in the enhancement of urban traffic management systems. There is ongoing potential to improve these models by combining real-time, multi-scale, and IoT-driven data streams. This would help to bridge the existing research gap and promote the advancement of the MSH-GAN model that was described in this study.

9.2 When GANs Hit the Road: The Gaps in Current AI Models

GANs have demonstrated their remarkable significance. Their generative powers, which include the generation of lifelike images as well as the synthesis of realistic voices, have almost become associated with the innovation of machines. On the other hand, traffic constitutes a unique landscape situation. It does not move in a sequential manner like audio or in a static manner like an image. Dynamic, interactive, multi-agent, and frequently turbulent are all things that describe it.

Applying standard GANs to the task of modeling urban traffic is like attempting to capture a moving cityscape in a still image. There is a persistent absence of a crucial component within the frame.

Generally speaking, GANs operate through a competition: one network, known as the generator, works toward the goal of producing data that is indistinguishable from real data, while the other network, known as the discriminator, works toward the goal of distinguishing between the two types of data. It is possible for the generator to fool the discriminator, which results in the mechanism becoming more effective. Facial recognition and handwriting analysis are two examples of domains that do exceptionally well with this paradigm since they are defined by the presence of homogenous data and confined context. There is neither uniformity nor absence of context in the traffic.

When it comes to the realm of traffic, the reality that needs to be modeled covers multiple dimensions. These scales include the individual decisions that drivers or pedestrians make, as well as the macro-patterns that are generated by the accumulation of thousands of those decisions at intersections, roundabouts, or barriers. Conventional GAN designs are primarily developed for the generation of data on a single scale. They have difficulty representing layered dynamics, which requires them to simultaneously comprehend emergent macro-level movement and micro-level behaviors.

In addition to this, GANs are extremely brittle. There is a substantial amount of documentation regarding training instability, mode collapse, and sensitivity to input distributions. This goes beyond a simple academic issue when it comes to situations involving transportation. An artificial neural network (GAN) that restricts its output to a single form of traffic flow, such as lanes that flow in a regular manner, does not take into account the whole range of urban mobility options. For the model to be useful for prediction or simulation, it is necessary for the synthetic data to precisely reflect all of the factors that contribute to delays, including rush-hour congestion, accident-related detours, and delays in school zones.

Coherence between space and time is another issue that needs to be addressed. The inherent capability of conventional GANs to understand that a car turning left at a specific crossing is likely to influence subsequent congestion is not present in these systems. There is a possibility that congestion on the principal arterial route will occur following the accumulation of traffic on a feeder road. Not only does traffic include recent happenings, but it also includes activities that are about to take place. When they are in their initial configuration, GANs usually fail to perform well in the forward-predictive dimension. Foresight and temporal synchronization are areas in which they are not natively meant to function.

One of the most restricting aspects of the majority of the currently available GANs is their consistent structure. A lone discriminator is responsible for determining authenticity, while a solitary generator is responsible for acquiring a universal pattern. However, the traffic in metropolitan areas is not at all consistent. Behavioral diversity, localized disturbance, and frequently asynchronous flow are all characteristics that it demonstrates. It is not possible for a single generator to simulate both the fluid acceleration pattern on a freeway and the unpredictable behavior of changing lanes at a congested intersection at the same time. When calibrated solely for general flow characteristics, the discriminator does not have the acuity to recognize localized anomalies. This is because it is not calibrated for the specific features of the flow.

Although we are making an effort to adapt traffic to the original GAN structure, we are unwittingly simplifying a problem that requires complexity. In many cases, the conclusion seems visually convincing, yet it is devoid of any behavioral significance. There is a possibility that a heatmap that has been developed will mimic traffic, with its pixel values possibly agreeing with historical averages. However, such outputs are not capable of conveying the semantic precision of traffic; they are unable to depict the basic causality, decision-making processes, or inter-agent dynamics that are characteristic of genuine urban mobility. Consequently, this brings us to a more significant worry. GANs are remarkable imitators when they are configured in their normal manner. In spite of this, imitation is not the same thing as modeling. To properly assist intelligent traffic management, a model must not only match actual traffic circumstances but also offer both explanatory and predictive skills. Only then can smart traffic management be effectively implemented. It is necessary for it to take into account not only the current status of traffic but also its origins, development, and reaction to changes.

For example, when a demonstration blocks a key artery, the traffic model needs to be able to comprehend the cascade effects. It must not only be able to duplicate a congestion pattern, but it must also be able to predict route options, changes in signal dynamics, and following congestion areas. In order to accomplish this, standard GANs are unable to provide the level of abstraction and awareness that is required. They are lacking in structure, order, and causality. They are fundamentally flawed.

As urban mobility systems become more dependent on data and sensors, our modeling approaches will need to develop in order to accommodate this increased dependence. It would not be sufficient to merely copy traffic patterns; rather, we need to construct models that simulate the behavior of the system, which requires architectural innovation.

It is precisely at this point that the concept of multi-scale hierarchical GANs comes into existence. This is not a rejection of GANs; rather, it is an evolution of them that has been reengineered to conform to the structural realities of urban traffic, to handle behavioral scales, and to integrate localized insights with broad system knowledge. Within the following sections, we will investigate this design, not as a matter of intellectual curiosity but rather as a crucial step toward the development of real-time, environmentally friendly, and intelligent transportation systems.

9.3 Reimagining Intelligence: The Architecture of MSH-GAN

During the process of developing any system, there will come a time when the solutions that have been adopted will be found to be insufficient. However, rather than merely providing additional reinforcement, it is necessary to do a reevaluation of the framework. Currently, the time has come to reevaluate the modeling of urban traffic, and multi-scale hierarchical GANs are a prime example of this approach.

It is possible that at first glance, MSH-GAN appears to be nothing more than an extension of the traditional GAN architecture. To speak in terms of architecture, it is. The opposite is true; its design concept is one of a kind. Through the process of deconstructing and reconstructing the traffic environment at various degrees of granularity, MSH-GAN is designed to comprehend structure. The primary focus of conventional GANs is typically on the replication of surface patterns. Not only does it imitate movement, but it also imitates the mentality that lies behind the surface.

It is possible to observe stratification in the building. In order to construct this structure, there are at least three generators, each of which is calibrated to a different scale of the traffic system. While one may concentrate on capturing the decisions made by individual drivers, such as lane changes, stops, and accelerations, another may analyze the patterns of traffic across entire neighborhoods or significant corridors. The functioning of these generators is interdependent, meaning that they influence and condition one another in a reciprocal manner while simultaneously recognizing patterns in both the individual components and the whole composition.

There is no element of luck involved in the assignment of tasks; rather, it has been designed to reflect the hierarchical structure of traffic. There is a high probability that incidents that involve a single car or intersection will result in systemic implications, such as the spread of congestion throughout an entire network. Instead of ignoring this fact, MSH-GAN was designed to take it into account and include it into its operational framework. The hierarchy of generators that a conductor possesses is paralleled by the manner in which the conductor interprets notations written by a soloist as well as the rhythms of the full ensemble. The only way to achieve harmony is through the combination of both components (integration).

All things considered, generation is but one component of the process that takes place. In an adversarial setting, the equivalent is a multi-class discriminator, which functions as a more sophisticated evaluator than conventional Google Adversarial Networks (GANs). Instead of providing a binary evaluation of authenticity, it differentiates between a wide range of traffic patterns, both in terms of their size and the sorts of traffic they involve. Through this, one develops the ability to differentiate between traffic that is likely to be on highways and traffic that is unlikely to be in urban environments. Because of this, its contextual awareness is improved, and it gives it the ability to direct the generators toward results that are more complicated and semantically precise.

The circumstance is considerably impacted by this design, which incorporates a large number of heads. As a result of its ability to facilitate the production of traffic scenarios that are both statistically accurate and behaviorally coherent, MSH-GAN offers advantages for predictive analysis. As an illustration, it may attempt to emulate the manner in which a sudden obstacle influences micro-level decisions such as braking and acceleration, while simultaneously forecasting macro-level implications such as queue spillbacks and delays that affect the entire network. This can be accomplished by simulating the outcomes that would result from such judgments. On the contrary, it is not the result of a random occurrence but rather the result of deliberate planning.

For the purpose of educating such a system, a project of this magnitude is required. One of the most important steps is the equilibration of a large number of loss functions. Synchronizing the dependencies that exist between the generators is an absolutely necessary step. The synchronization of feedback loops across a wide range of scales is necessary in order to avoid divergence. The advantage, on the other hand, is a system that comprehends the dynamics of traffic rather than only focusing on the superficial characteristics of traffic. Not only does the upcoming image provide a preview of its actual substance, but it also provides a glimpse into the story that will be told.

It is important to note that MSH-GAN is not limited to static learning, making this observation relevant. The integration of data collected in real time from the Internet of Things makes adaptability possible. It is able to adjust its behavior in reaction to new information, taking into account the changes in patterns that occur throughout time periods such as the monsoons, festivals, or construction. This is something that can be accomplished by the functionally adaptive components. MSH-GAN's generating powers are further enhanced by additional characteristics like as memory, spatial-temporal depth, and the knowledge of several realities that are occurring simultaneously.

As a consequence of this, the model goes beyond its function as a simple simulation instrument. It performs the role of a consultant by acting as a computer representation of the traffic logic of the city. It has the ability to evaluate efforts, forecast outcomes, and make adjustments to tactics in real time. The availability of such an engine provides policymakers and traffic engineers with the opportunity to optimize signal designs, boost emergency response routes, implement proactive rerouting during peak hours, and provide individuals with tailored mobility estimates.

Within the context of MSH-GAN, what was initially conceived of as a development of GANs evolves into a paradigm shift with reference to the conceptual orientation. The system provides predictions that are inconsistent; rather, it creates an ecosystem of interconnected predictors, each of which is uniquely adapted to its particular situation and functions as a cohesive organism. An AI system that comprehends the city itself, as opposed to only the data, is a notion that has the potential to be considered the most powerful.

9.4 The City in Layers: Micro and Macro-level Generators

It is possible that if you were to observe traffic from a rooftop, you would notice waves of cars moving through the area, with some areas being congested and others being fluid. As soon as you go down to street level, however, the picture shifts. The decisions that individuals make are the ones that have the most impact in this situation: a quick change in lane, a pedestrian crossing with hesitation, or a motorcycle trying to squeeze through spaces. The truth of urban movement is comprised of these two layers, which are the panoramic and the personal. Both are what MSH-GAN was designed to see.

There are two different kinds of generators that are at the core of its architecture: micro-level and macro-level generators. It is not simply that they are different in size; they are also distinct in how they are seen. The fact that they view the city through glasses that are complementary to one another is what makes MSH-GAN so powerful with their collaboration.

9.4.1 Micro-level Generators: Learning the Pulse of the Street

In its most refined form, urban traffic can be described as a dance performed by agents such as vehicles, pedestrians, and bicycles, each of which has its own intentions and behaviors. The micro-level generators that are used in MSH-GAN are educated to comprehend these fine-grained operations. The below are all characteristics that are modelled by these vehicles.

- Individual vehicle trajectories,
- Acceleration and deceleration decisions,
- Gap acceptance and lane change behaviors,
- Interactions with pedestrians at crosswalks or jaywalking zones.

A pothole that induces swerving, a flashing pedestrian signal that slows turning vehicles, or a sudden downpour that affects driving techniques are all examples of conditions that can lead this layer to be more sensitive to specific conditions.

The contextual framing of these generators is what gives them their extraordinary force. They do more than merely imitate behavior; they comprehend the factors that contribute to it. A vehicle that is slowing down is not just a data point; it may also be an indication of signal anticipation, merging behavior, or the presence of pedestrians. This level of thinking, which is represented in probabilistic form through adversarial training, enables the micro-model to reflect urban traffic with a level of realism that is difficult for standard models to mimic.

9.4.2 Macro-level Generators: Capturing Systemic Flow

When you zoom out, a new pattern becomes apparent. In this day and age, the question of who breaks where is less important than the question of how flows grow, disseminate, and disintegrate. The MSH-GAN model incorporates macro-level generators to simulate these phenomena.

- Congestion formation and dissipation across zones,
- Influence of traffic signals across networked intersections,
- The ripple effect of one bottleneck on adjacent corridors,
- Temporal rhythms: morning peaks, afternoon lulls, weekend inversions.

Rather than dealing with density, volume, and timing in isolation, these generators deal with all three. Due to the fact that the macro level does not appear out of thin air, its outputs are influenced by the trends that occur in the micro layer. This is the result of a series of cumulative micro-decisions. Although a single late brake may not generate congestion, ten thousand late brakes do produce congestion.

Through the hierarchical modeling of these flows, MSH-GAN has the ability to establish a connection between individual behaviors and the dynamics of the system. Because of this, the macro-generators are not blind to the micro-data; rather, they take in the abstracted patterns that are generated by the micro-generators and simulate how those patterns scale and interact with one another.

Component	Micro-level generator	Macro-level generator
Focus	Lane-level behavior, individual vehicle dynamics	Corridor-level traffic flow, congestion evolution
Input Data	Speed, acceleration, proximity, signal phase	Density maps, travel time, aggregated flow counts
Output Format	Short-term trajectories, lane pressure maps	Regional congestion forecasts, peak-hour flows
Time Horizon	Seconds to a few minutes	15–30 minutes and beyond
Resolution	Per-vehicle	Per-road segment or traffic zone
Use Cases	Rerouting, pedestrian conflict resolution	Traffic redistribution, load balancing

9.4.3 The Synchronization Mechanism

The manner in which these two layers continue to be synced is one of the most sophisticated features of MSH-GAN. It is not as easy as going from the micro to the macro level. Influence can be thought of in both directions. Micro-predictions can be informed by macro patterns, and micro-events can bubble up to modify macro-flows. Both of these things are possible.

An example of this would be a forecast of significant congestion in the macro layer, which might potentially influence the micro-generator to imitate driving behaviors that are more aggressive or evasive, respectively. On the other hand, a pattern of unpredictable lane shifts in the micro layer may indicate the beginning of congestion, which would then encourage the macro-generator to adjust its flow estimates.

Through the utilization of shared latent regions and conditioning mechanisms, this interaction is facilitated, which in turn enables the model to function in a cohesive manner. To put it another way, the parts communicate with the entire, and the whole makes sure to listen.

9.4.4 Real-world Implications

This multitiered structure offers a number of concrete advantages. When it comes to emergency response planning, micro-level outputs can effectively imitate the manner in which individual drivers either yield to emergency vehicles or fail to yield to them. In the meantime, outputs at the macro level forecast the influence that this will have on the traffic density in the region. When it comes to optimizing signal timing, micro-level models are able to make predictions about how changes in light cycles affect individual acceleration patterns, and macro-level models evaluate how such changes spread throughout the grid.

Not only do the results produce more accurate forecasts, but they also produce simulations that can be put into action. These simulations are the kind that city planners and transport engineers can rely on to create, test, and refine interventions before putting them into practice in the real world.

It is a fundamental realization that cities are not merely physical environments; rather, they are systems of intent and emergence. MSH-GAN embodies this fundamental realization by separating the micro and macro dimensions of transportation while simultaneously connecting them. In order to create a good model of them, we need to pay attention to both the details and the drift.

9.5 Listening to the City: Real-time IoT Data Integration

The data that a model takes in is the only thing that gives it life, regardless of how smart the model is. A lack of data is no longer an issue in this day and age of smart cities. This is accomplished through the use of embedded road sensors, GPS pings from automobiles, signal control logs, weather monitors, and surveillance cameras. The streets are continually communicating with one another. Being able to hear the city is not the difficulty. It is the ability to comprehend it in the present moment and to react with intelligence. This is where the integration of MSH-GAN with the infrastructure of the IoT becomes a distinguishing feature.

In the past, urban data was typically delivered in the form of daily reports or hourly averages. Immediacy, however, is needed in today's modern city. The onset of congestion can occur in a matter of minutes, accidents can alter flow patterns in a matter of seconds, and changes in the weather can almost instantaneously transform a continuous stream into a jam. In order to properly manage traffic in the modern day, prediction alone is not sufficient; perception must be achieved in real time.

IoT source	Feature captured	Used by	Purpose in MSH-GAN
Traffic Cameras	Vehicle density, movement trails	Micro and Macro Generator	Real-time vehicle tracking
GPS from Vehicles	Position, route choice	Micro Generator	Trajectory simulation, behavioral context
Speed Sensors	Flow velocity, queue lengths	Macro Generator	Flow estimation, bottleneck prediction
Weather Feeds	Rain, visibility, temperature	Macro Generator	Traffic model modulation, demand variation
Signal Controllers	Phase timing, delays	Micro Generator	Signal approach simulation
Incident Reports	Road blocks, accidents	Macro Generator	Detour planning, congestion estimation

The MSH-GAN system was developed with this realization in mind. It acts as a live interface between Internet of Things data streams and traffic modeling, in contrast to models that are static or that are developed in batches. By continuously taking in inputs such as vehicle speeds, traffic density, signal status, and event feeds, it is able to generate a running understanding of the traffic situation over both space and time.

Integration starts at the periphery of the system. In order to reduce the latency of transmission, data from roadside devices, vehicle telemetry, and environmental sensors are processed either locally or regionally. They are then aligned and fed into the MSH-GAN architecture, where the micro-generators utilize them to calibrate fine-scale behaviors and the macro-generators alter broader forecasts based on developing flow patterns. This process is repeated until the micro-generators and macro-generators have reached the desired results.

Let us take a straightforward example. A particular borough is the location where the rainfall starts. At this point, the sensors in the vehicle begin to detect decreased speeds and more sudden brakes. Despite the fact that signal timings have not altered, should they? In the case of a conventional system, it is possible that such insights might never reach the control center in a timely manner. In MSH-GAN, the data is sent directly into the model, which then causes adaptive predictions to be made. These predictions include the forecasting of delays, the simulation of alternative routes, and the provision of anticipatory guidance to emergency services.

It is not only a matter of input-output mechanics here. Indeed, the model is capable of reconditioning itself. The IoT data assists in correcting and retraining its short-term knowledge, making it more resistant to aberrations and edge cases. Not only does it learn from pretrained distributions, but it also learns from live variations, which helps it develop a kind of intuition that is more street-level.

MSH-GAN is distinguished not only by the amount of data it employs, but also by the degree to which it combines different forms of data in a meaningful way:

- Flow assumptions are shaped by numerical data obtained from speed sensors.
- Categorical data, such as accident reports or event alerts, has the ability to modify behavior forecasts that are specific to a location.
- Visual data obtained from cameras can be preprocessed to generate density maps or estimates of lane occupancy, allowing for more precise implementation of traffic models.

The combination of these layers is what enables the model to react not only to movement but also to the context in which it is being used.

The operations of the city are revolutionized by this context-aware modeling. MSH-GAN is able to forecast increases in pedestrian density during a big event, model the influence that these surges have on vehicle flow, and recommend dynamic signal adjustments across crossings that are affected by the surges. In the event of a city marathon, it is able to simulate the way in which closures spread to neighboring neighborhoods and direct navigation apps accordingly.

It is essential to note that these forecasts are not isolated occurrences; rather, they represent a component of a feedback loop. For the purpose of refining its internal weights and expectations, the system compares its simulations to subsequent data and makes adjustments accordingly. Not only does it update, but it also transforms.

MSH-GAN, in its most fundamental form, functions as a kind of digital nervous system for the city. It does not perform these functions in discrete phases but rather as a continuous, synchronized function. It senses, models, predicts, and adjusts. Every single piece of information evolves into a whisper that directs its evaluation. With each new traffic signal comes a new opportunity to gain knowledge.

This capacity not only causes the system to be more precise, but it also causes it to be more resilient. It is able to adjust to ambiguous weather conditions, unanticipated detours, and unanticipated incidents. In the city's rhythm that is constantly shifting, MSH-GAN pays close attention and reacts as if it were a part of the metropolis.

9.6 Understanding the Why: Hierarchical Modeling and Contextual Awareness

It is never the case that a traffic jam is only the result of a collection of automobiles. A number of factors, including decisions, responses, indicators, climate, events, and even accidents or ambiguity, all play a role in determining the outcome of the situation. A mixture of local factors has prompted this response, which is a systemic response. This level of complexity requires a model that goes beyond merely mapping out the positions of the cars in order to be effectively managed. They need to understand the reasons that are driving their actions. Hierarchical modeling and contextual awareness are the two components that serve as the basis for MSH-GAN at this point in the program's development.

Data is commonly viewed as a two-dimensional layer in typical traffic models. This layer is where input features are provided, and it is also where expected patterns are formed. Cities are not able to function on a plane that is two-dimensional. With regard to the influence they wield, they function in layers:

- At the **bottom layer**, sensors are tasked with providing fundamental parameters such as the number of vehicles, their speeds, and the amount of time they must wait.
- The **next layer** captures patterns which may include delays during peak hours, blockages in lanes, and congestion clusters.
- A **higher layer** embeds causality, which includes problems caused by the weather, malfunctioning signals, and the number of pedestrians.
- And above that still, lies **intent and policy**, the issues of emergency protocols, event management, and mobility goals are included in the considerations.

This multifaceted perspective on the environment is encapsulated within the MSH-GAN architecture. Because of its hierarchical architecture, which imitates and organizes different sizes of flow, knowledge is organized through chains of cause and effect and is structured through these chains. The model is able to learn not only the behavior of traffic but also the ways in which context influences that pattern of behavior.

Imagine a traffic intersection that is suffering an unexpected increase in the amount of time it takes for vehicles to arrive when they are expected to arrive. It is possible that a typical model will recognize and highlight the oddity. In contrast, MSH-GAN performs its interpretation by taking into account the entire system. The behavior of the microlayer is investigated, particularly with regard to the suddenness of the stops made by automobiles. It looks at macro patterns in order to determine whether or not this congestion is confined to a specific area or is spreading throughout the area. The final step involves the incorporation of contextual information, such as the presence or absence of rain. Is there a problem that has been reported in the region? Is there a public event that could potentially affect the driving conditions in this area?

This vertical synthesis of information transforms detection into **understanding**.

Detection is transformed into comprehension by the utilization of this vertical synthesis of information. For example, take into consideration the weather. Not only does MSH-GAN recognize rain as a factor that is associated with slower speeds, but it also acknowledges that the type of precipitation, the time of day, and the geographical location all have different effects on the outcomes. A light drizzle in the morning might not be enough to delay traffic, but a heavy deluge in the evening close to schools might make it far more difficult to commute. Not only does the model take into account deeper context, but it also takes into account fundamental linkages.

When dealing with high-stakes and dynamic occurrences, such as accidents, it is essential to maintain the same level of attention. Not only is it necessary for the system to determine the deceleration that occurs as a consequence of a collision, but it must also evaluate the ensuing consequences, which may include the closing of lanes, the creation of congestion due to detours, and the strain placed on alternate routes. Furthermore, because its layers are interconnected, the model is able to imitate not just the primary effects but also secondary and tertiary changes in flow patterns. This is because the model is able to replicate all of these changes.

Not only does this consciousness extend beyond reacting to actual stimuli like the weather or accidents, but it also exists beyond that. MSH-GAN is also capable of learning from behavioral cues, such as patterns of reluctance at particular turns, sites with frequent honking, and intersections that are marked by weak yield compliance. These complex inputs, which are frequently ignored in standard models, lend support to the phenomena of "urban intelligence" that is displayed by the system.

As a consequence of this, the model is beneficial not only for the purpose of creating projections but also for the purpose of making management decisions easier. The question that should be asked of city planners should go beyond just asking, "What would happen if we put this signal plan into action?" Nevertheless, the question that needs to be asked is, "What kinds of behaviors and responses would this change generate throughout the entire system?" Due to the fact that it encodes both details and context, MSH-GAN is able to respond to such requests.

It is possible that the most exciting part of it is the generalizability of the contextual options it offers. In spite of the fact that it is confronted with unusual circumstances, such as a flyover that has just been inaugurated or an intersection that has been reconfigured, the model may make use of the structural logic that it has gained in order to generate predictions that are both credible and flexible. It does not rely heavily on the repetition of data; rather, it generates a varied representation of the dynamics of the metropolis.

The result of this is that MSH-GAN has progressed beyond the role of a simple modeling tool and is now functioning as a collaborative partner in the process of designing and improving mobility support systems. Because it acknowledges the significance of movement, it improves traffic in ways that go beyond simple motion. In contrast, meaning differentiates between merely being aware of facts and having genuine comprehension.

9.7 Thinking at the Edge: Decentralized Computation for Faster Response

Cities are too quick to think about the far future. The decisions that are made in control rooms located in far-off locations are not quick enough to be implemented when an ambulance is stalled behind a sea of traffic or when a firetruck needs to reroute immediately. Taking effective action in real time requires intelligence that is both real-time and localized. At this point, decentralized edge computing transforms from merely being an enhancement in performance to becoming an absolute requirement. And MSH-GAN was designed to function at this edge of the network.

MSH-GAN is designed to enable the distribution of its intelligence, in contrast to typical AI models, which necessitate the use of centralized servers and pipelines that are hosted in the cloud. It is possible to deploy its components across several junctions, within roadside units (RSUs), and even in vehicles that are connected to one another. In addition to processing localized Internet of Things data, running inferences, and carrying out micro-level predictions, each instance operates in a semi-autonomous manner while simultaneously remaining synced with bigger macro-level patterns through shared learning.

It is a notion that has been around since the beginning of biology that this architectural choice reflects: intelligence is most effective when it is close to action.

Similar to how the spinal cord, rather than the brain, is responsible for controlling the speed of reflexes in the human body, traffic reflexes, such as rerouting after a blockage, need to be triggered at the location where they begin to take place.

Live sensor data, including video feeds, vehicle speed logs, signal statuses, and environmental variables, are gathered by MSH-GAN modules that are deployed at the edge of the network. They make use of the micro-level generator in order to replicate immediate behaviors and friction points, which allows them to foresee how a jam would manifest itself or how the behavior of lanes might change in the following few seconds. It is possible for the local models to develop without becoming blind because the macro-level context, which is educated on broader traffic knowledge, is synchronized sporadically from a higher-layer cloud or citywide node.

It is essential to strike this balance between autonomous decision-making and synchronized learning. It enables the system to be responsive without increasing the likelihood of fragmentation. Not only does every edge node comprehend the flow of traffic in its immediate vicinity, but it is also cognizant of the role that its segment plays in the wider city puzzle.

The benefits are numerous and varied:

- **Decreased Latency**: Decisions are made at the location where data is created, eliminating the need for round-trip delays to remote servers.
- **Fault Tolerance**: ensures that even in the event that central systems fail, the functioning of localized control remains uninterrupted.
- **Scalability**: The ability to add more nodes without overwhelming a central AI system.
- **Bandwidth Efficiency**: Only aggregated insights or updates are required to travel back to the core system, as opposed to raw high-frequency data.

Imagine a network of junctions that are taken care of by edge devices that are powered by MSH-GAN. In a crucial lane, one of the nodes identifies a trend of slowing down. It does this by employing micro-behavior simulation in order to project a likely buildup. Without having to wait for an instruction, it immediately makes the necessary adjustments to the green-time allocation of the incoming signal. At the same time, it notifies surrounding nodes about the anticipated delay, which prompts them to make adjustments to the flow that is further upstream. Within milliseconds, all of this takes place, and there is no need for centralized coordination to be a burden.

When it comes to contingency situations, the same rules apply. An ambulance that is connected to the internet sends out its location and route in real time. Edge nodes along its path replicate the behaviors of drivers in the surrounding area and have the ability to dynamically change signal priorities in order to clear the way. The answer is not only quick, but it is also coordinated and self-consistent. This is because each decision is taken locally, but it is simultaneously aligned worldwide.

It goes without saying that this kind of decentralization also brings about difficulties, such as version synchronization, local hardware limits, and model update consistency. However, modular deployments were taken into consideration when designing the architecture of MSH-GAN. It is possible for lightweight inference versions of the model to function on edge GPUs or on ARM devices that have been tuned. Nodes are able to interact with one another and learn from one another through the use of protocols that allow for incremental updates and experience sharing.

According to this line of thinking, the city itself functions as the computer. In today's world, streets are no longer merely passive infrastructure; rather, they are active agents. No longer does the traffic system wait for insight to arrive; rather, it generates it, lives in it, and acts on it **at the source**.

This change, from centralized planning to decentralized intelligence, is more than just a breakthrough in technology. It embodies a new philosophy, which is that the future of mobility will not be controlled from above, but rather will be **co-created by distributed intelligence that is based on context, proximity, and purpose**.

9.8 Measuring Intelligence: Evaluating the Performance of MSH-GAN

Providing reassurance is not the only thing that an outstanding model accomplishes. The ability to deliver is something that they possess. In the complex world of urban traffic, where precision and time are the determining factors between smooth flow and congestion, performance is a sensible concern that needs to be taken into consideration. Within the parameters of this discussion, performance is the determining factor rather than a theoretical concern pertaining to evaluation. It is eventually necessary for MSH-GAN to demonstrate that it is not just capable of thinking effectively but also of acting proficiently, despite the architectural ambitions and contextual complexities it possesses. Yes, it does in fact.

When compared to conventional AI traffic models, such as classic GANs, LSTMs, and shallow hybrid systems, MSH-GAN consistently displays improved fidelity, increased generalization, and rapid adaptability to situations that are representative of the real world. As a result of its comprehensive evaluation in simulated environments, synthetic benchmarks, and real-time urban feeds, it outperforms earlier models in a variety of dimensions.

9.8.1 Fidelity: Ensuring the Real Truth Is Not Lost

One of the most important metrics for evaluation is **data fidelity**, which refers to the degree to which the output of a model is similar to the actual traffic data that has been seen. On the other hand, pattern matching is not the only component that contributes to the development of realistic traffic modeling. It is the purpose of this endeavor to determine the **truths of behavior**. An example of a model that generates traffic levels that appear realistic but fail to accurately describe genuine microbehaviors is an example of a model that creates illusions.

MSH-GAN improves upon this earlier development by doing an **analysis of traffic patterns** that are semantically relevant on both the micro and macro levels. It is not as simple as aligning the amounts of vehicles; rather, it represents emergent processes that are occurring beneath the surface. In addition to determining whether or not the system is realistic, its multi-class discriminator is also modified in order to validate the structural coherence of the system. When compared to the baselines of conventional GANs, the ultimate result was a 118.17% improvement in data fidelity. This was verified using parallel simulations that were validated against actual ground truth data acquired from networked devices.

In practical terms, this means:

- Projections of congestion are more correctly associated with the actual queue lengths.
- It is possible for flow models to provide a more realistic representation of the pressure that is applied to each individual lane during peak hours.
- There is a strong correlation between delay estimations and the observed signal timings and throughput ratios.

9.8.2 Accuracy: Predicting the What and the When

The capability of accurately predicting both the what and the when is a measure of precision.

Depending on the particular layer for which the prediction is created, it is common practice to evaluate the accuracy of predictions by employing metrics such as mean absolute error (MAE), root mean square error (RMSE), or classification F1-scores. This evaluation is done within the context of the standard operating procedure. By creating outputs that are not only point forecasts but also **sequence-consistent** and **behavior-aware**, MSH-GAN regularly outperforms standard time-series and sequence models. This is accomplished by producing outputs that generate outputs.

Long short-term memory (LSTM) systems have the ability to properly forecast the average flow of traffic along a route. On the other hand, MSH-GAN is able to forecast:

- The beginning of flow breakdown,
- The length of time it will last,
- The driving behaviors that will have been prevalent throughout the event.

The outcomes of its projections are consequently more complete, more practical, and more aligned with the knowledge that operators or adaptive systems require on a regular basis.

It has been demonstrated that MSH-GAN can:

- Reduce the RMSE in short-term forecasts within a 15–30-minute interval,
- Improve its response to sudden disruptions like as accidents and changes in the weather,
- Recalibrate as quickly as possible following abnormal data spikes.

9.8.3 Responsiveness: Learning as Conditions Evolve

There is never a stationary city. In addition to the redesign of flyovers, the introduction of brand new flyovers and the revision of signal laws are currently taking place. Numerous AI models have shown a decrease in accuracy as a result of this transition, necessitating either retraining or manual recalibration, respectively. The architecture of the MSH-GAN makes continuing customization easier.

The model has:

- No requirement to undergo a comprehensive retraining of the system in order to make adjustments to the local weights,
- Capability to recalibrate the dynamics that exist between the generator and the discriminator in response to new circumstances,
- Capability to simulate counterfactuals in order to evaluate adaptive procedures before they are put into action.

Considering this, MSH-GAN is not only a predictor, but it is also a dynamic system that is able to acquire knowledge over time rather than the information decreasing with it.

9.8.4 Simulation Utility: Planning Before Acting

When it comes to traffic management, the ability to anticipate unforeseen occurrences is just as important for city planners as the ability to react in real time. When it comes to this particular field, MSH-GAN shines. There are a number of tasks that can be accomplished through the generation of coherent and contextually aware traffic simulations. These tasks include:

- Evaluating new signal plans under virtual conditions,
- Assessing the ramifications of road closures prior to their execution,
- Estimating pedestrian-vehicle conflicts in mixed-use areas,
- Predicting the influence of weather-sensitive traffic management.

As a result of its capacity to generate **high-fidelity synthetic futures**, urban designers are able to move away from making assumptions and instead collect data. This enables them to simulate before implementation and evaluate before spending.

9.8.5 Robustness: When the Unexpected Happens

In spite of the fact that it receives very little attention, resistance to the unexpected is in fact one of the most important strategies. In studies in which noise was added into sensor feeds, simulated partial data loss occurred, or nonstationary behaviors were implemented (such as rerouting due to protests), MSH-GAN displayed a graceful decline, maintaining its utility despite fluctuations in input quality. This was the case even when the input quality fluctuated.

The durability of the system is due to a number of factors:

- Hierarchical design, which makes it easier to incorporate macro-level patterns in order to address micro-data deficiencies,
- The generative modeling framework, which has the ability to propose credible continuations,
- Ongoing feedback loop with real-time data, which makes it possible to quickly recalibrate the system.

The promise of AI is not that it will deliver perfection, but rather that it will provide a constructive imperfection that will improve over time, react to reality, and assist humans in making judgments that reflect increasingly accurate information. Not only does MSH-GAN fulfill that need, but it also improves upon it to an even greater extent. Additionally, this action places it as both a technological achievement and a practical collaborator in the evolution of smart cities, which is a significant accomplishment.

9.9 From Control to Care: MSH-GAN and the Future of Smart Cities

Transportation in urban areas is no longer solely concerned with infrastructure; rather, it is also concerned with **intelligence, sustainability, and equity**. It is possible to gain a second in terms of productivity, safety, or even survival for every second that is spared in traffic. Every gram of carbon dioxide that is prevented from being released into the atmosphere is a step toward making the planet habitable. Every route that is optimized makes cities more humane, more robust, and more responsive to their visitors. This bigger tale is finally served by MSH-GAN, which has a great deal of computational depth. Not only does it aim to make cities smarter, but it also aims to make them **kinder and wiser**.

9.9.1 Sustainable Mobility: From Emissions to Efficiency

Congestion in traffic not only wastes gasoline but also wastes resources and opportunities. As a result, it exacerbates urban pollution, raises stress levels, and wastes time on a massive scale. Even a slight enhancement in flow, when implemented across the entire city, can result in substantial savings in the following areas:

- Greenhouse gas emissions,
- Airborne particulates,
- Noise pollution,
- Fuel consumption.

Through **proactive decongestion**, MSH-GAN contributes to the achievement of this goal. It functions as a sustainability amplifier by predicting bottlenecks before they become more severe, changing flows in a dynamic manner, and enabling better modal distribution (for example, separating freight from commuter traffic during sensitive hours). In other words, it embeds **ecological intelligence into everyday transportation**.

Furthermore, it is able to model alternate design scenarios, such as the possibility of pedestrianizing this corridor. What if we offered incentives to commuters who arrive at off-peak times? How about we reroute the freight while the schools are in session? In addition to providing answers to these questions, MSH-GAN also assists in testing the ramifications of these questions before they are implemented.

9.9.2 Emergency Responsiveness: Making Every Second Count

The influence that real-time, predictive traffic modeling has on **emergency services** is one of the most tangible benefits related to this type of modeling. The implications can be fatal if an ambulance or fire engine is delayed because of clogged intersections or drivers who are ignorant of their immediate surroundings.

MSH-GAN enables systems in:

- Detection of emergency vehicle trajectories in advance,
- Simulation of the optimal clearance courses in real-time,
- Reconfiguration of signals to construct dynamic corridors,
- Communication of impending right-of-way alterations to other vehicles.

Instead of being carried out by centralized orders, all of this is accomplished by distributed intelligence that is implanted at the edge, which is where speed and context are of the utmost importance.

9.9.3 Citizen-centric Planning: From Data to Dignity

Cities are not machines; rather, they are ecosystems in which people live. Traffic is not only about automobiles; it is also about people, including workers, students, parents, and senior citizens. Some of the ways that MSH-GAN assists in the development of systems that better serve people are:

- Providing planners and policymakers with insights into travel time variability across neighborhoods,
- Providing fairer routing for underserved zones with poor transit access,
- Simulating how interventions affect **walkability** or **bike safety**,
- Providing tools to engage communities in data-informed planning.

The technology that is used to represent signal flow can also be used to model human fairness using the following: Are locations with low incomes more likely to be affected by congestion than other areas? Which populations are more likely to have lengthier commutes as a result of obsolete routing? In addition to assisting in the surface of these patterns, MSH-GAN can also support a more egalitarian redesign.

9.9.4 Adaptation and Governance: A Living Urban System

It is necessary for urban traffic networks to develop to accommodate not only development but also change. Projects involving infrastructure, public events, changes in climate, and emerging mobility trends (such as electric scooters or autonomous vehicles) are all examples of things that provide a threat to the fixed paradigms of the past.

Due to the fact that MSH-GAN is:

- Modular,
- Context-aware,
- Continually adaptive,

...it has the ability to continuously adapt to new circumstances and keep up with the ever-changing urban narratives. As a result, it becomes a collaborator in the formulation of policy, a device for governance, and a platform for experimentation.

Imagine living in a city where commuter plans are revised on a weekly basis rather than once every ten years. Where possibilities of "**what if**" are tested within hours, rather than after citizen complaints have been received. In this context, judgments are not made based on inertia but rather on data, simulation, and predictive insight respectively. In preparation for this transformation, MSH-GAN builds the groundwork.

Scenario	MSH-GAN simulation capabilities	Outcome
Peak Hour Congestion	Predicts choke points before saturation	Signal plan optimization
Emergency Vehicle Clearance	Reconfigures local signal phases along the route	Reduced emergency response time
School Zone Behavior	Models pedestrian-vehicle interactions in constrained environments	Safety-aware flow modulation
Weather-induced Disruptions	Adjusts flow patterns under reduced visibility or road friction	Dynamic rerouting and demand redistribution
Event-based Surges (e.g., stadium)	Simulates pre/post-event flow amplification	Temporary lane prioritization or routing

9.9.5 A Philosophy of Coexistence

Over and above all else, what MSH-GAN stands for is a new philosophy: urban intelligence as a shared ecosystem that is responsive to its surroundings. It is no longer a collection of instructions that are imposed from the top down and are driven by rules; rather, it is a system that perceives, learns, and reacts in a manner that is reflective of the rhythm of life itself.

Planners are able to look into the future, not via the use of an oracle, but through the use of calculation. Because of this, residents are able to navigate the city with a higher level of assurance. Systems are able to fail gracefully and recover intelligently, thanks to this capability. Furthermore, by doing so, it provides support to cities that are not only intelligent but also just, sustainable, and profoundly aware of the human reason for which they exist.

9.10 Looking Ahead: The Road Beyond MSH-GAN

When it comes to cities that never stop moving, traffic never really stops; rather, it only changes form, transforming itself according to peak hours, events, seasons, and unexpected situations. A task that is both one of the most difficult and consequential aspects of modern urban living is the management of this flow that is always shifting. The chapter that started out as a discussion of multi-scale hierarchical GANs is, in reality, a plea to rethink the way in which we model, manage, and navigate through our cities.

MSH-GAN is not merely a technological improvement; rather, it represents a paradigm shift in terms of thought. It is a reflection of a transition from control to comprehension, from reaction to prediction, and from uniform rules to contextual adaptation. It is a model that acknowledges the complexity of the situation and replies with layered abstraction that is intelligent.

Throughout this chapter, we have traced the road from the limitations of traditional traffic systems to the gaps in early AI models to the promise and performance of MSH-GAN. MSH-GAN is a system that not only simulates flow but also understands it in levels, including micro and macro, temporal and spatial, and behavioral and systemic.

Nevertheless, this is not the conclusion of the adventure. Nevertheless, it is merely the basis. Continuing to be curious, creative, and careful will be necessary in order to navigate the road that lies ahead, which is full of chances as well as problems.

9.10.1 Federated and Privacy-preserving Models

Concerns over the privacy of sensitive data are growing as cities become more linked. Additionally, it is possible that future iterations of MSH-GAN would incorporate federated learning approaches. This would enable several city zones or jurisdictions to train models locally and share just model updates, rather than raw data. Not only would this safeguard the data of citizens, but it would also make it possible to make innovations that are scalable and collaborative across areas.

9.10.2 Integration with Reinforcement Learning (RL)

Integrating MSH-GAN with reinforcement learning could be the next step forward, despite the fact that it is particularly effective at generation and simulation. The predictive simulations of MSH-GAN could provide RL agents with the information they need to actively test signal policies, routing strategies, or incident reactions, thereby learning the most effective actions to take in a closed-loop feedback environment. The result would be that cities would be transformed into self-optimizing systems, capable of enhancing themselves not only through design but also through experience.

9.10.3 Multi-agent and Human-aware Traffic Models

Not only does traffic not flow in a consistent manner, but it is also made up of agents who have objectives, limitations, feelings, and unpredictability. It is expected that multi-agent GANs will be used in the future of intelligent traffic modeling. Here, many types of entities, such as drivers, cyclists, pedestrians, and autonomous vehicles, will be modeled with behaviors that are distinct from one another but interact with one another. It is possible to improve both realism and responsiveness by incorporating human consciousness, which can be accomplished through behavioral modeling, psychology, or even game theory.

9.10.4 Diffusion Models and Beyond-GAN Architectures

Recent developments in diffusion-based models provide fresh promise in capturing long-range dependencies, uncertainty, and fine-grained information. While GANs have been at the forefront of the revolution in generative modeling, these models have also presented new opportunities. In subsequent rounds of traffic models, it may be possible to hybridize both designs, so as to combine the speed of GANs with the robustness of diffusion systems. This is especially true in urban settings that are neither linear nor complicated.

9.10.5 From Research to Deployment: Bridging the Gap

The main value of MSH-GAN resides in its deployment, despite the fact that it is quite sophisticated. Collaboration between data scientists, policymakers, transport authorities, civic technologists, and the general public is required in order to make the transition from the prototype to the impact on a city-wide scale successful. The development of interfaces, simulated sandboxes, visual tools, and layers of explainable AI will be essential in order to put insights into action.

9.10.6 A Message to the Reader

To have read this far indicates that you are not merely a student of AI or transportation; rather, you are a member of the generation that is responsible for designing the **next era** of mobility. **Curious minds who dare to think in systems** and ask questions that bridge the gap between code and consequence will be the ones to build the future. Perfect models will not be the ones to build the future.

An example of such a bridge is MSH-GAN. On the other hand, a great number more are required. The hope is that this chapter will not only provide information but also motivation. to think of systems that are able to adapt. Cities have a sense of air. Streets that are attentive. And technology that, at every step, creates space for humanity to advance in a manner that is more intelligent, more rapid, and with a more significant purpose.

References

1 Goodfellow, I., Pouget-Abadie, J., Mirza, M., Xu, B., Warde-Farley, D., Ozair, S., Courville, A., and Bengio, Y. Generative adversarial nets. In *Advances in Neural Information Processing Systems*, pages 2672–2680, 2017.

2 Xu, J., Tang, S., and Yang, H. KI-GAN: Knowledge-informed generative adversarial networks for traffic flow prediction. *Applied Sciences*, 11(22): 10956, 2021.

3 Kumar, A., Nema, P., and Srinivas, K. SpectraGAN: GAN with spectral normalization for robust traffic modeling. *Symmetry*, 13(5): 779, 2021.

4 Zhang, Y., Li, Q., and Gao, H. Graph-based generative adversarial networks for traffic data prediction. *International Journal of Intelligent Systems*, 38(1): 59–76, 2023.

5 Wang, Y., Zhao, L., and Liu, Y. Enhancing traffic flow prediction with GAN and graph convolutional networks. *IEEE Transactions on Intelligent Transportation Systems*, 24(9): 2238–2247, 2023.

6 Liu, X., Chen, R., and Zhang, J. GAN-based traffic prediction considering external weather conditions. *Journal of Advanced Transportation Research*, 8(3): 142–155, 2023.

10

IoT and AI Integration in Traffic Management

J. Steffi[1,], K. Merriliance[1], and Mary Immaculate Sheela Lourdusamy[2]*

[1] *Department of Computer Applications and Research Centre, Sarah Tucker College (Autonomous), Affiliated to Manonmaniam Sundaranar University, Tirunelveli, Tamil Nadu, India*
[2] *Department of Computing and Engineering, School of Engineering Technology and Applied Science (SETAS), Heritage Christian University College (HCUC), Accra, Ghana*

*Corresponding author: stefficsc10@gmail.com

10.1 Introduction

10.1.1 Urbanization and Traffic Challenges

The rapid increase of urban populations has brought about an exponential growth in the wide variety of automobiles on roads. This surge in vehicular traffic contributes to continual site visitors jams, longer trip times, and better degrees of noise and air pollutants [1]. Existing infrastructure and traditional site visitors manipulate mechanisms, which depend heavily on static sign timing and manual monitoring, are now not adequate to manipulate dynamic urban website visitors conditions.

10.1.2 Limitations of Traditional Traffic Management

Traditional website online traffic manage structures are typically reactive and limited in scope. They lack the capability to evolve in actual-time to changing visitors situations, sudden events such as injuries or street closures, and fluctuations in site visitors quantity.

These boundaries bring about inefficient traffic float, improved gasoline consumption, and extensive delays that have an effect on the satisfactory of life and monetary productiveness in cities.

10.1.3 Emergence of Smart Traffic Management

To address those challenges, the idea of smart traffic management has emerged, leveraging the strength of modern virtual technology. The integration of IoT and AI allows a shift from static to dynamic site visitors manipulate models [2].

These technologies facilitate real-time facts collection, superior analytics, and automated choice-making methods that adapt site visitors controls based on stay situations.

10.1.4 The Role of IoT and AI

IoT era entails deploying a network of sensors and related devices throughout urban infrastructure. These gadgets continuously gather facts on site visitors flow, vehicle speeds, weather situations, and avenue occupancy. AI algorithms technique this facts to derive insights, predict future situations, and suggest or execute moves that include adjusting signal timings or rerouting visitors [3].

10.1.5 Objectives and Benefits

The major goal of combining IoT and AI [4] in website online site visitors systems is to boost town mobility by reducing congestion, lowering environmental effect, and increasing avenue defense. Other advantages include increased public shipping coordination, faster emergency response, and data-pushed infrastructure planning. These systems help to advance sustainable and smart cities.

10.2 Role of IoT in Traffic Management

10.2.1 IoT in Urban Traffic Systems

IoT is a collection of devices connected to each other, gathering, sending, and processing information. In traffic, IoT is the digital nervous system that facilitates continuous, real-time communication among vehicles, infrastructure, and control centers. Through connectivity, it is possible to manage road networks more responsively and intelligently.

10.2.2 Key IoT Components in Traffic Infrastructure

Traffic management [5] using IOT uses an array of sensors and equipment installed on strategic points within a city (Figure 10.1):

- **GPS (Global Positioning System)**: The singular vehicle logs speed for travel as well as instructions and mob reports.

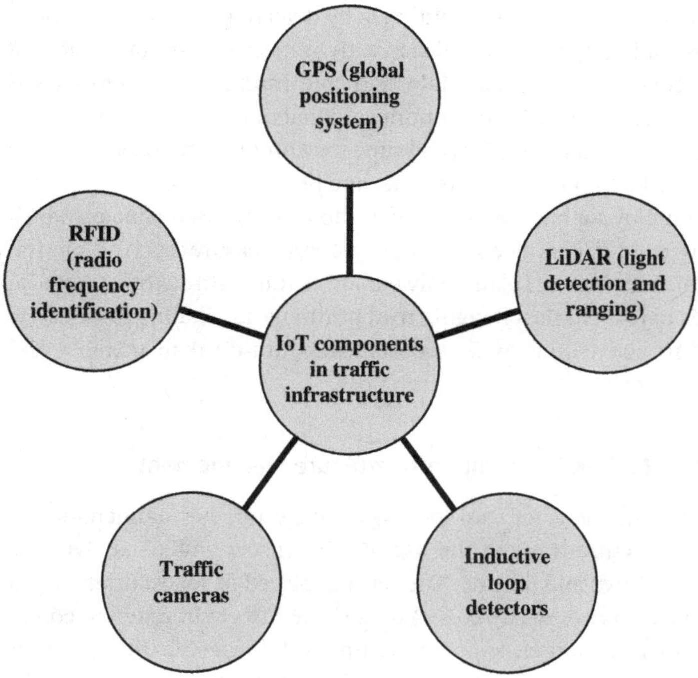

Figure 10.1 IoT components in traffic infrastructure.

- **RFID (Radio Frequency Identification)**: Identification, tracking, and movement of vehicles, mainly toll and access to restricted areas.
- **LiDAR (Light Detection and Ranging)**: 3D delivery and vehicle detection delivery and is best in autonomous vehicle management.
- **Traffic Cameras**: Support video monitoring for detection, traffic monitoring, and police enforcement.
- **Inductive Loop Detectors and Magnetometers**: Find the presence and number of vehicles on intersections and on the freeway.

10.2.3 Data Collection and Real-time Monitoring

Collecting real-time data [6] is the basis of adaptive and responsible control in intelligent traffic management systems. IOT sensors together with site visitors cameras, inductive loops, magnetometer, GPS trackers, and weather sensors constantly measure facts about car waft, pace, density, road surface, and even pedestrian moves. For example, GPS vehicle data facilitates travel patterns and congestion level monitoring in urban areas, while traffic cameras and loop detectors measure the number of vehicles and queues passing over intersections.

Environment sensors also provide useful data by detecting weather conditions such as rain, fog, and temperature, which greatly impact the driving style and road safety [7]. Pedestrian movement data, gathered from infrared sensors and computer vision systems, are utilized to optimize signal timings for pedestrians to cross safely. All of these data are fed in real time to centralized management systems or decentralized edge processors for real-time processing.

These systems employ sophisticated algorithms to identify discrepancies, adjust traffic light timing, send drivers alerts, or trigger control measures such as removal of traffic. Real-time monitoring [8] not only enhances traffic throughput, but also allows authorities to react to the changing road positions, facilitating more safety and efficiency. This active strategy is important for future urban mobility and development of smart city.

10.2.4 Predictive Maintenance and Infrastructure Management

The future maintenance is a forward-looking strategy for the maintenance of urban traffic infrastructure through the use of IoT sensors and AI analytics to predict premature failures and decline. The sensors placed in roads, bridges, and traffic control devices continuously collect data related to environmental conditions, structural stress, vibration, temperature ups and downs. Real-time data is analyzed through the AI algorithm to detect abnormal patterns and to predict the potential wear and tear or damage.

For example, a sensor may identify the growing vibration on a bridge, suggesting structural weakness. Instead of waiting for obvious damage or failure, the maintenance crew can step quickly to repair, prevent expensive breakdown, and maintain public safety. In the same vein, tracking the traffic light system and road conditions allows authorities to solve problems such as defective signs or pits before obstructing traffic flows.

This clever, records-based approach minimizes unplanned downtime, decreases the cost of preservation, and maximizes the lifespan of infrastructure property. It additionally protects street users from disruption by acting as protection at off-height times. Generally, predictive protection of the usage of IoT and AI converts city infrastructure management from a reactive to a preventive framework, facilitating the smooth running of clever cities and improving the road protection and reliability of commuters.

10.2.5 Vehicle-to-infrastructure (V2I) Communication

A vehicle-to-infrastructure (V2I) [9] is a building block of the communal clever transport structures, providing the real-time, wireless exchange of facts between automobiles and roadways infrastructure. This bidirectional conversation enables motors to talk with visitors signals, avenue signs, lane markings, and one-of-a-kind infrastructure capabilities, improving visitors performance and safety.

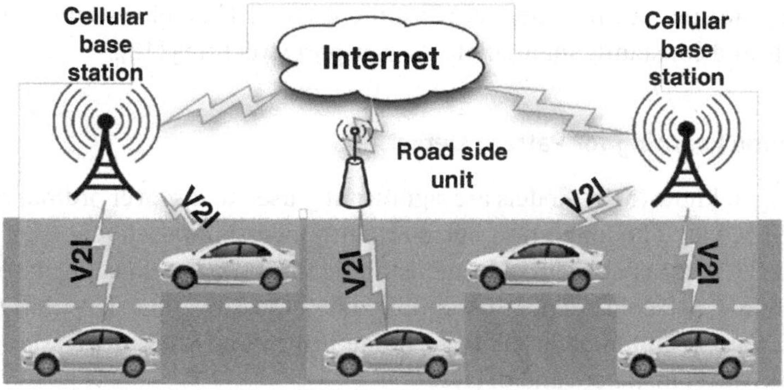

Figure 10.2 Vehicle-to-infrastructure (V2I) communication.

V2I structures make use of technology like dedicated short-range communications (DSRC) and cellular vehicle-to-everything (C-V2X) to permit this interaction. For example, a car coming up to an intersection can receive signal timing information, allowing the driver or autonomous system to modulate speed in response, eliminating sudden stops and enhancing traffic flow. In turn, cars can send information back to infrastructure, like speed and position, supporting traffic management and congestion studies.

The benefits of V2I communication are multifaceted:

- **Improved Safety**: Real-time warnings regarding unsafe conditions, such as slippery roads or accidents up ahead, enable drivers to exercise caution.
- **Traffic Optimization** [10]: Intelligent traffic signals can adapt based on real-time traffic patterns, reducing delays and lessening congestion.
- **Environmental Benefits**: Smooth traffic flow results in decreased fuel usage and lower emissions, making the environment more sustainable.

Figure 10.2 [11] shows the V21 communication. Implementing V2I communication requires investment in infrastructure upgrades, including the installation of roadside units (RSUs) and the integration of communication protocols. As urban areas evolve into smart cities, V2I technology plays a crucial role in creating a connected, efficient, and safer transportation ecosystem.

10.3 AI Applications in Traffic Optimization

10.3.1 AI-observation in the Traffic System

Artificial intelligence (AI) performs an important function for analyzing and understanding the large amounts of facts generated through IoTs in the network of urban visitors [12]. AI converts raw sensor data into action-rich insights that help manipulate techniques in adaptive, real-time visitors. This allows cities to more efficiently and constantly manipulate its transport structures [13].

10.3.2 Machine Learning for Pattern Detection

Machine getting to know (ML) models are significantly used to discover ordinary site visitors styles, pick out congestion hotspots, and understand each day commuting behavior. These fashions learn from historic statistics to are expecting visitors flow, top hours, and coincidence-susceptible zones. Classification algorithms help become aware of automobile kinds and lane utilization, while regression models estimate travel instances and delays.

10.3.3 Deep Learning for Object and Scenario Recognition

Deep learning fashions, along with convolutional neural networks (CNNs), allow excessive-accuracy reputation of automobiles, pedestrians, and street signs. These fashions are crucial in complicated environments like busy intersections or pedestrian-heavy zones. They decorate the performance of autonomous systems and laptop vision programs used in actual-time video surveillance and traffic tracking.

10.3.4 Reinforcement Learning for Dynamic Signal Control

Reinforcement gaining knowledge of (RL) algorithms [14] dynamically regulate traffic sign timings based on evolving site visitors conditions. By receiving feedback from the environment, these systems learn to reduce the length of the queue, reduce the wait, and improve float to visitors. RL-Mote performs with predetermined rules, adapting to unexpected adjustments.

10.3.5 Future Staging Analysis and Traffic Forecast

Predictive Analytics [15] uses timing collection analysis and long-term short-term memory (LSTM) networks and website such as Hybrid AI fashion to advocate for optimal routes to overload and go in for visitors as LSTM networks and visitors.

These models integrate actual-time sensor statistics with historic traits to assume traffic constraint-America and enable proactive traffic management.

10.3.6 AI in Automated Decision-making Systems

AI capabilities are the center decision-making engine in clever traffic systems. It processes numerous inputs, which includes weather, roadwork schedules, and occasion calendars to supply coordinated choices [16]. This allows proactive measures such as rerouting website visitors, sending alerts, or enhancing sign priorities throughout emergencies.

10.3.7 Key Benefits of AI in Traffic Optimization

- Real-time Adaptability: Enables immediate responses to dynamic situations.
- Improved Road Safety: Early detection and intervention help lessen collision fees.
- Operational Efficiency: Optimizes useful resource allocation, together with signal control and emergency response.
- Data-driven Planning: Provides insights for long-term infrastructure and policy development.

AI empowers site visitors management structures with the intelligence to now not just reply, but also to anticipate and adapt. By combining more than one AI strategy, cities can achieve extra performance, protection, and sustainability in their transportation networks. Machine learning algorithms come across styles and predict congestion developments, while reinforcement mastering optimizes signal timings in dynamic site visitors environments. Deep mastering models improve vehicle and pedestrian type in complex situations. Predictive Analytics estimates excessive traffic zones and proposes preemptive course modifications. AI consequently serves as the decision-making center, enabling faster and more correct responses to visitors fluctuations.

10.4 Smart Traffic Signals and AI-driven Control Systems

10.4.1 Introduction to Smart Traffic Signals

Smart traffic signals [17] take dynamic advantage of real-time data and AI algorithms to manage traffic lights. Unlike the traditional systems that work with a certain schedule, these intelligent signals adapt their cycles based on real traffic flow conditions, which improves the general efficiency of urban traffic networks.

10.4.2 Integration with IoT Sensors

IoT sensors, such as inductive loop detectors, magnetometer and camera-based systems, are distributed at intersections to collect live traffic data. This data includes the calculation of vehicles, vehicle type, queue length, and pedestrian movement. The sensors [6] continuously transfer this information to checking devices, which are able to make informed decisions in each signal phase.

Figure 10.3 [6, 17] shows the elements of the smart traffic management system.

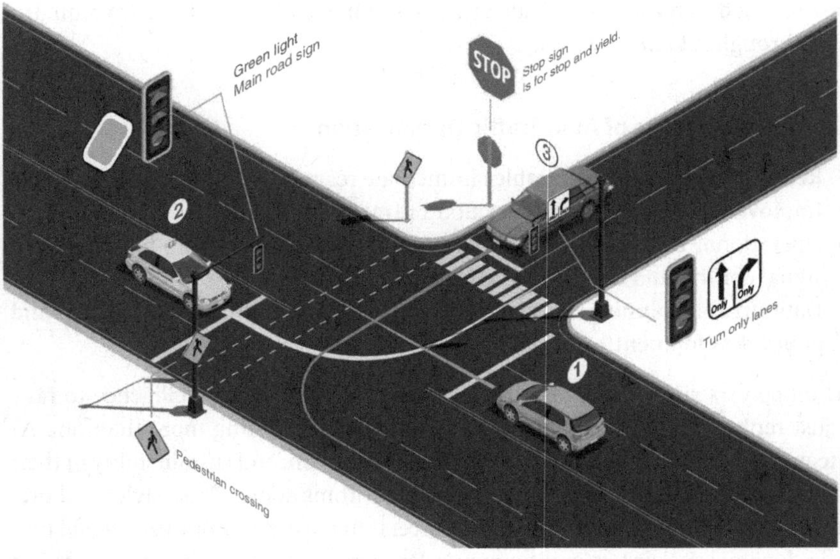

Macrovector/Adobe Stock Photos

Figure 10.3 Elements of a smart traffic management system.

10.4.3 Use of Reinforcement Learning for Signal Optimization

Reinforcement learning (RL) is an essential AI technique used to adapt to signaling timing. RL models consider each intersection as an environment and learn the best signal phase actions through testing and errors. Over time, these models identify optimal patterns that reduce waiting time, improve traffic flows, and vehicles reduce passive duration.

10.4.4 Real-time Adaptive Signal Control

Smart traffic alerts can respond to immediate deviations consisting of visitors waves, accidents, or street boundaries. It is achieved by analyzing live data and adjusting light cycles accordingly. For example, if an abnormally high number of vehicles is detected in one direction, the duration of the green light can be temporarily expanded to clear the amount.

10.4.5 Multi-interest Coordination

The AI system can coordinate signal behavior in multiple intersections to ensure even the vehicle's progress, known as the "Green Wave" traffic flow. This reduces stop-and-go driving, which not only increases the traffic liquidity but also cuts fuel consumption and emissions.

10.4.6 Emergency Vehicle Prioritization

Smart signal V2I can identify emergency vehicles through communication or acoustic sensors and override normal traffic cycles to provide a clear way. It ensures quick response time for ambulances, fire trucks, and law enforcement.

10.4.7 Benefits of AI-driven Traffic Signals

- Minimized Traffic Delays: Dynamic cycle adjustment reduces unnecessary waiting time.
- Low Emissions: Reduction of fuel consumption decreases and greenhouse gas is produced.
- Better Traffic Safety: The liability of truth helps prevent accidents at intersections.
- Scalable Infrastructure: The AI model can be scaled to manage urban or regional traffic networks.

Smart traffic signal systems provide examples of how AI and IoT integration convert reactive traffic management into a future, efficient, and responsible system that adapts to the ever-increasing conditions in urban dynamics. The reinforcement is able to learn these systems with time to learn the optimal signal patterns over time, reduce the wait, and reduce bottleneck. Integration with IoT sensor allows for dynamic decisions, where signal changes are not only based on the cycles of time, but also on live data from the intersection. These systems minimize the traffic delays, fuel intake, and car emissions.

10.5 Incident Detection and Emergency Response

10.5.1 Introduction to Incident Detection

Urban traffic networks are particularly vulnerable to accidents, roadworthy vehicle failures, or unsafe road conditions. Efficient and timely detection and response to incidents are necessary for the reduction of disruption, public safety, and successful traffic movement [18].

10.5.2 AI-driven Computer Vision Systems

Computer vision powered by means of AI is a key era for real-time detection of incidents. Such systems scan stay video feeds from site visitors cameras using deep mastering models like YOLO (you only look once) and Faster R-CNN.

These models are able to detect and classify matters and activities such as visitors injuries, abandoned cars, and pedestrian offenses.

10.5.3 Automated Alert Systems

Once a incidence is identified, AI systems can trigger automated alerts and ship them to site visitors facilities and emergency services. This automation extensively reduces reaction instances and allows mobilizing assist correctly.

10.5.4 Smart Rerouting and Traffic Adjustment

AI algorithms can at once examine neighboring traffic situations and observe rerouting strategies to divert vehicles away from the affected region [19]. This is no longer the most effective permit for emergency cars to reach the scene faster but additionally prevents congestion from building up at the incident website on-line.

10.5.5 Integration with Emergency Services

Traffic management systems integrated with emergency services can provide real-time information on road conditions and expected arrival times. This allows ambulance and health provider crews to select the shortest and safest routes, enhancing the effectiveness of emergency responses.

10.5.6 Benefits of AI-enabled Emergency Response

- Quicker Response Times: Robotic detection and communication simplify emergency processes.
- Improved Public Safety: Early warnings and intelligent rerouting reduce secondary accidents.
- Operational Coordination: Enhanced coordination among traffic authorities and emergency devices.
- Data Collection for Future Planning: Records of incidents enable the improvement of traffic rules and emergency preparedness strategies.

10.6 Public Transport Enhancement with IoT and AI

10.6.1 Overview of Public Transport Challenges

Public transportation systems [12] often struggle with inefficiencies which include fixed routes, inconsistent schedules, overcrowding, and negative adaptability to actual-time call for. These troubles discourage use, boost private automobile reliance, and make contributions to urban congestion and pollutants.

10.6.2 IoT-based Real-time Fleet Monitoring

IoT gadgets permit for round-the-clock tracking of public transport vehicles including buses, trains, and trams. GPS and onboard sensors deliver real-time place, pace, occupancy, and automobile health statistics. All these records are vital for tracking delays, direction compliance, and operational efficiency.

10.6.3 Route and Schedule Optimization Using AI

Artificial intelligence algorithms process historical and real-time data to provide optimized routes and scheduling. Reinforcement learning and genetic algorithms are employed in dynamically resetting bus intervals, minimizing waiting times, and directing resources to areas of maximum demand. Predictive analytics assists in forecasting demand by time of day, events, and weather conditions.

10.6.4 Passenger Information Systems

IoT and AI-driven platforms support intelligent passenger information systems that provide real-time updates on arrival times, delays, and crowding levels via mobile apps and digital signage. This will increase a person's pleasure and self-belief in public shipping.

10.6.5 Multimodal Transport Integration

AI facilitates easy coordination among diverse modes of transport like buses, metros, and journey-stocks. AI-powered integrated journey planners provide optimized travel tips with the aid of joining actual-time availability, direction optimization, and passenger convenience.

10.6.6 Maintenance and Safety Management

Artificial intelligence techniques sense facts through sensors for predictive security of public vehicles. By predicting future breakdowns, it reduces interruption of service. Security is also improved through real-time video surveillance and automatic alerts to any issues or offenses on board.

10.6.7 Sustainability and Emission Reduction

Public shipping enhanced by using AI and IoT contributes to lower emissions by lowering reliance on personal automobiles. AI can prioritize electric-powered or low-emission fleet scheduling and identify greener routes.

10.6.8 Key Benefits of AI and IoT in Public Transport

- Improved Reliability and Punctuality: Real-time tracking and AI optimization reduce delays.
- Improved User Experience: Timely feedback and smart routing improve user confidence.
- Efficient Resource Utilization: AI optimizes transit asset deployment.
- Environmental Benefits: Encourages environmentally friendly urban mobility through lowered emissions.

By leveraging advanced monitoring, real-time analysis, and intelligent scheduling, IoT and AI are revolutionizing public transportation to become a more efficient, sustainable, and dependable form of urban mobility.

10.7 Environmental and Sustainability Benefits

10.7.1 Reduction in Greenhouse Gas Emissions

One of the maximum sizeable environmental blessings of IoT and AI-driven visitors control systems is the reduction in greenhouse gas (GHG) emissions [17]. Traditional site visitors congestion results in prolonged idling, common acceleration and deceleration, and longer journey instances—all of which increase carbon dioxide (CO_2), nitrogen oxides (NO_x), and particulate depend emissions. Smart visitors alerts and AI-based routing lessen stop-and-cross conditions, streamline automobile drift, and minimize gas intake. This now not most effectively cuts emissions from inner combustion engine vehicles but additionally extends the battery variety of electrical motors by using lowering strength-losing visitors delays.

10.7.2 Promotion of Public and Shared Mobility

Through optimization of public shipping overall performance, dependability, and splendor, AI and IoT systems drive the shift away from personal car use. Public shipping turns into a handy choice with the aid of optimized route efficiency, real-time positioning, and incorporated multimodal capability. Intelligent traffic systems, furthermore, can prioritize buses and trams at intersections, minimize their waiting times, and optimize scheduling efficiency. This modal change ends in fewer cars on the roads, reducing average emissions and decreasing the carbon footprint of city regions.

10.7.3 Energy Efficiency of Traffic Infrastructure

AI-supported traffic control systems can also reduce energy intake for traffic infrastructure. For instance, IoT-powered adaptive lighting fixtures can decrease or flip off streetlights when there are little site visitors.

Also, AI can optimize the running hours of site visitors lighting fixtures and different roadside infrastructure to reduce wasteful electricity utilization. All those measures factor toward an electricity-efficient metropolis.

10.7.4 Better Air Quality and Reduced Noise

Less congestion and smoother visitors equate to decreased emissions of pollutant gases including CO, NO_2, and $PM_{2.5}$, which might be adverse to human fitness and the environment. AI-powered monitoring gadgets can display air quality in actual time and offer actionable intelligence for pollution management measures. Moreover, less idling of motors and fewer surprising brakes translate into reduced urban noise pollution, making the town a healthier and more livable region.

10.7.5 Support for Sustainable Urban Policy and Planning

IoT and AI produce high-definition facts regarding individuals traveling, natural outputs, and infrastructure use. Such data contribute toward long-term sustainable metropolis planning, allowing metropolis planners to plan green corridors, maximize pedestrian areas, and impose low-emission zones.

AI can also be used to evaluate the effect of policy measures (e.g., congestion charging or car bans) earlier than they are delivered to assess their environmental effect.

10.7.6 Promotion of Green Vehicles

Smart traffic structures may additionally assist infrastructure for electric and coffee-emission vehicles. For example, AI may want to decide the most suitable locations to put in electric-powered car charging stations by studying utilization styles. In addition, visitors control structures can deliver preferential remedy, which includes devoted lanes or signal precedence, to EVs and other smooth delivery modes.

10.7.7 Circular Economy and Smart Waste Reduction

IoT in visitors infrastructure includes sensor-enabled gadgets which can be designed for lengthy-time period durability, upgradability, and reusability. AI can help expand the lifestyles cycle of those devices thru predictive maintenance and failure forecasting. As a result, fewer digital additives end up in landfills, assisting a greater circular and environmentally responsible approach to city technology deployment.

10.8 Challenges and Future Trends

10.8.1 Challenges in IoT and AI-driven Traffic Systems

10.8.1.1 Data Security and Privacy Issues

The enormous IoT sensor and AI network gather gigantic pools of real-time data, such as vehicle locations, routes traveled, and potentially even user identifiers. This is very much a privacy concern, as well as a matter of user consent and potential abuse. IoT devices can also be hacked to bring about visitors disruption, presenting safety and safety risks. Robust encryption, get entry to controls, and regulatory compliance (e.g., GDPR) are crucial to ensure statistics integrity and person privateness.

10.8.1.2 Maintenance and Infrastructure Costs

Intelligent traffic systems for an entire city entail massive initial capital expenditure in cameras, speech exchange structures, computer devices, and sensors.

Public coffers are also burdened with maintenance, calibration, updates, power, and hardware expenses at regular intervals. Regular updates of software and technical staff for maintenance, operation, and troubleshooting are also necessary.

10.8.1.3 Legacy System Integration

Most cities continue to use conventional traffic management systems, which can be incompatible with contemporary AI or IoT-based systems. Smooth integration between legacy infrastructure and future systems can be technically challenging and costly. Phased implementation plans and interoperability standards are needed to fill this gap.

10.8.1.4 Data Quality and Sensor Reliability

The reliability of traffic decisions by AI relies on the data quality from IoT sensors. Weather, electromagnetic interference, and the sensor getting old may also affect records accuracy. Defective or mis calibrated sensors may additionally bring about suboptimal selection-making and decreased machine dependability.

10.8.1.5 Ethical and Social Implications

AI systems that make actual-time choices for visitors may also be a source of ethical troubles—like giving priority to specific routes or styles of vehicles, or using surveillance devices that could tread on civil rights. Building transparency, accountability, and fairness into algorithmic choice-making is vital to maintain public self-assurance and make certain equitable access.

10.8.2 Future Trends in Smart Traffic Management

10.8.2.1 Integration with 5G Networks

The deployment of 5G technology holds the promise of ultra-low latency and high-bandwidth communication, which can significantly improve IoT device performance. 5G will offer quicker and more reliable facts exchange among motors, infrastructure, and manage centers—enabling extra responsive and scalable real-time analytics and manage.

10.8.2.2 Autonomous and Connected Vehicles (CAVs)

The increase of self-sustaining vehicles (AVs) and vehicle-to-the whole thing (V2X) linked vehicle technology will be key to destiny visitors environments [19]. V2X communique, in which cars communicate with infrastructure, other cars, and pedestrians, will enable site visitors systems to better manipulate vehicle movement and accomplish that in a safer way.

10.8.2.3 Edge and Fog Computing

Rather than transmitting all information to central servers, edge and fog computing enable data processing near the source (e.g., at traffic lights or roadside units). This decreases latency, enhances reliability, and saves bandwidth. It allows for quicker decisions, like turning on an inexperienced light or sending an emergency alert.

10.8.2.4 AI-driven Urban Planning

In the future, AI will no longer simply coordinate actual-time site visitors, but can even play a position in long-time period urban making plans.

City planners will be able to use predictive modeling and simulation tools to optimize road design, zoning, and public transportation systems based on traffic flows and population growth projections.

10.8.2.5 Green and Sustainable Mobility

Smart traffic management will increasingly enable green mobility measures—like route optimizing electric vehicles, low-emission zone management, and promoting cycling or walking by means of traffic-calming. AI will feature in planning sustainable cities with a focus on nature and habitability.

10.8.2.6 Digital Twins and Simulation

Digital twins, or virtual representations of real-world traffic systems, will be employed to model traffic conditions and evaluate policy modifications or system changes prior to application. This permits for greater efficiency and more secure traffic planning without interfering with real avenue users.

10.8.2.7 AI Regulation and Governance

As AI turns into an increasing number of included into public infrastructure, governments will enforce stricter policies to shield equity, transparency, and accountability. Policies might include regular auditing of AI selections, express regulations for facts utilization, and moral norms for surveillance and automation.

References

1 Sundararajan, V. and Podgorelec, V. Predictive modeling in traffic management systems using IoT and AI. *Journal of Transportation Research Part C: Emerging Technologies*, 115: 415–427, 2020. https://doi.org/10.1016/j.trc.2020.05.010.

2 Ranjan, R. and Lee, C. Autonomous vehicle communication in smart traffic systems. *IEEE Transactions on Vehicular Technology*, 68(4): 3581–3590, 2019. https://doi.org/10.1109/TVT.2018.2886446.

3 Jiang, S. and Sun, Z. Data analytics in traffic incident detection: AI approaches and their impact. *Journal of Transportation Safety & Security*, 11(6): 563–576, 2019. https://doi.org/10.1080/19439962.2018.1501687.

4 Li, L. and Xie, L. Data-driven approaches to traffic flow prediction with IoT and AI. *Journal of Transportation Engineering, Part A: Systems*, 146(1): 04019023, 2020. https://doi.org/10.1061/JTEPBS.0000246.

5 Alonso, A. and Sánchez, J. Smart traffic lights optimization using reinforcement learning algorithms. *Journal of Artificial Intelligence Research*, 63: 123–139, 2018. https://doi.org/10.1613/jair.5727.

6 Zhang, X. and Chen, T. AI in public transport systems: A comprehensive review. *Transportation Research Part A: Policy and Practice*, 131: 109–123, 2019. https://doi.org/10.1016/j.tra.2019.06.003.

7 Mokhtari, M. and Shahandashti, S. Artificial intelligence applications in traffic flow optimization. *Journal of Traffic and Transportation Engineering*, 4(5): 323–331, 2017. https://doi.org/10.1016/j.jtte.2017.06.002.

8 Kim, M. and Park, S. Artificial intelligence for intelligent transportation system: Applications and challenges. *Transportation Research Part B: Methodological*, 121: 91–106, 2019. https://doi.org/10.1016/j.trb.2019.02.002.

9 Singh, R. and Chauhan, S. Artificial intelligence for traffic optimization: A review of algorithms and future prospects. *Computers and Industrial Engineering*, 118: 319–333, 2018. https://doi.org/10.1016/j.cie.2017.12.021.

10 Li, B. and Jin, Y. Exploring the impact of IoT on the environment: Traffic emissions reduction through smart systems. *Environmental Impact Assessment Review*, 86: 106477, 2021. https://doi.org/10.1016/j.eiar.2020.106477.

11 Liu, F. and Li, Z. The role of artificial intelligence in traffic flow optimization. *Journal of Transportation Engineering, Part A: Systems*, 146(7): 04020040, 2020. https://doi.org/10.1061/JTEPBS.0000248.

12 Zhao, L. and Yang, H. AI-based traffic incident detection and its application in real-time control systems. *IEEE Transactions on Intelligent Transportation Systems*, 21(5): 2101–2112, 2020. https://doi.org/10.1109/TITS.2019.2928347.

13 Zhou, L. and Xu, W. The role of AI and IoT in sustainable traffic management systems. *Sustainable Cities and Society*, 53: 101874, 2020. https://doi.org/10.1016/j.scs.2019.101874.

14 Wang, J. and Zhang, H. Smart traffic signals and IoT in future urban mobility. *Transportation Research Part C: Emerging Technologies*, 105: 146–161, 2019. https://doi.org/10.1016/j.trc.2019.06.011.

15 Zhao, Y. and Yang, Y. Smart traffic management using internet of things and artificial intelligence. *IEEE Internet of Things Journal*, 6(3): 456–468, 2019. https://doi.org/10.1109/JIOT.2018.2874965.

16 Bai, S. and Xu, Y. Leveraging IoT for smart public transport management in urban areas. *International Journal of Sustainable Transportation*, 14(6): 463–472, 2020. https://doi.org/10.1080/15568318.2020.1796163.

17 Shia, J. and Kim, J. AI in traffic signal control and incident management. *Journal of Traffic and Transportation Engineering*, 5(1): 11–19, 2018. https://doi.org/10.1016/j.jtte.2018.05.001.

18 Yang, X. and Zhang, W. A review of IoT-based intelligent transportation systems. *International Journal of Transportation Science and Technology*, 9(4): 287–298, 2020. https://doi.org/10s.1016/j.ijtst.2020.09.003.

19 Ghosh, A. and Banerjee, S. IoT and AI in traffic management: Innovations and future directions. *Computers, Environment and Urban Systems*, 76: 71–84, 2019. https://doi.org/10.1016/j.compenvurbsys.2019.02.009.

11

Intelligent Urbanism: AI and Big Data-driven Approaches to Planning, Design, and Transportation

R. Saradha

Department of Computer Science, SDNB Vaishnav College for Women, Chennai, Tamil Nadu, India

Corresponding author: drsaradhaganesh@gmail.com

11.1 Introduction

Since the tremendous growth in global urbanization, which is anticipated to reach around 70% by 2050, cities have been under increasing economic, social, environmental, and governmental pressure [1]. Scholars, urban planners, and politicians are currently engaged in a heated debate about COVID-19 pressures, climate change externalities, population growth mismanagement, housing instability, environmental degradation, and insecurities related to the water-food-energy nexus [2]. With the use of real-time data-driven planning, AI helps planners and urban designers create dynamic urban landscapes. Using real-time data, AI can assist in further integrating sustainable practices into the urban environment. AI can help urban planners identify routes that support more equitable public transportation, better traffic control, and more intelligent, efficient utilities. AI can help urban designers create responsive and efficient urban settings that benefit residents. Although the use of technology in managing cities dates back to the late 1950s and early 1960s, its potential has only recently been broadly acknowledged [3]. With the rapid progress in data collection technologies [4], achieving the Sustainable Development Goals, especially those closely associated with urban development [5], is becoming increasingly feasible. Currently, the adoption of both new and established disruptive technologies has been accelerated by the deployment of big data gathering sensors that track changes in the use of land, commuting patterns, property ownership, and consumption of energy.

Traffic management solutions driven by AI are essential for reducing congestion and enhancing the effectiveness of the road network [6]. In order to forecast traffic patterns and dynamically modify traffic signals, sophisticated machine learning algorithms employ contemporaneous information from cameras, sensors, and GPS devices. These systems contribute to decreased carbon emissions, better fuel economy, and fewer delays. Furthermore, commuters can choose the quickest and most effective routes with the help of AI-driven navigation apps, cutting down on travel time and lowering the stress that comes with everyday commuting.

Predictive analytics is another essential element of AI-driven traffic management. AI models may estimate areas of congestion and provide alternate routes for commuters by examining both current and historical traffic statistics. Transportation authorities can take preventative steps like rerouting cars, changing lane priorities, and sending emergency response teams to accident-prone regions thanks to these findings. Similar predictive analytics [7, 8] are used by AI-driven navigation apps like Google Maps and Waze to provide users the best possible route recommendations depending on the traffic situation both now and in the future.

Emerging as a game-changing technology, AI has the potential to completely reinvent urban mobility by improving the sustainability, efficiency, and intelligence of transportation systems [9]. Cities may optimize public transportation, improve traffic management, and increase citizens' overall mobility by using AI-driven solutions.

AI-powered buses and shuttle services are examples of autonomous public transportation solutions that are being developed to offer economical and effective mobility options in urban areas [10, 11]. These developments help to improve accessibility and lessen traffic, especially for senior persons and those with impairments.

There is a lack of research that simultaneously addresses big urban data analytics and AI tools within the urban context. As a result, there is a need for a structured framework grounded in existing literature to evaluate the combined influence of these technologies on design and urban planning. To address this gap, a conceptual model has been proposed to explore the ways in which the integration of AI and urban big data has shaped planning practices amid ongoing urban transformation. The outcome of this framework is illustrated in Figure 11.1, presenting a typology that categorizes the application of these technologies in supporting urban development.

The paper is divided into six parts. The introduction is followed by a summary of earlier works that highlights research problems and characterizes the gap in the existing literature that this study fills. The background section is a survey of the literature, with an emphasis on AI-based techniques and big data analytics.

Figure 11.1 The future of urban planning: AI and technology.

The methodology for the study is covered in the third section. The following section discusses the techniques and tools used in AI for urban planning. The following examines data sources and the various AI-based technologies utilized in urban analytics. Various uses of AI-based tools and urban big data are analyzed and evaluated in the same area with regard to the impact of these technologies on city planning and design. The research questions and the state-of-the-art discussed at the beginning of the study are used as a lens through which to examine the key findings in the Results Section. Six main areas where these technologies can help the planning process can be identified thanks to it. Finally, cognitive conclusions, planning practice recommendations, and future application trends outline the essential components for AI and big data-based analysis to more successfully reach policymakers.

11.2 Literature Background

Smart city models are produced by technologies like AI, cloud computing, and the Internet of Things. To accomplish intelligent and intelligent urban management, use advanced information methods to assess a variety of areas, including social security, urban transportation, urban planning, and people's living policies. Then, respond quickly and in a timely manner to these sectors.

Many Chinese cities are currently developing while dealing with issues such as resource depletion, traffic congestion, air and waste pollution, and water contamination. It has hampered the city's sustainable growth by negatively affecting residents' livelihood and health. Promoting shifts in citizens' behavior is essential for tackling these challenges and fostering a sustainable and harmonious urban environment [12]. Therefore, advancing smart city initiatives carries significant real-world importance in achieving these goals. Image recognition technology, which falls under the category of human work intelligence skills, has been extensively applied in many industries during the smart city construction phase. The image recognition system that was developed with the aid of technology is capable of recognizing faces, fingerprints, gaits, and other features. These technologies support the advancement of information technology in our nation. Thus, the following characteristics primarily represent the particular application of ML and AI technologies in smart cities.

Although AI is a commonly used word, most people, even urban planners, do not fully understand it. The idea of incorporating intelligence into machines and systems dates back to the sixteenth century. Computer scientists Allen Newell, John McCarthy, Herbert Simon, and Cliff Shaw were the first to use the term "AI" in 1956 [13, 14]. AI hadn't yet been sufficiently conceptualized to have a common definition back then. According to some informal definitions, AI is the ability of highly adaptable, general-purpose systems to achieve goals in a range of unpredictable settings through self-directed learning [15]. AI can be broadly understood as the ability of machines to replicate human cognitive functions and behaviors involved in tasks such as learning, reasoning, planning, perception, communication, and knowledge processing [16, 17].

Humanity entered the Anthropocene after the Industrial Revolution because of the growing effects that all sizes of the ecosystem are being impacted by human activity [18]. Cities and human settlements are growing more complicated than ever at the same time. Until the 1960s, when the study of cities began to take off, scholars were unaware of this intricacy [19]. Furthermore, a number of complexity theory applications to urban planning appeared in the 1990s [20–22]. Urban environments should no longer be perceived as static constructs defined merely by buildings and roads. Instead, they must be understood as dynamic systems shaped by continuous interactions among human activities, commercial functions, and the movement of energy and materials.

Furthermore, big data mining applications have been increasingly common in urban planning and studies in recent years. Utilizing new data types to improve system performance and fully utilize its real-time nature is made possible by urban big data mining. This is accomplished by extracting new insights from existing data sources and extrapolating trends. However, these fresh perceptions might also be helpful for studies in urban planning.

Moreover, advancements in urban systems can be significantly enhanced through data-driven city planning that leverages urban big data analytics and supports real-time decision-making. Urban big data, also referred to as geographic or geospatial big data—consists of large-scale, location-specific datasets collected from diverse sources such as sensors, mobile applications, and satellite imagery. This wealth of information enables the design of advanced and integrative research methodologies that influence urban policy-making and support the implementation of intelligent, data-oriented strategies and development plans. The use of high-frequency datasets facilitates continuous data mining and the detection of large-scale spatial and temporal patterns across cities.

The growing availability and sophistication of AI tools have made it possible to carry out analytical tasks that were previously unattainable, such as object recognition and classification in data-limited environments, including studies of informal urban settlements [23] and the digital preservation of cultural heritage [24]. Additionally, AI enhances conventional analytical models, such as urban growth simulations, by enabling a deeper understanding of the underlying complexity of urbanization processes. These advancements allow researchers and planners to better utilize the potential of big data collected from multiple heterogeneous sources [25].

It is crucial to explore the interconnected roles of AI and big data analytics, as their integration has been shown, according to Allam and Dhunny, to enhance urban livability and foster the development of more interconnected, sustainable, and economically feasible cities. A growing body of urban research supports the idea that AI-assisted big data analytics holds significant promise in delivering benefits such as adaptive urban systems, predictive planning capabilities, increased energy efficiency, improved quality of life (QoL), and enhanced accessibility [26–28].

Data-driven innovations, particularly those powered by AI, are also informing the next generation of geographic information system (GIS) platforms by enabling the fusion of diverse data inputs into coherent analytic frameworks. AI-based methods are now widely applied in urban research scenarios that demand high spatial and temporal resolution—such as real-time analysis of pedestrian movements and intelligent traffic flow monitoring. These tools enable a transition from aggregate and static data interpretations to dynamic and granular modeling of spatial processes, thereby allowing urban researchers to capture individual-level variations and micro-level regional dynamics [29]. For instance, Faust et al. [30] utilized a combination of multi-agent systems and cellular automata to model and forecast patterns of urban expansion. Table 11.1 presents the ten leading smart cities globally, according to the Smart City Index 2023.

Table 11.1 Top ranked smart cities of the world 2023. The rankings were given based on the following dimensions: 1) Service Innovation, 2) Urban Intelligence, 3) Urban Sustainability, 4) Urban Openness, 5) Infrastructure Integration, 6) Urban Innovativeness, 7) Collaborative Partnership and 8) Smart City Governance.

Country	City	Rank
Switzerland	Zurich	1
Norway	Oslo	2
Australia	Canberra	3
Denmark	Copenhagen	4
Switzerland	Lausanne	5
United Kingdom	London	6
Singapore	Singapore	7
Finland	Helsinki	8
Switzerland	Geneva	9
Sweden	Stockholm	10

Source: Smart city index 2023 (Smart City-Observatory Web, 2023).

11.3 Methodology

The development of a conceptual framework to assess the impact of emerging technologies on urban planning and design addressed the primary objective of this study—namely, to explore how AI tools and urban big data can support city development strategies. This framework was constructed through an integrative systematic review of existing literature on the application of AI and big data in urban planning and design. It helped identify key criteria for evaluating the effectiveness of these tools, particularly in terms of data availability and accuracy, as well as the adaptability and scalability of the technologies involved. The review also enabled the classification of six major application areas where big AI-driven data analytics are being utilized in urban contexts. To further investigate the role of these tools in promoting urban transformation, the study examined a range of current practices and applications across different domains. Key assessment parameters included the depth and scope of the analyses, the specific types of data and AI tools employed, their influence on planning and design processes, and the associated strengths and limitations.

This research follows the systematic review methodology outlined by Fernández-Güell and González López [31], which emphasizes a structured, repeatable, and scientific approach aimed at minimizing bias through a comprehensive literature search and maintaining a transparent audit trail of methods and findings. The integrative review approach, recognized as one of the most inclusive forms of scholarly evaluation, allows for the combination of both theoretical and empirical studies—enabling a holistic understanding of the topic. Unlike prior studies that focused solely on big data applications within a specific national context, such as China, this study offers a broader perspective by examining the global use of AI-based tools in urban big data analytics.

The following steps served as the foundation for this integrated systematic literature review's approach, which was adjusted for the many disciplines of study. (1) Identifying the problem; (2) searching the literature; (3) evaluating and analyzing the findings; and (4) presenting the results.

To respond to the research questions in the introduction, the issue needs to be identified. The author reviewed academic publications on the use of AI-based technologies and big data analytics in urban planning and design as part of his literature review. Search phrases "URBAN/CITY/CITIES" and "ARTIFICIAL INTELLIGENCE" were employed to locate the included articles in the Web of Science Core Collection, which formed the basis of the initial literature corpus. We looked for those terms in the abstracts, titles, and article titles. The latter was more thorough because it included numerous irrelevant searches and used the terms "BIG DATA" and "URBAN/CITY/CITIES" in the second literature inquiry, even though the most significant sources could be discovered in both of the previously mentioned searches.

This essential information was examined from several angles during the data assessment process. Owing to their wide distribution, primary sources were grouped based on several factors that were pertinent to this study, including the research center, the publication period, the type of article, the data type, the AI-based approaches employed, and the article style (experimental, review, and theoretical). The final sample for this integrative review consisted of theoretical publications (4), reviews (14), and empirical research (64). About 9.7% of the publications were only released before 2010. Volunteered geographic information data and mobile phone (including social media data), search engines, GPS, point-of-interest data, sensor data (including drone, satellite, and urban sensor data), data from civic and governmental equipment, and new sources of massive amounts of government data are the main data types used.

Finding the potential and limitations that could help or hinder the application of AI and big data in the development of urban practices was the first stage in the data analysis process. The literature corpus's review articles and experimental investigations acknowledged the benefits and drawbacks of utilizing different AI-based urban big data analytics. This study took into account the flexibility and duplicability of the instruments, as well as the availability and reliability of the data.

Further research helped define the format of the results tables, categorize the design and planning implications, and determine the benefits and drawbacks of each application area for AI-based urban big data analytics. Using the research questions that were posed at the start of the study as a guide, the main conclusions are discussed at the end of the paper. In order to assess the potential of using AI-related tools for urban big data analytics in city planning and design, as well as the role of AI-based tools in influencing policies to support urban change, the author identified six key areas where these tools can help in the planning process. Last but not least, a summary of the main ideas of massive data, analytical findings, and practice recommendations—identifying the main domains for AI and big data-based analysis—were created.

11.3.1 Types of AI-based Tools for Urban Planning

With the introduction of planning support systems (PSS), the usage of quantitative planning models has seen a renaissance in the twenty-first century. Various sorts of models have been implemented by planners, including

1. Urban Models on a Large Scale: DRAM/EMPAL, MEPLAN, TRANUS, and UrbanSim are a few examples of these models. For urban regions, these models may anticipate future land uses, housing, job patterns, and resident travel habits.
2. Models Based on Rules: Rule-based approaches offer explicit guidelines and requirements for land development. The California Urban Futures (CUF) model, Community Viz, and INDEX are a few examples of these models.
3. Models Using Cellular Automata (CA): Areas are divided into smaller cells using CA models, and each cell is given a "state" like "developed" or "undeveloped." Based on a set of predetermined criteria, these models represent how states change over time. CA urban models one example of a CA model utilized in urban planning is DINAMICA SLEUTH.
4. ABMs or Agent-based Models: ABMs replicate alterations to the built environment, including land usage, brought on by intricate interactions among several agents, including local residents, companies, governmental organizations, and property developers. These simulations represent the buying and selling of land in a specific market. ABMs and CA models are examples of ABMs, which are direct applications of AI in urban planning, as they come from the discipline of AI.

With the help of these numerous modeling techniques, planners have access to powerful tools for comprehending and simulating urban dynamics, enabling them to examine a range of scenarios and weigh the possible effects of various planning actions. Processes for planning analysis and decision-making have been significantly impacted by the development of various planning models and new technologies. In reality, planners have broadly embraced a number of PSS models. Additionally, today's planners frequently employ GIS and statistical methods to pinpoint spatially related urban problems like congestion and pollution, fostering spatial thinking.

The planning profession has not been able to effectively utilize PSS models and spatial technology due to a number of issues, notwithstanding their value. These constraints include data restrictions, complexity, and the presence of "wicked" planning scenarios that challenge standard planning methods. Accordingly, AI has a better chance of overcoming the challenges that PSS models and GIS present, particularly when it comes to handling "super wicked" planning concerns.

11.3.2 New Approaches to the Complexity of Urban Planning

Since then, nearly every facet of urban life—such as utilities and infrastructure, transportation systems, pollution, and crime, to mention a few—has adopted complexity-based approaches to urban concerns. Deeper understanding of the existence and consequences of structural interdependencies across urban environments has also been facilitated by the particular conflict between local, specialized causes and dynamics and the generic nonlinear effects typical of all urban environments [32].

Despite their academic context, most quantitative methods based on systems theory are challenging to apply in a collaborative process with stakeholders from different professional backgrounds because they require a high level of technical expertise and have limited capacity to represent complex problems. When these conditions arise, planners should be able to interpret a city's complexity in a more qualitative and user-friendly manner, create more robust conceptual frameworks. A more comprehensive manner while simultaneously facilitating the efficient participation of local stakeholders along the organizing procedure. The new conceptual framework should therefore satisfy four essential criteria.

First and foremost, planners must be able to convey the complexity of cities by representing all significant local stakeholders, sectoral systems, and functional relationships in a thorough and complete way. By graphically representing all the factors in a situation and the roles of various stakeholders, conceptual maps and process diagrams can assist in organizing complicated occurrences.

Second, when examining complicated urban challenges, planners should create new communication technologies that make it easier for local stakeholders to participate in collaborative planning. These tools need to make it possible for urban stakeholders to handle disagreements and conflicts while also looking for common ground.

Third, in order to create comprehensive future visions for the city, planners should incorporate foresight tools into the planning process. They would be able to act in concert with the dynamic urban environment thanks to these tools. The continual reconfiguration of actor-networks and city systems as a result of ongoing cultural, economic, technical, and environmental changes is what we mean when we use the term "co-evolution."

Fourth, rather than attempting to address clearly defined problems using conventional quantitative tools, planners should build novel research methodologies capable of managing fuzzy interrelated difficulties with the use of qualitative instruments and collective intelligence. These strategies must respond nimbly to agendas that are founded on an evolving process of collective intelligence.

In short, a new model framework built on the earlier assumptions could offer conceptual assistance to spatial planners who are now finding it difficult to develop a comprehensive grasp of a city's structure and operations as they change in an unknowable environment.

11.3.2.1 New Framework for Conceptualization

Conceptualizing the city as an evolving functional ecosystem capable of adaptability and self-organization in response to context changes is one way to address the diversity, uncertainty, and complexity that are inherent to modern cities with others [33]. Additionally, as part of the planning process, foresight tools should be used to support the systemic approach, assisting planners and stakeholders in looking beyond their immediate issues.

As a result, a city can be thought of as a complex ecosystem of interconnected components or pieces that provide similar functions, where human activities are connected through communications and transportation and interact dynamically as the system changes and adapts through time in a particular socioeconomic and physical setting. In other words, a city is created by a variety of independent decisions made over time by a large number of participants who are closely related to one another. Any geographical or structural change to one component of this ecosystem has the potential to influence the other components as well.

Concept mapping was employed to describe the intricate workings of a modern metropolis. A conceptual model was created that summarized and depicted the urban ecology as a collection of several interconnected subsystems. This updated framework includes Urban Transportation as both a vital enabler and an integrative infrastructure component (Figure 11.2).

Subsystem of urban demand. The center of the model is occupied by this subsystem, which is made up of residents, economic stakeholders, societal institutions, and tourists. To be able to work and live in a city in a comfortable

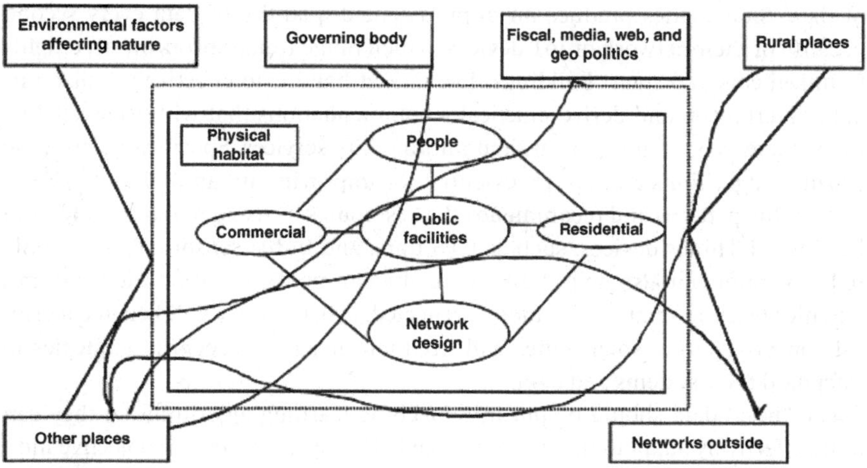

Figure 11.2 Framework of urban ecosystem functionality with integrated urban transportation.

environment, all demand segments impose a variety of demands on the services, resources, and infrastructures offered by the urban subsystems. The city's sociological, environmental, political, and economic subsystems work to understand and meet demand segments' needs from the supply side. Each subsystem is characterized in terms of its resources, services offered, operating agents, and technology employed.

External Trends: The total urban system is exposed to external change factors that have an impact on its functional balance, such as population changes, economic cycles, technology advancements, or environmental impacts. Studies of the future offer a variety of methods for spotting and evaluating trends of every description. By taking a systemic approach, local stakeholders can find common ground for cooperation and come to a greater knowledge of the underlying causes of urbanization.

11.3.3 Data Sources Supporting AI-based Urban Analysis

11.3.3.1 Role of Data Analytics and AI in Smart City

In the context of interconnected urban landscapes known as "Smart Cities," the advent of Internet of Things (IoT) technology has ushered in a new era of plentiful data. The enormous amount of data produced by IoT devices has become both a promise and a danger as smart cities continue to develop. Better decision-making and more efficient city governance are now feasible because of the development of data analytics and AI. These are strong tools that can interpret the abundance of data. Smart cities produce an unprecedented quantity of data every second because of their network of IoT devices, which range from sensors on streetlights to linked cars and smart buildings. The task at hand is to effectively utilize this data, interpret it, and derive practical recommendations that will boost productivity, increase QoL for citizens, and improve city services. Specifically, machine learning algorithms and AI are essential for improving urban mobility. AI predicts traffic patterns and recommends the best routes by using real-time data from Internet of Things devices, such as GPS units and traffic sensors. This not only reduces traffic but also helps build more efficient public transportation systems. As cities become smarter and more connected, cybersecurity will become a critical concern. Future smart cities will need advanced cybersecurity strategies to safeguard their systems and data.

The "best" data should be put into machine learning algorithms so they can more effectively adapt to the specified goal [34]. Depending on the objective indicators and the questions that need to be answered, these data on urban planning come from a variety of sources. Numerous data sources that frequently correspond with the topics being studied have been employed by researchers. These sources include the most traditional urban research methods (surveys and sensing) as well as the most advanced, contemporary, and widespread data. Today's city problems, like "sustainable planning, smart city, digitalization, resilience, developing better strategies and studying the impact of new urban development projects," can now be solved more intelligently thanks to the development of urban data that can train potent machine learning algorithms. Sensor-based technology and institutional surveys are typically the sources of urban data. According to our review of the literature, 33% of machine learning research employs hybrid data, 9% use surveys, and 58% use data that comes only from sensors.

11.3.3.2 Data Sources

11.3.3.2.1 Remote Sensing

The most crucial stage for the majority of urban data is sensing, which is accomplished by onboard sensors, which are carried in a mobile device or placed elsewhere on the device to cover an area of a smaller spatial scale [33]. The area the mobile device covers and the accuracy of the data it gathers determine its speed. More than 58% of the urban data needed to train machine learning algorithms comes from sensors, with 33% coming from surveys and hybrid sensing, according to surveys.

Large volumes of very varied sensor data are produced now, and this amount is predicted to increase rapidly. The reason this method is dubbed "remote sensing" is that the objective is typically geographically far. It is most frequently employed to gather data about cities. Sensor-based data sources can be categorized or classified using the transportation technologies that are employed,

Table 11.2 Comparative summary of urban data sources for ML modeling.

Category	Device/source	Spatial coverage	Temporal data	Cost level	Data type
Sensors	Satellite/radars	Global	Yes	Low	Vector
Sensors	Drone/plane	Partial	Yes	High	Vector, Attribute
Sensors	Bikes/cars/motos	Partial	Yes	Low	Attribute
Sensors	Mobile devices	Partial	Yes	Low	Attribute
Sensors	Fixed devices (cameras, street lights)	Partial	Yes	High	Vector, Attribute
Survey and institutional	Social networks/media	Global	Yes	Low	Attribute
Survey and institutional	Crowd-sourcing	Partial	No	Low	Vector, Attribute
Survey and institutional	Interviews	Partial	No	High	Attribute
Survey and institutional	Institutional statistics	Partial	No	High	Vector, Attribute

which show the spatial extent of the region covered and the degree of accuracy/data precision.

We categorize data types according to the carrier device: radars/satellite, plane/drone, ground transportation (cars, e-scooters, scooters, and bikes), fixed devices, ubiquitous mobile devices, social media and social networks, interviews, crowd-sourcing, and institutional statistics. Table 11.2, which evaluates their (a) complexity, (b) geographic coverage, (c) temporal information, (d) free and open accessibility, (e) cost, and their data structure—that is, (a) vector or (b) attribute—describes them and discusses their benefits and drawbacks. Cost and complexity are evaluated as having either low or high spatial coverage (partial or global), whereas accessibility and temporal information are evaluated as having a yes-or-no Boolean function.

11.3.3.3 Urban Survey and Statistical Information

Although surveys are the main tool used by urban-planning experts to get new data, they also employ a variety of more inventive techniques. ICT and social interaction are being incorporated into these techniques more and more. Citizen actions, whether individual or collective, are crucial to the gathering and dissemination of this kind of information. Consequently, we categorize this type of data as follows: statistical data from an institution's survey; social network data (also referred to as web sensing); interview data; crowd-sourcing data [35], and so forth. These data are frequently employed in isolation or in conjunction with sensor data, which raises the degree of accuracy and dependability. Furthermore, compared to sensory data, they are easier to gather and process and more opportunistic.

11.3.3.4 Urban Big Data Sources

Massive information is required to expedite rational advancement. The application of BDA has now caused a change in the pattern of world development. Because BDA offers an open, ongoing development process, defects and gaps in specific activities of many organizations can be avoided [36]. When it comes to solving issues pertaining to sustainable urban development, BDA are particularly pertinent and significant. Furthermore, big data is crucial to maintaining the capabilities of many sectors in relation to urban growth.

The key big data sources, including individual, public, and sensor device sources, are shown in the following table. To be more precise, big data can be found in a number of additional places, such as social media platforms, public transportation systems, waste management systems, websites run by commercial companies, and government websites [37]. In essence, these resources are constantly employed in the creation and upkeep of numerous industries, including urban growth, which is ultimately prosperous and sustainable.

Major sources	Identified specific data sources
Public sources	• Websites belonging to different governments • private organizations • open repositories • Banking system transactions and • Surveillance systems.
Individual sources	• Social media network • data obtained via mobile access • data obtained via internet access • data obtained from internet service providers (ISPs)
Sensor devices	• Information from closed-circuit television (CCTV) cameras, public transportation systems • waste management systems, and • community entrance auto-detection systems

11.3.4 Feasibility of AI-related Tools in Urban Design

According to UN estimates, there will be 10.4 billion people on the planet by the year 2100. Two out of every three people will live in cities by 2050 as a result of the possibilities, conveniences, and services they offer. Given that all of these people will require access to food, water, clean infrastructure, public spaces, and, most importantly, housing, this places even more strain on metropolitan regions. To handle this population expansion, projections indicate that more than two billion dwellings would need to be built by the end of the twenty-first century. Urban sprawl, which has its own set of environmental and socioeconomic issues, expands with cities. Sustainable urban development must make sure that, in the face of climate change, new and updated housing options are constructed to support healthy communities and give equal weight to the needs of people and the environment. As a result, cities will need to be developed more quickly and densely, but only after fulfilling a number of demanding requirements. This is the only method to prevent the unfavorable, frequently disregarded repercussions of unchecked hyper densification, which damage urban development's reputation.

Densification can take many forms, and as each city or location is different, a customized strategy is needed. Simultaneously, we must minimize urban sprawl by making the most of the little land that exists in our current cities. Although skyscrapers are frequently linked to densification, more subtle solutions are more appropriate for intricate urban environments. Imagine more human-scale, compact, mixed-use neighborhoods that blend higher and lower scales and densities based on the streetscape and urban quality. These include transit-oriented developments and the 15-minute city, which make use of the compact urban layout to enable people to easily walk or bike to services, public transportation hubs, or places of employment. Concurrently, infill, Brownfield, and adaptive reuse initiatives are now given priority in order to optimize existing urban space and revitalize underutilized districts.

AI and data-driven digital tools are pushing the boundaries of innovation and have radically changed how architects design and conceptualize the built world. With the aid of AI, they may capitalize on the fact that computers are best at accessing and processing vast volumes of data and coming up with numerous answers to even the most difficult issues. With the use of technology, any project can have a more efficient design process, fewer mistakes, more time, and less guessing and effort. Overall, this research has found various topics that are associated with data-driven sustainable smart cities, such as technology, applications, competencies, infrastructure, and data sources. The data collected from multiple sources shows how different themes lead to distinct forms of sustainable, data-driven smart cities. The degree to which the objectives of sustainable development are supported in the administration, planning, building, and operation of the city is one of the primary characteristics that set these models apart. This includes the usage of various components to enhance and promote sustainability as well as the state of technical advancement at the moment.

The outcomes of the two cases under investigation form the basis of the integrated framework (Figure 11.3). Consequently, its core elements are based on

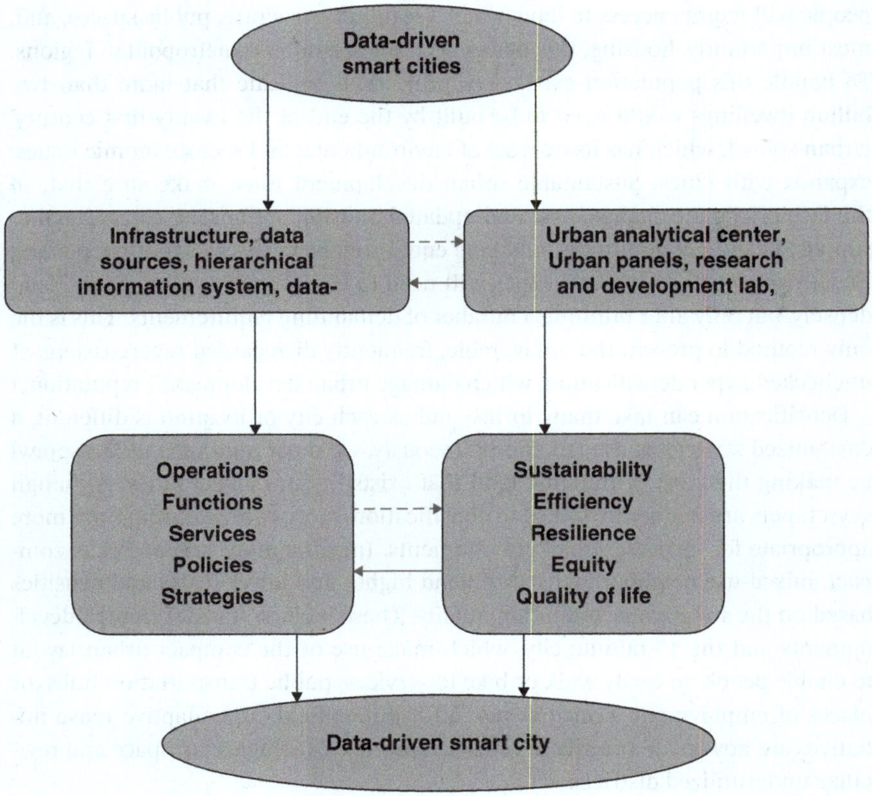

Figure 11.3 A theoretical paradigm for sustainable smart cities.

the sustainable and data-driven strands of the recently emerging idea of smart urbanism. As such, it attempts to methodically capture the essential components of the environmentally friendly, data-driven smart city. The parameters used to establish such a city fall into four major categories: technology, competencies, practices and processes, and intelligence and sustainability. This framework is also a conceptual structure intended to act as a roadmap for developing various kinds of data-driven, sustainable smart cities.

By giving them insights and recommendations based on data analysis, these new data-driven technologies support architects and urban planners, enhancing their judgment and intuition. For example, algorithms that employ machine learning can be used to estimate the amount of energy required to run a building or to track and forecast foot traffic. By using ideas such as generative design, parametric design, and AI can be used to interpret data entered by the architect and provide a variety of high-performing design options. Architects are able to fine-tune their preferred design solutions to satisfy community demands, utilize available space, and minimize environmental effect.

11.4 Results and Discussion

Even though urban planning is still a relatively new field, research so far shows that big data and AI-based techniques are useful in a variety of planning areas. Based on this assessment of the literature, the author assessed the potential of applying AI-related urban big data analytics tools to support city planning, design, and transportation. Six key areas were found where these tools can support the planning process. These include the following fields:

- Big data analytics applied to urban areas: large-scale urban modeling, large volumes of data can be used for analysis, both in terms of the number and magnitude of observations (e.g., picture interpretation), thanks to AI-based methods like artificial neural networks. The use of individual attributive data in complex systems techniques, such as agent-based models, is becoming more and more prevalent
- High velocity and frequency—provide the means to undertake nearly real-time observation, which aids in defining urban change—instead of waiting for the periodic publishing of data from official administrative sources in order to capture urban processes like fast urbanization [38, 39].
- Utilitarian, frequently a liquid meaning of the review region—the limits among districts and regions are frequently obscured, and the development of "delicate spaces" forestalls examinations that are restricted to director limits from catching various peculiarities, for example, versatility or normal work markets. Metropolitan enormous information can uphold different techniques arranged for "delicate in the middle of between the formal, legal sizes of government [40], from regional master plans to multiregional development systems.
- High granularity: Given that the data is often geocoded, massive data can do more than just enable quantitative analyses of urban morphology with more precision and resolution [6, 41]. They license inside and out examinations, yet additionally give a more extensive point of view to uncover the elements behind complex metropolitan cycles and primary examples.
- Collaboratively Sensing the City: User-generated content is being utilized more and more as a source of information for urban participatory practices to gather information, generate ideas, and even develop solutions for the problems that have been found [42, 43].
- Empirical Urban Research: Using Big data and AI broadens the range of potential analyses to determine how to quantify phenomena like urban

image and spatial quality, leading to evidence-based decision-making [40, 44]. Effectively integrating quantitative research with a qualitative approach is beneficial, as is quantifying emotions, behaviors, or happiness [45].

The examinations presented in this paper affirm that metropolitan large information and man-made intelligence-based devices can assist in molding metropolitan strategies with supporting metropolitan change. Without a doubt, the range of potential research has expanded, and methods that were previously only used in other domains can now be used to examine and enhance the built environment. For instance, picture recognition can be used to evaluate and enhance public spaces by identifying urban features like cars and trees. Additionally, the use of new information sources such as GPS logs, cell phone data, focal points, and chipped-in geographic data enables scientists to conduct much more research on topics like the "beat of the city." The scope of investigations can also be increased by connecting preexisting information sources, such as cell phone information, with web-based entertainment information. Information-based urbanism considers dynamic assets as the executives of metropolitan resources and framework, to help the preparation of stronger metropolitan designs and social consideration in versatility. As certain wellsprings of enormous information are gathered progressively and not restricted to the limits of managerial units, there is the likelihood to acquaint another methodology with arranging metropolitan regions in light of useful associations, as examinations need not be restricted to factual office information, which is delivered two times every year and collected to authoritative units. Furthermore, acquiring geo-coded information considers fine-grained examinations of the assembled climate, which, thus, empowers better ID of clients' necessities. Metropolitan enormous information and man-made intelligence-based apparatuses support the production of plans that, rather than presenting the last vision, show potential situations of improvement and assessments that permit the appraisal of and plan for metropolitan change.

The literature mapping study demonstrates that the application of big data analytics based on AI tools to city planning can, in many cases, more accurately describe the complexity of a city's functional and spatial structure and aid in the successful management of urban change. This is because AI-based tools enable more precise research on urban dynamics, which serves as a foundation for analyzing the city as flows of people, energy, and goods rather than as a planned, static structure. Geolocated data makes it possible to examine the variability of space and offers precise urban studies that demonstrate the methods used to bring about urban transformation. An increased level of refinement and more accurate empirical research are made possible by AI-based big data analytics tools. They can help with dynamic, even real-time, evaluation in addition to improving the accuracy and precision of conventional spatial planning analyses. Furthermore, the frequency of data is a major factor in determining the potential applications of their various kinds. The strategic value of big data in urban development lies in its capacity to connect two fundamentally different time scales of urban processes: the fast-changing short-term dynamics and the slower long-term processes associated with traditional planning and governance [10]. This contrast often creates a tension between the immediacy of high-frequency data and the persistent nature of deep-rooted urban challenges. These characteristics significantly increase the proximity of big data analytics to the concepts of urban transformation and the intricacy of city systems. Consequently, traditional planning methods, which frequently overlook temporal fluctuations and rely largely on static data, can be supported by big data analytics based on AI-related tools.

Simultaneously, despite the fact that the advances related to man-made brainpower and enormous information can possibly deliver various up-sides of the metropolitan texture, they ought not to be indiscriminately embraced. Innovation should be coordinated into the cultural texture and be created to answer the requirements of metropolitan tenants. Also, given the portrayal predisposition of virtual entertainment information, cell phone information, and chipped-in geographic data, these procedures can't fill in for exemplary metropolitan examinations. To empower a comprehensive way to deal with planning and arranging, there is a need to coordinate those information sources and join them with other more conventional strategies for metropolitan evaluation. However, there are still a number of issues with big data analytics based on AI-related tools, such as those pertaining to big data accessibility and accuracy as well as the limitations of various AI-based tool types that prevent this type of analytics from completely replacing traditional urban planning analyses. Big data application in design and planning can, in terms of technological advancement, considerably help conventional planning techniques and provide conditions for creativity; nevertheless, because of its limitations, it can only complement conventional urban studies and cannot, in any way, replace them.

11.5 Conclusion

More use of machine learning is possible anytime and anywhere in smart cities, enhancing people's comfort. Planning and establishing states for future development are provided by more machine learning algorithms. Additionally, IoT devices are enabled to enhance the methods. The traffic dataset is analyzed and predicted using these smart city technologies and advancements, which are built on different software. By categorizing the cars, these records offer extensive traffic analysis.

Economic, social, behavioral, biological, and physical principles provide the theoretical foundation for analyses of urban systems, enabling the simulation of intricate interactions, flows, movements, and diffusion patterns. In contrast, the rapidly developing field of data science frequently adopts a purely empirical methodology without taking into account the social, psychological, economic, and regional planning theories [43] that surround urban research. Urban big data and AI-based tools are used in the nexus of those two approaches to facilitate the analysis of patterns found, knowledge discovery, empirical justification, and formulation of hypotheses about urban phenomena and trends. As this study affirms [46], to get this going, there is a need to recover and extricate data from unstructured or extremely voluminous floods of information, and further to reconfigure and structure large information through information planning procedures for it to meet the info necessities of existing or arising metropolitan demonstrating approaches.

The theoretical foundation of big data analyses is crucial, according to Kandt and Batty [47], since there is a lot more leeway in how big data and AI analyses are interpreted than, say, survey data. Thus, in the practice of big data analytics, theoretical reasoning and contextualization play a much greater yet even more elusive role. The significance of the cognitive processes involved in comprehending the patterns revealed in big data is confirmed by the examined publications [47, 48]. Furthermore, encouraging open scientific elements and greater disciplinary integration [49] might guarantee that the potential applications of big data and AI-based research in urban planning are more widely acknowledged.

We can determine potential application trends for the use of AI-based urban big data analytics tools in urban design and planning based on the results of the study that was conducted. As processing and computer power rise and pretrained ANN algorithms become more widely available, deep learning models may find greater use in urban analytics. This tendency can also be strengthened by cloud-based services, which in some situations enable easier access to data at a fraction of the cost and availability of computer system resources, particularly the data storage and processing capacity. It is reasonable to anticipate that the technologies being utilized in other domains will eventually find their way into urban planning, meaning that we will likely witness a wider range of AI-powered urban analytics tools.

The majority of evolutionary computing models employed in urban planning to date are applications of supervised and unsupervised learning, although this may also concern models of reinforcement learning, which can be utilized for optimal decision-making in complicated contexts. As a result, employing predictive models for interdependency modeling based on precise input data might enhance longer-term scenario planning. We anticipate that digital twin technologies will continue to advance and play an increasingly significant role in both urban management and urban planning. Although AI-based tools won't be able to resolve urban planning issues on their own, new technologies can be incorporated into current information systems to address urban issues in a more thoughtful and efficient manner. We may anticipate that, in addition to the already-rising worries about data security and safety, new networks, protocols, and security-enhancing systems will be launched, and that the question of ethical responsibility and regulation will become even more prominent. Simultaneously, most existing models concentrate on single sector domains like transportation or pollution, which necessitate the introduction of integrated tools that connect various urban layers by exchanging data and introducing models that enable greater system interoperability.

11.5.1 Future Direction of Studies

Although AI has a lot of exciting and captivating possibilities, businesses must plan carefully before allocating resources to it. It's critical to get ready for change, and some disruption is normal. Because we are used to our routines, it could be difficult to convince staff members to embrace AI. This article explains why it's critical to have a solid grasp of the benefits AI can provide for a company and to strategically consider any possible drawbacks. After implementing AI, it's critical for an organization to monitor its performance and assess the benefits and advantages of its use in order to compile a thorough understanding of the ways in which AI helps the company, its employees, and its stakeholders. Additionally, when new data and analytic formats for AI applications develop, planners should be mindful of the possibility of ethical issues surrounding the use of AI throughout the adoption process.

Lastly, the introduction of tools that can greatly improve decision-making, optimize resource allocation, and raise general livability is what the future of AI promises to bring to urban planners. By forecasting urban expansion, assessing traffic patterns, and supporting the planning of sustainable infrastructure, the incorporation of AI can result in smart, responsive, and sustainable cities. The ethical use of data, protecting privacy, and guaranteeing fair access to AI's advantages are important factors to take into account. Urban planners must constantly participate in interdisciplinary discussions, work with AI specialists, and involve the public as we transition to this AI-assisted future. This will help to guarantee that AI-based urban planning is open, responsible, and most importantly, meets the requirements of everyone.

The suggested framework, which outlines the primary applications of AI-based tools for urban big data analytics in recent studies, can be useful for both urban theorists developing new techniques for evaluating urban transformation and planning practitioners interested in these tools. The conceptual contribution of this study is to describe how AI-based tools for urban big data analytics might be used to model urban evolution. In the process of analyzing the applications of

AI-based tools and big data in urban settings, this study can serve as a foundation for future comparative analyses between cities and regions, highlighting the key issues that impede the use of these tools in local contexts. Because different countries or even cities have different access to the data, it is important to do studies that account for geographical differences in the potential applications of big data. An analysis of this kind can assist in recognizing the diversity of methods and, in particular, location-specific regulations related to the collection, dissemination, management, and storage of data.

The next crucial step is to investigate the possibility of modeling the resilience of urban structures and assisting with the regenerative design and planning of cities through the use of urban big data analytics based on AI-related tools. This might be a significant area of research in the future. Intelligent systems have the potential to optimize resource utilization in urban settings; however, to achieve this, ecologically sound indicators must be identified. This enables ongoing built environment monitoring and unlocks the potential of big data analytics based on AI-related tools.

It is necessary to evaluate the potential of AI-based tools to support decision-making in specific sectors, like energy or transportation, but also to address more complex issues, like the resilience and well-being of urban dwellers, as most revised papers concentrate on sectoral solutions or analyses. Since AI can be a useful tool for modeling the dynamics and heterogeneity of urban space, which in turn helps to model urban change, such an approach could undoubtedly aid in evaluating the usability and reliability of urban big data analytics and AI-related tools for multidisciplinary research and urban planning practice.

References

1 Perveen, S., Kamruzzaman, M., and Yigitcanlar, T. Developing policy scenarios for sustainable urban growth management: A Delphi approach. *Sustainability*, 9(10): 1787, 2017. 10.3390/su9101787.

2 Berawi, M. The role of industry 4.0 in achieving sustainable development goals. *International Journal of Technology*, 10(4):644–647, 2019. https://ijtech.eng.ui.ac.id/article/view/3341.

3 Huang, Z. AI in urban planning: 3 ways it will strengthen how we plan for the future, 2021. https://www.ura.gov.sg/Corporate/Resources/Ideas-and-Trends/AI-in-Urban-Planning.

4 Zhou, M., Ma, Y., Tu, J., and Wang, M. SDG-oriented multi-scenario sustainable land-use simulation under the background of urban expansion. *Environmental Science and Pollution Research*, 29(48): 72797–72818, 2022. https://doi.org/10.1007/s11356-022-20904-9.

5 Sanchez, T., Shumway, H., Gordner, T., and Lim, T. The prospects of artificial intelligence in urban planning. *International Journal of Urban Sciences*, 2022. https://doi.org/10.1080/12265934.2022.2102538.

6 Chen, G. and Wan Zhang, J. Intelligent transportation systems: Machine learning approaches for urban mobility in smart cities. *Sustainable Cities and Society*, 107: 105369, 2024.

7 Abduljabbar, R., Dia, H., Liyanage, S., and Bagloee, S. A. Applications of artificial intelligence in transport: An overview. *Sustainability*, 11, 189, 2019.

8 Tao, X., Cheng, L., Zhang, R., Chan, W. K., Chao, H., and Qin, J. Towards green innovation in smart cities: Leveraging traffic flow prediction with machine learning algorithms for sustainable transportation systems. *Sustainability*, 16(1): 251, 2023.

9 Allam, Z. and Dhunny, Z. A. On big data, artificial intelligence and smart cities. *Cities*, 89: 80–91, 2019.

10 Yan, Z., Jiang, L., Huang, X., Zhang, L., and Zhou, X. Intelligent urbanism with artificial intelligence in shaping tomorrow's smart cities: Current developments, trends, and future directions. *Journal of Cloud Computing*, 12: 179, 2023. https://doi.org/10.1186/s13677-023-00569-6.

11 Anyoha, R. The history of artificial intelligence, 2017. https://sitn.hms.harvard.edu/flash/2017/history-artificial-intelligence.

12 Herath, H. and Mittal, M. Adoption of artificial intelligence in smart cities: A comprehensive review. *International Journal of Information Management Data Insights*, 2(1): 100076, 2022. https://doi.org/10.1016/j.jjimei.2022.100076.

13 Legg, S. and Hutter M. A collection of definitions of intelligence. In Goertzel, B. and Wang, P., editors, *Advances in Artificial General Intelligence*, pages 17–24. IOS Press: Amsterdam, 2007.

14 Corea, F. AI knowledge map: How to classify AI technologies, 2018. https://www.forbes.com/sites/cognitiveworld/2018/08/22/ai-knowledge-map-how-to-classify-ai-technologies.

15 Frankenfield, J. What is artificial intelligence (AI)?, 2022. https://www.investopedia.com/terms/a/artificial-intelligence-ai.asp.

16 Crutzen, P. J. The "Anthropocene." In Ehlers, E. and Krafft, T., editors, *Earth System Science in the Anthropocene*, pages 13–18. Berlin/Heidelberg: Springer, 2006.

17 Batty, M. *The New Science of Cities*, Cambridge, MA: MIT Press, 2013.

18 Allen, P. M. Cities and regions as evolutionary complex systems. *Journal of Geographical System*, 4: 103–130, 1997.

19 Portugali, J. *Self-Organization and the City*, Berlin: Springer, 1999.

20 Batty, M. *Cities and Complexity: Understanding Cities with Cellular Automata, Agent-Based Models, and Fractals*. MIT Press: Cambridge, MA, 2005.

21 Ibrahim, M. R., Haworth, J., Cheng, T. URBAN-i: From urban scenes to mapping slums, transport modes, and pedestrians in cities using deep learning and computer vision. *Environment and Planning B: Urban Analytics and City Science*, 48: 76–93, 2021.

22 Mager, T. and Hein, C. Digital excavation of mediatized urban heritage: Automated recognition of buildings in image sources. *Urban Plan*, 5: 24–34, 2020

23 Grekousis, G., Manetos, P., and Photis, Y. N. Modeling urban evolution using neural networks, fuzzy logic and GIS: The case of the Athens metropolitan area. *Cities*, 30: 193–203, 2013.

24 Ownsend, A. M. *Smart Cities: Big Data, Civic Hackers, and the Quest for a New Utopia*. W. W. Norton & Company: New York, NY, 2013.

25 Kourtit, K., Nijkamp, P., and Steenbruggen, J. The significance of digital data systems for smart city policy. *Socio-Economic Planning Sciences*, 58: 13–21, 2017.

26 Batty, M. Urban analytics defined. *Environment Planning B: Urban Analytics City Science*, 46: 403–405, 2019.

27 Hao, J., Zhu, J., and Zhong, R. The rise of big data on urban studies and planning practices in China: Review and open research issues. *Journal of Urban Management* 4: 92–124, 2015.

28 Rienow, A., Stenger, D., and Menz, G. Sprawling cities and shrinking regions-forecasting urban growth in the Ruhr for 2025 by coupling cells and agents. *Erdkunde*, 68: 85–107, 2014.

29 Cook, D. J., Mulrow, C. D., and Haynes, R. B. Systematic reviews: Synthesis of best evidence for clinical decisions. *Annals of Internal Medicine*, 126: 376–380, 1997.

30 Faust, K. M., Abraham, D. M., and DeLaurentis, D. Coupled human and water infrastructure systems sector interdependencies: Framework evaluating the impact of cities experiencing urban decline. *Journal of Water Resources Planning and Management*, 143(8): 04017043, 2017.

31 Fernández-Güell, J. and González López, J. Cities futures. A critical assessment of how future studies are applied to cities. *Foresight*, 18: 454–468, 2016. https://doi.org/10.1108/FS-06-2015-0032.

32 Jordan, M. I. and Mitchell, T. M. Machine learning: Trends, perspectives, and prospects. *Science*, 349, 255–260, 2015.

33 Kamel Boulos, M. N., Resch, B., Crowley, D. N., Breslin, J. G., Sohn, G., Burtner, R., Pike, W. A., Jezierski, E., and Chuang, K. Y. S. Crowdsourcing, citizen sensing and sensor web technologies for public and environmental health surveillance and crisis management: Trends, OGC standards and application examples. *International Journal of Health Geographics*, 10: 1–29, 2011.

34 Thakuriah, P., Tilahun, N. Y., and Zellner, M. Big data and urban informatics: Innovations and challenges to urban planning and knowledge discovery. In

Thakuriah, P., Tilahun, N., and Zellner, M., editors, *Seeing Cities Through Big Data*, pages 11–45. Springer International Publishing: Berlin/Heidelberg, 2017.

35 Kaginalkar, A., Kumar, S., Gargava, P., and Niyogi, D. Review of urban computing in air quality management as smart city service: An integrated IoT, AI, and cloud technology perspective. *Urban Climate*, 39: 100972, 2021.

36 Baeza, R. A. A methodology for urban sustainability indicator designs. *TEMA Journal of Land Use Mobility and Environment*, 11: 285–303, 2018.

37 Aschwanden, G., Wijnands, J. S., Thompson, J., Nice, K. A., Zhao, H. F., and Stevenson, M. Learning to walk: Modeling transportation mode choice distribution through neural networks. *Environment and Planning B: Urban Analytics and City Sciences*, 48: 186–199, 2021.

38 Allam, Z. and Allam, Z. Urban chaos and the AI messiah. In *Cities and the Digital Revolution: Aligning Technology and Humanity*, pages 31–60. Springer Nature: Basingstoke, 2019.

39 Sun, Y. and Shao, Y. W. Measuring visitor satisfaction toward peri-urban green and open spaces based on Social Media Data. *Urban Forestry Urban Greening*, 53: 126709, 2020.

40 Ghahramani, M., Galle, N. J., Duarte, F., Ratti, C., and Pilla, F. Leveraging artificial intelligence to analyze citizens' opinions on urban green space. *City and Environment Interactions*, 10: 100058, 2021.

41 Haqbeen, J., Sahab, S., Ito, T., and Rizzi, P. Using decision support system to enable crowd identify neighborhood issues and its solutions for policy makers: An online experiment at Kabul municipal level. *Sustainability*, 13: 5453, 2021.

42 Bazzan, A. L. C., Do Amarante, M. D., and Da Costa, F. B. Management of demand and routing in autonomous personal transportation. *Journal of Intelligent Transportation Systems*, 16: 1–11, 2012.

43 Filomena, G., Verstegen, J. A., and Manley, E. A computational approach to 'The Image of the City'. *Cities*, 89: 14–25, 2019.

44 Thakuriah, P., Tilahun, N., and Zellner, M. Big data and urban informatics: Innovations and challenges to urban planning and knowledge discovery. In *Proceedings of the NSF Workshop on Big Data and Urban Informatics*, pages 4–32. Chicago, IL, USA, 11–12 August 2014.

45 Christodoulou, S., Deligianni, A., Aslani, P., and Agathokleous, A. Risk-based asset management of water piping networks using neurofuzzy systems. *Computers Environment And Urban Systems*, 33: 138–149, 2009.

46 Fan, J., Han, F., and Liu, H. Challenges of big data analysis. National Science Review 1: 293–314, 2014.

47 Wellmann, T., Lausch, A., Andersson, E., Knapp, S., Cortinovis, C., Jache, J., Scheuera, S., Kremerg, P., Mascarenhasa, A., Kraemera, R., Haase, A., Schug, F., and Haase, D. Remote sensing in urban planning: Contributions towards ecologically sound policies? *Landscape and Urban Planning*, 204: 103921, 2020.

48 Bambara, G., Peyras, L., Felix, H., and Serre, D. Developing a functional model for cities impacted by a natural hazard: application to a city affected by flooding. *Natural Hazards and Earth System Sciences*. 15, 2015. https://doi.org/10.5194/nhess-15-603-2015.

49 Bibri, S. The leading data-driven smart cities in Europe: Their applied solutions and best practices for sustainable development. In: *Advances in the Leading Paradigms of Urbanism and their Amalgamation. Advances in Science, Technology & Innovation*, pages 227–258. Springer: Cham, 2020. https://doi.org/10.1007/978-3-030-41746-8_9.

12

AI-driven Public Transport Solutions

Shantanu Bindewari, Prakhar Consul, Hilal Ahmed Shah, Basab Nath, and Mansi Trivedi*

School of Computer Science Engineering, ILLM University, Greater Noida, India

*Corresponding author: bindewarishantanu@gmail.com

12.1 Introduction

Artificial intelligence (AI) has not only arisen, but it is one of the most transformative technologies of the twenty-first century, transforming industries and changing the way societies operate. Some of the industries most highly affected by the use of AI include the transportation industry. The transportation system of the world, encompassing the road systems, rail, air travel, and maritime transaction, is undergoing a huge reformation due to the incorporation of AI technologies [1]. This change is not just about automation but also a deep change to how mobility is visualized, planned, and experienced.

The current state of affairs concerning the way our world moves has a number of problems to resolve. Dense population in urban areas, rampant vehicular density, and traffic congestion are some of the effects of urbanization. The traditional systems, which were made to cater to lower volumes and less complex modes of traveling, are failing to fulfill the needs of modern life [2]. As a result, the passengers are left with prolonged commutes, variable travel periods, and security issues. Such obstacles require a paradigm shift in the process of planning, operating, and maintaining the transportation systems. AI comes as a promising solution providing intelligent systems capable of analyzing large amounts of data, learning patterns, and making real-time decisions. Adaptive traffic signals through to predictive maintenance, AI is enhancing more responsive, efficient, and safe transportation networks. Not only is the integration of AI increasing the efficiency of operation of transportation systems, but it is also immensely improving the user experience [3, 4].

Over the last few years, there has been a boom in the development and integration of AI-driven solutions across different forms of transport. For example, AI algorithms are now utilized in optimizing logistics routes, thereby lowering both fuel consumption and delivery time. In our cities, AI-powered platforms help commuters receive up-to-the-minute traffic notifications, alternative route suggestions, and customized travel advice [5]. The autonomous vehicles under the powers of AI are being piloted and rolled out in controlled settings, promising to change the face of personal and public transport. In addition, the safety concerns are being addressed through AI's important role. Computer vision systems are being used in cars to identify obstacles, to read traffic signs, and to observe the driving pattern. These technologies play a crucial role in the reduction of accidents and the safety of the roads [6, 7]. To the same extent, public transportation systems are using AI for surveillance, identification of incidents, and facilitation of emergency responses. Digital makeover is also crossing over to the ticketing and fare collection process. Old ticketing counters and human checks are being phased out to AI-propelled kiosks and mobile applications that automate ticket generation and validation [8]. Not only do these systems make the process more efficient, but they also reduce human interaction, which became strongly relevant during COVID-19 (Figure 12.1).

Another noteworthy application is in the realm of predictive maintenance. By analyzing data from vehicle sensors, AI can predict component failures before they occur, allowing for timely maintenance and reducing downtime [9]. This approach not only enhances vehicle reliability but also reduces maintenance costs and extends the lifespan of assets. In addition to technological advancements, the implementation of AI in transportation raises important questions related to ethics, data privacy, and regulatory compliance. As AI systems collect and analyze vast amounts of personal and operational data, it is essential to ensure that these processes are transparent, secure, and aligned with legal standards [10, 11]. Addressing these

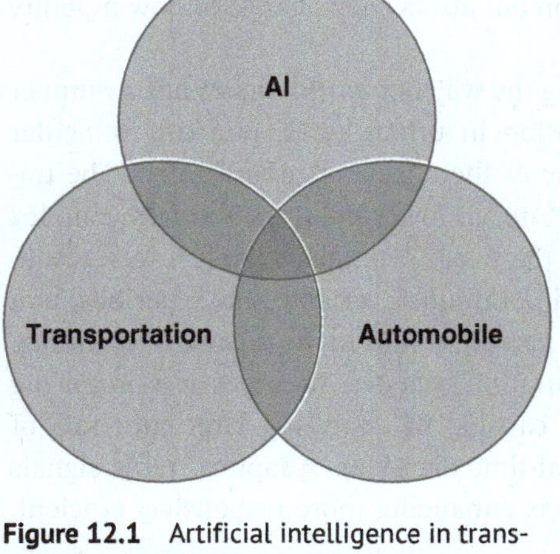

Figure 12.1 Artificial intelligence in transportation and the automobile industry.

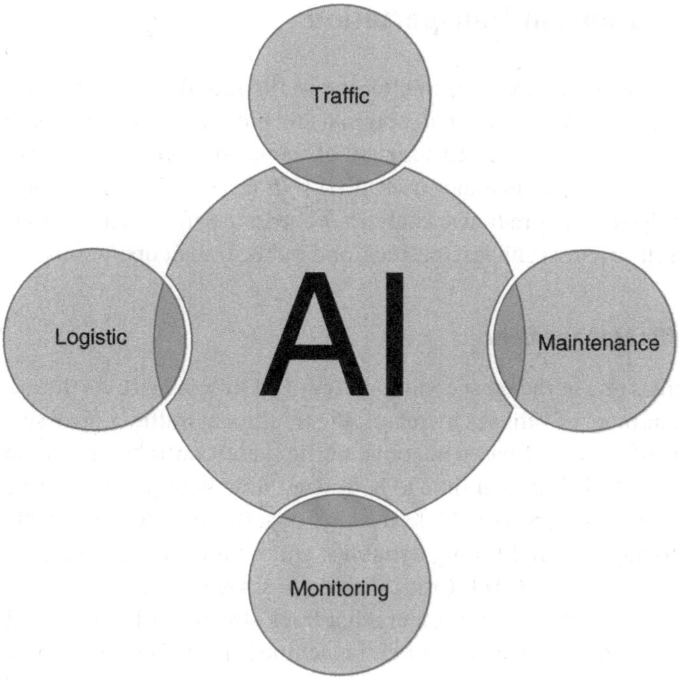

Figure 12.2 Operation management using AI in automobile and mobility.

concerns is critical for building public trust and ensuring the sustainable adoption of AI technologies (Figure 12.2).

More than that, the effects of AI go beyond the immediate gains of efficiency and safety. It can change the face of urban planning, environmental sustainability, and economic productivity [12, 13]. With smarter transportation systems, AI helps to achieve a reduction of carbon emissions, cleaner air, and cities that are more livable. It also provides new economic opportunities in the form of posting jobs in the field of technology development, system integration, and data analysis.

This chapter explores the numerous applications of AI in the transport sector, including its applications in traffic control, route optimization, ticket systems, safety, and maintenance, as discussed by the authors [14, 15]. Each section analyzes certain use cases, technological frameworks, implementation strategies, and real-life case studies. The challenges of AI deployment and ethical considerations are also shared in the chapter, along with future trends that are likely to define the future course of intelligent transportation systems. This chapter aims to provide a comprehensive overview of how AI is redefining mobility and to offer insights into the path forward for stakeholders in the transportation ecosystem.

12.2 AI Applications in Transportation

Transportation systems' structure and functioning at the foundational level are being reengineered by AI. It is no longer an optional tool but has become the spine of decision-making, operations, and development of transportation [16]. This section provides greater detail on the major uses of AI in transportation and ranges from traffic control systems to predictive analytics, vehicle-to-infrastructure (V2I) communications, self-driving vehicles, logistics, and public transit operations.

12.2.1 AI in Traffic Management

Traffic management is one of the most common uses of AI in transport. As the cities expand and the number of vehicles increases, the traditional traffic control systems are no longer sufficient. AI-based adaptive traffic signal control systems use camera, sensor, and GPS data in real time to optimize traffic light periods. These systems study such parameters as vehicle density, pedestrian flow, and traffic patterns in order to adjust signal timing dynamics and, as a result, reduce congestion, time, and emissions [17, 18]. Cities such as Los Angeles and Singapore have established AI-run traffic control centers that track streams of live data and make split-second decisions to regulate traffic. A machine learning algorithm may identify anomalies like accidents or stalled vehicles and automatically reroute the traffic or send emergency services (Figure 12.3).

12.2.2 Predictive Maintenance and Diagnostics

AI allows predictive maintenance because it uses historical and real-time data from the sensors in the vehicles to create patterns that lead to the failure of

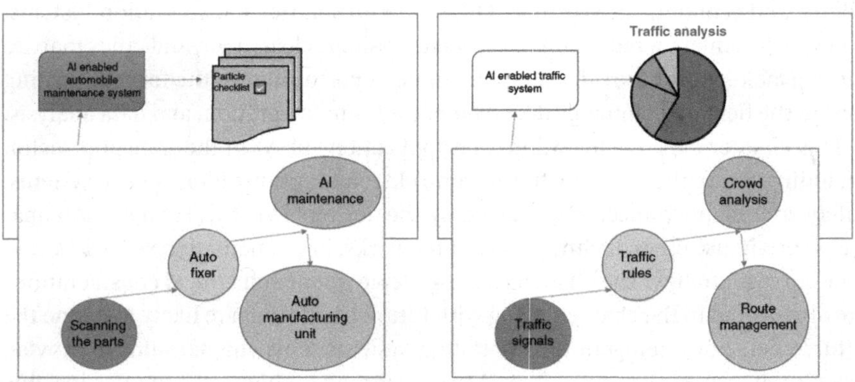

Figure 12.3 AI in traffic management.

Figure 12.4 Predictive maintenance using AI in the automobile sector.

their components. The transportation agencies and the fleet operators can use the machine learning models and predict when an auto part is likely to fail and schedule maintenance work [19]. Such a predictive method reduces unplanned downtime, improves safety, and lowers operational costs. For instance, the AI systems can detect the vibration, temperature, and pressure in engine parts and send alerts before fatal failures happen. Airlines apply AI for analyzing the telemetry data of aircraft to foresee a need for maintenance and avoid expensive delays, and ensure the safety of passengers (Figure 12.4).

12.2.3 AI in Logistics and Fleet Management

Logistics is another domain where AI's impact is profound. Companies use AI to optimize delivery routes, reduce fuel consumption, and improve customer satisfaction. Algorithms analyze traffic conditions, delivery windows, vehicle load, and driver schedules to generate the most efficient route plans. AI also plays a critical role in demand forecasting and inventory management. Retailers and logistics firms use predictive analytics to determine the optimal stock levels and anticipate demand spikes, ensuring timely delivery and cost-effective supply chain management [20]. In fleet management, AI helps monitor driver behavior, vehicle usage, and maintenance schedules. Telematics systems integrated with AI provide fleet managers with insights into fuel efficiency, route performance, and vehicle health, allowing them to make data-driven decisions.

12.2.4 Autonomous Vehicles

One of the most revolutionary uses of AI in the field of transportation could be found in autonomous vehicles development. AI technologies like computer vision, sensor fusion, deep learning, and natural language processing are largely used by these vehicles. Through AI, the vehicle can know its environment, make driving decisions, and communicate with other road users in real-time. Autonomous vehicles rely on a number of LIDAR, radar, cameras, and GPS for collecting data about the surroundings. AI analyzes this data to identify barriers, locate traffic signs, as well as to forecast pedestrians' and other vehicles' behavior. Companies such as Tesla, Waymo, and Uber are front-runners in the development of self-driving cars that would help minimize human error, which is one of the key causes of road accidents. Even though completely autonomous vehicles are still being tested, semi-autonomous features, which are adaptive cruise control, lane-keeping assistance, and automatic emergency brakes, are already implemented on modern cars, thus making the vehicles safer and more comfortable.

12.2.5 Public Transport Optimization

There has been a growing use of AI by public transportation systems for efficient delivery of services and customer satisfaction. AI applications include dynamic scheduling, real-time tracking, predictive arrival times, and fare optimization. Transport authorities analyze usage patterns based on ticketing systems, GPS, and mobile apps' data to alter services accordingly. AI also increases multimodal transport by bringing together various modes of transport—buses, trains, bicycles, and ride-sharing into a single unit. Commuters can organize entire journeys with little waiting times and ideal stopovers, thus enhancing the user experience. Additionally, the use of AI is applied in analyzing the passengers' feedback and the data retrieved from social media to identify service gaps and enhance communication with the public. Processing tools in natural language provide ways of interpreting the public sentiment and engaging better.

12.2.6 Infrastructure Monitoring and Smart Cities

AI applications can be applied not only to monitoring and managing transportation infrastructures but also to guarantee that finance is available for the operation of trains. By means of drone surveillance, computer vision, and sensor data, AI systems can monitor the condition of roads, bridges, and tunnels in real-time. Maintenance tasks can then be scheduled depending on urgency and resource availability. In smart cities, AI connects transportation systems to other urban networks, including energy grids and communication networks, so as to perform seamlessly [21]. Data from traffic is compiled with pollution, weather forecasts, and emergency alerts to come up with complete urban mobility plans. AI also helps in planning infrastructure by simulating the flow of traffic through various scenarios for city planners to plan the roads and transport systems more efficiently.

12.2.7 Emergency Response and Incident Management

AI systems are important to the identification and management of transportation emergencies. Machine learning models can watch video streams as well as sensor readings, spot accidents, fires, or suspicious conduct in real time. AI systems can issue alerts to the authorities, redirect traffic, and initiate emergency procedures once an incident is identified. Smart traffic systems apply AI to offer alternatives and communicate with the drivers through dynamic signboards or mobile apps. When it comes to natural disasters or large-scale events, AI assists in emergency evacuations and route optimization of emergency vehicles.

12.2.8 Environmental Monitoring and Sustainability

AI supports environmental sustainability since it helps achieve more efficient and eco-friendly transportation. AI algorithms optimize engine performance, track emissions, and give suggestions for changes in driving habits that will minimize fuel consumption. The few benefits of AI to electric cars are in battery control and forecasting the range. Also, AI helps to facilitate green logistics strategy implementation, such as a combination of delivery and alternative modes of transport use [22, 23]. Urban planners apply AI models to measure the effect of the transport projects on the environment and to support sustainable mobility orientation.

12.3 Introduction to AI in Automation and Ticketing

There has been a significant effect on the automation of service processes across different sectors owing to the transformation of AI; transportation is a prominent beneficiary. The deployment of AI to the automation of transportation and ticketing systems not only made them more efficient from an operational point of view but also transformed the user experience into something faster, safer, and more personalized. From self-service ticket kiosks to smart fare systems and automatic transit operations, AI is improving reliability and efficiency on every touchpoint. With the rapid urbanization and rise in demand for public and private transport services, available manual ticketing systems and operations are no longer sufficient. The need to automate through the use of AI stems from issues such as long queues, human error, fare evasion, and inefficiency in calculating fares. The modern AI systems are created to address such problems using real-time data examination, machine learning, and the prediction of user behavior (Figure 12.5).

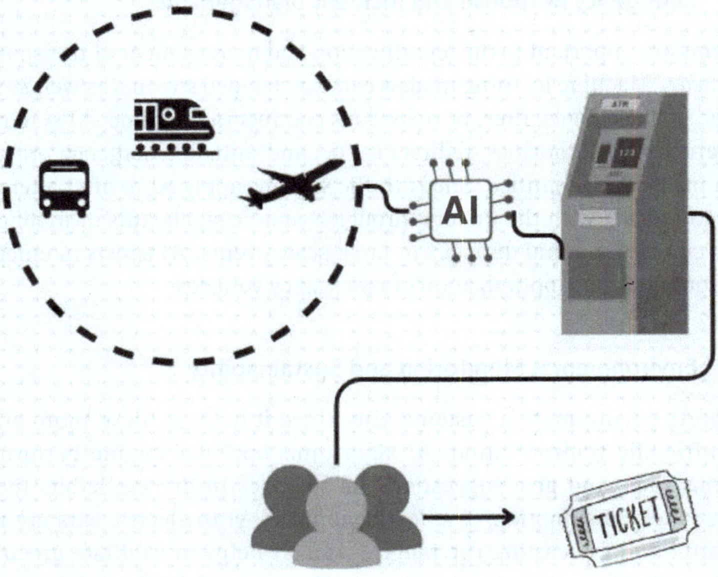

Figure 12.5 Ticket machine using AI in public transport.

12.3.1 AI in Ticket Generation and Validation

Some of the modern AI technologies, like optical character recognition (OCR), natural language processing (NLP), and biometric authentication, are currently at the core of the modern ticketing systems. Machine-based kiosks and mobile ticketing software use AI algorithms to recognize the users, verify the travel details, and assign tickets within dynamic pricing frameworks [24]. Smart ticketing systems also include scanning of QR codes and facial recognition, and command by voice, whereby automatic entry and exit in various forms of transport is possible without human contact. Such innovations improve the comfort of clients, eliminate bottlenecks on ticket counters, as well as boarding points. One of the announcements is the integration of AI with the digital wallets and ride-hailing apps so that travelers can purchase tickets, top up the fare card, and validate the rides through one mobile interface. This is a merging of services that brings a single transportation ecosystem that is efficient and satisfaction-providing to the users.

12.3.2 AI-based Ticket Collectors and Validators

AI is also changing the way fare collection is conducted. Intelligent ticket collectors and validators come with AI-based image recognition, sensor fusion, and cloud enablement. These systems can authenticate and find duplicates of tickets and detect irregularities such as the expiries, or the wrong fares. AI-enabled fare gates are increasingly being seen in metros and trains around urban systems. These gates can respond to crowd density, reroute passengers when congestion builds up, and even detect suspicious behavior with the help of built-in video analytics. AI algorithms can also monitor boarding patterns and alighting patterns in order to revise transit schedules and fine-tune route planning.

Exterior to the physical systems, virtual ticket collectors like AI chatbots support ticket bookings, travel updates, and complaint resolution among their users. These bots are learning from past data and user behavior and are quick to respond and precise.

12.3.3 Self-handling Ticketing Systems

Self-handling ticketing systems that work on the basis of AI are created with the aim of reducing human involvement. These systems come with such capabilities as predictive ticket recommendations, fare top-up automation, and individualized travel guidance. AI models examine users' previous travel history, travel preferences, and real-time inputs (e.g., delays, no seats available) to suggest best-priced ticketing options. These systems have been adopted by the airports and long-distance bus terminals to achieve a smooth boarding process. Passengers can do self-check-in using biometric data, print tickets, and open doors to gates without serving staff [25]. These innovations cut wait times and enhance scalability in operations, particularly in the peak periods. The use of AI in crowd estimation only escalates the self-handling systems. Systems can make better use of passengers' influx and travel behavior in preemptively maintaining loads and dynamically allocating resources, and to avoid overcrowding.

12.3.4 AI-enabled Women Safety Alarm Systems

The inclusion of safety aspects, especially for the vulnerable segments of society, such as women, is one critical area of utilizing AI in the field of transportation ticketing and automation. AI-based alarm systems are integrated in ticketing applications, ride-hailing systems, and public transport vehicles. These systems apply tracking of location, behavior pattern analysis, and audio-video live monitoring to find anomalies. Once a safety threat is detected—either by the user or automatically using behavioral indicators—the system can, in turn, send the alerts to the surrounding security personnel, sound vehicle alarms, and alarm emergency contacts. In some cities, AI-enabled emergency kiosks in transport hubs enable passengers to get immediate help. These systems include video streaming, two-way communication, and panic button features, among others, and timely intervention in any case of crisis as discussed by the researchers [26, 27].

12.3.5 Integration with Smart Mobility Platforms

AI-based ticketing and automation systems are part and parcel of the concept of smart mobility platforms. These platforms bring together all the transport services, such as buses, metros, cabs, bikes, and rental cars together under one interface. AI coordinates smooth relationship between such services with instantaneous updates, dynamic pricing, and route optimization. A smart mobility app allows a passenger to make a multimodal journey, get AI-made ticket bundles, and pay in an integrated wallet. AI studies traffic, weather, and the availability of public transport to suggest the most optimal modes of transport. The governments and transit agencies are collaborating with the tech firms to roll out such platforms, particularly in smart city programs. These integrations promote environmentally-friendly traveling and resource optimization, thereby making the advantages of AI-facilitated ticketing spill over.

12.3.6 AI in Dynamic Pricing and Revenue Management

Dynamic pricing in terms of transportation is another field where AI has achieved significant contributions. AI algorithms make changes in fares based on demand, time of the day, availability, and user history, among other variables. This model is more popular among airlines, ride-hailing services, and rail operators to achieve the highest levels of occupancy and revenue. Machine learning models use a record of past and real-time data to predict demand trends and optimum prices. By so doing, these systems guarantee affordability to the passengers in low demand seasons and at the same time, optimize operators' profits during their high demand seasons. In addition, the AI tools also observe fare evasion patterns and fraud detection. By correlating user behavior with purchase history, AI can highlight the anomalies and help the authorities to take remedial measures.

12.3.7 Challenges and Considerations

Notwithstanding the AI potential in automation and ticketing, some challenges arise, and they include the following. Issues of data privacy and user consent are important, especially if biometric information and travel history are involved. The secure storage and transparent decision-making process in AI are important for public trust. Interoperability and standardization of systems are also barriers. Due to the use of different technologies, which might not be compatible, the integration of the different transit systems is a challenge. Strong APIs, one universal system of ticketing, and collaborative ecosystems are essential to deal with these challenges [28]. In addition, though automation decreases human intervention, it creates job replacement and the digital divide anxiety. Inclusive design, reskilling programs, and equitably ensuring that AI does not accentuate divisions among demographic groups are essential in relation to AI benefits distribution.

12.3.8 Case Studies and Real-world Implementations

Some of the AI-accelerated automation and ticketing systems in several global cities have been splendidly successful: London's Oyster login and contactless card system uses AI to monitor usage of travel behavior by the user and change fare capping. Singapore's land transport authority (LTA): Performs facial recognition and crowd analytics for metro ticketing. Indian Railways piloted AI ticketing kiosks with facial verification and translation of language capabilities. New York's MTA integrates the use of AI toward optimizing ticket prices depending on passenger flows and congestion. These are some excellent examples of how AI can transform the provision of responsive, secure, and user-centric transport services.

12.4 AI for Safety and Security

As the sphere of transportation is becoming interrelated and data-driven, safety and security have become the number one priority. AI is currently gaining more traction in the effort to boost safety measures, to monitor real-time conditions, and take precautions against accidents or criminal activities [29]. By utilizing its abilities in data analysis, computer vision, natural language processing, and sensor fusion, AI is changing the safety horizon of both public and private transportation systems.

12.4.1 AI in Surveillance and Monitoring

Intelligent surveillance is a prominent AI application in the field of transportation safety. We see today in buses, trains, at stations, and in airports the use of AI-powered cameras for facial recognition, license plate identification, and behavioral analysis. These systems may also include the use of machine learning to identify unusual activity, unauthorized entry, or left-out items. Also, for instance, they use computer vision to study passenger body language and movement which in turn may indicate aggressive intent or unstable behavior. AI in these settings reports to security, and at the same time streams live video, which in turn enables for prompt response.

12.4.2 Real-time Incident Detection

AI systems are also used for on-time incident detection on roads and highways. Smart infrastructure cameras and sensors gather information about the speed of vehicles, lane departure, wrong-way driving, and crashes. AI processes this data to find anomalies and report them to the emergency services immediately. Predictive algorithms can forecast accidents that have not yet taken place because of danger patterns, i.e., erratic driving behaviors or bad weather circumstances. Such a proactive approach not only reduces the response time but can also assist in deploying preventive measures such as dynamic speed limits and rerouting traffic.

12.4.3 AI for Emergency Management

In emergencies, time is critical. AI supports emergency response with real-time route optimization for ambulances and fire trucks, and police vehicles. GPS reads along with AI algorithms points to the most expedient route, taking in viewpoints of traffic jams, construction sites, and roadblocks. In addition, AI-powered dispatch systems prioritize calls depending on how severe they are and where they're made from, making it easier to coordinate agencies. In mass transit systems, AI observes the pattern of the flow of passengers and assists evacuations by determining the safe exits and paths.

12.4.4 Women and Vulnerable Passenger Safety

AI safety systems for women and vulnerable passengers are becoming more and more incorporated into the transport infrastructure. Mobile apps and in-vehicle systems provide AI-driven panic buttons, real-time tracking, and automatic provision of information to emergency contacts. Behavioral analysis tools can identify distress indicators using their tone of voice, face expressions, and movement, and this can trigger alerts despite situations where users are unable to manually report an issue. Some metro systems have dedicated kiosks for AI where women can ask for help in a discreet way.

12.4.5 Preventive Maintenance for Safety

Mechanical malfunctions are one of the major reasons for transportation accidents. AI-powered predictive maintenance systems analyze the past data and sensor data combined with usage patterns to predict the occurrence of faults on vehicles before they become dangerous [30]. For instance, the machine learning models can be used for the prediction of brake wear-out, engine failures, or unusual tire pressure. This data assists transportation operators in scheduling timely repairs and preventing breakdowns to provide reliability of the fleet, thereby avoiding accidents due to mechanical failure.

12.4.6 Cybersecurity in AI-enabled Systems

The complete integration of the AI systems into transportation calls for guaranteeing the security of these systems from attacks by cyber individuals. AI helps in monitoring network traffic, detecting suspicious usage, and data leaks. Machine learning algorithms are able to recognize threats such as denial-of-service attacks, phishing attacks, or unauthorized access to data. With the ability to learn from new threats, the AI improves the IT infrastructure and guarantees that the sensitive data of the passengers is secured.

12.4.7 AI Use in Driver Aid and Autonomous Cars

AI is used in Advanced Driver Assistance Systems (ADAS) and autonomous vehicles to guarantee their operational safety. The list of functions performed by AI is an endless one, ranging from lane-keeping assist to adaptive cruise control, collision avoidance, and pedestrian detection. Autonomous vehicles rely on AI in order to process data from LIDAR, radar, cameras, and GPS, which come in at split seconds to make decisions. These systems can determine the movement of pedestrians and other vehicles, thus minimizing the occurrence of accidents. In addition, AI continues to learn from driving situations through which it enhances vehicle responses.

12.4.8 Ethical and Legal Implications

While AI significantly improves safety, it also raises ethical and legal concerns. Misidentification in facial recognition, bias in behavioral analytics, and the accountability of AI-driven decisions are areas of active debate. Transportation authorities must establish ethical guidelines, ensure algorithmic transparency, and involve public stakeholders in decision-making. Legal frameworks should define the responsibilities of AI system developers, operators, and regulators in the event of system failure or misuse.

12.5 Dynamic Route Optimization Systems

With increasing global urbanization and an ever-increasing number of vehicles on roads, dynamic route optimization has become necessary to minimize traffic congestion, optimize fuel usage, and improve commuters' experience. The old routing approaches are not sufficient in handling the real-time complexities of transportation. AI provides dynamic, adaptive, and context-aware routing solutions that revolutionize how both public and private transport systems operate.

Dynamic routing may be defined as the process of constant analysis and subsequent changes in the travel path depending on real-time data (e.g., traffic flow, road closures, weather, and passenger demand). Unlike static routing, which refers to preconfigured routes, dynamic systems are able to adjust in real-time through machine learning, predictive analytics, and the data received from the IoT sensors. AI algorithms like reinforcement learning and deep learning allow simulating and assessing millions of potential routes to choose the efficient one according to the current and future conditions. This practice makes sure that vehicles bypass traffic snarl-ups, reduce travel time, and enhance on-time performance.

12.5.1 AI in Real-time Traffic Monitoring

AI is very important in dealing with real time traffic monitoring. Computer vision and AI-based pattern recognition methods are used to process collected data retrieved from GPS devices, traffic cameras, mobile apps, and smart traffic lights. Systems such as Google Maps and Waze use AI to examine the trend of congestion and give alternative routes. More developed systems incorporate the use of predictive modeling in order to predict future traffic situations and prompt active rerouting to guarantee smoother traffic distribution.

12.5.2 Personalized Route Recommendations

The modern AI systems provide customized route recommendations according to users' choices, travel history, and time-specific information—appointments, weather prediction, and the fuel level. Using mobile apps, AI may offer the routes that would fit in with a driver's habits or current requirements—for example, faring clear of toll roads or low-emission zones [14, 19]. For commercial logistics and delivery services, AI designs routes according to the service-level agreements, lowers the fuel costs, and fits with the customer delivery windows. This customization enhances operational efficiency as well as the satisfaction of customers.

In public transport, AI adjusts schedules for buses, trams, and metros dynamically, depending on the density of the passenger flow, traffic conditions, and the schedule of events. Real-time data gathered from the fare system, vehicle tracking, and social media forecast the spikes in demand. Some of the cities have installed AI-based demand-responsive transport (DRT) systems that redirect vehicles in real time to meet the changing needs of the passengers. These systems shorten waiting times, increase vehicle usage, and provide a more flexible choice of commuting, especially in low-density areas.

12.5.3 Environmental and Economic Benefits

Dynamic route optimization plays a significant role in environmental sustainability by limiting idle time and consumption of fuel, as well as vehicle emissions. Eco-routing models with AI on top are concerned with finding routes that use the least energy or emit the lowest level of pollutants. Economically, optimized routes reduce the costs of operations for logistics providers, public transport agencies, and ride-hailing firms. Businesses enjoy improved delivery reliability, lower costs of fuel, and more efficient fleet management.

12.5.4 AI and Emergency Response Routing

The AI systems are crucial for emergency response, and time-sensitive routing is important. With the help of the live traffic data and infrastructure information, AI can give ambulances, fire services, and police the best route to their destinations. In addition, AI can mimic clearance strategies of traffic and coordinate signal prioritization to warn nearby drivers to clear routes for the emergency vehicles. This ability saves lives and reduces disruption in highly stressful situations [24]. Related to the benefits of dynamic routing systems, their implementation is problematic. The accuracy of data, latency, and integrations between platforms may affect the performance of AI algorithms. Inconsistent infrastructure and data standards in different cities make scalability difficult. Issues regarding privacy are also raised when the location and patterns of travel of the users are constantly being examined. Cybersecurity quality, data practices transparency, and compliance with the regulations must be ensured.

12.6 AI in Public vs. Private Transportation

The transport systems across the world are significantly changing fast due to AI. From improving the public transit routes to facilitating the use of self-driving cars, AI solutions are building smarter, safer, and more efficient mobility networks. This essay examines to what extent AI can be useful in both public and private transport, with most of the focus on use cases in public transit, how smart mobility is achieved by combining AI with ride-sharing and autonomous vehicles, and case studies that provide real-world examples.

12.6.1 Rise of AI in Transportation

Inclusion of AI in transportation is a part of the general trend toward intelligent urban infrastructure. Under the mounting city pressure caused by rapid urbanization, population increase, and environmental concerns, AI offers the tools for the optimization of traffic, reduction of emissions, enhancement of safety, and UX. Overall, AI in transportation includes such technologies as: Machine learning for predictive analytics

- Computer vision for vehicle navigation
- Natural language processing for customer service
- Reinforcement learning for dynamic routing
- Sensor fusion in autonomous systems

While AI impacts both public and private sectors, the use cases and implementation strategies vary significantly across these domains.

12.6.2 Use Cases in Public Transit

Public transportation systems—such as buses, trains, trams, and subways—are essential for urban mobility. AI plays a vital role in enhancing the efficiency, reliability, and accessibility of these systems.

A. **Predictive Maintenance**
   AI-driven predictive maintenance systems use sensor data and machine learning to anticipate failures in buses, trains, and infrastructure. For example, vibration sensors on train wheels can detect anomalies before they lead to breakdowns. This reduces downtime, improves safety, and lowers maintenance costs.

   Example: The Chicago Transit Authority uses predictive analytics to monitor escalators and elevators, ensuring uninterrupted service for passengers with mobility needs.
B. **Real-time Passenger Information**
   Using real-time data from GPS, ticketing systems and social media to produce up to date information about transit, AI monitors. AI-powered apps can inform the users regarding the delays, provide alternative routes, or come up with an estimated time of arrival. Example: Transport for London (TfL) applies AI to notify commuters on train delay, congestion, and other options.
C. **Optimized Fare and Fraud Detection**
   With the help of AI, transit authorities can optimize affairs in accordance with the patterns of the passenger behavior and recognize fraudulent activity through pattern recognition. Example: NY's MTA applies AI to recognize fare evasions on the basis of turnstiles' usage trend.
D. **Accessibility Enhancements**
   Computer vision and voice assistants enhance the accessibility for old and disabled people. AI instruments are able to help blind passengers navigate stations utilizing cameras and instructions in real time.

12.6.3 Smart Mobility in Ride-sharing and Autonomous Vehicles

The private sector, particularly in ride-sharing and autonomous vehicles, is driving innovation in smart mobility. AI is central to the development of these services.

A. AI in Ride-Sharing Platforms

Companies like Uber, Lyft, and Didi Chuxing rely heavily on AI for a range of operations:

a. Dynamic pricing: Algorithms adjust prices based on supply and demand.
b. Matching algorithms: AI pairs passengers with the closest drivers.
c. Route optimization: Real-time traffic data is used to find the fastest routes.
d. Demand prediction: Machine learning forecasts ride demand across different locations and times.

AI not only optimizes logistics but also enhances safety through driver behavior analysis and customer support automation.

B. Autonomous Vehicles (AVs)

Autonomous vehicles represent the pinnacle of AI in transportation. These systems rely on a combination of:

a. Computer vision for obstacle detection and lane following
b. LIDAR and RADAR for 3D mapping and object recognition
c. Deep learning for decision-making in complex environments
d. Sensor fusion for situational awareness

Companies like Waymo, Tesla, and Cruise are leading efforts in self-driving technology. Although full autonomy (Level 5) is not yet mainstream, semi-autonomous vehicles with AI-driven features such as adaptive cruise control and automatic emergency braking are increasingly common.

C. Shared Autonomous Mobility

The convergence of ridesharing and AVs will enable fleets of shared autonomous vehicles (SAVs), which can reduce vehicle ownership, lower emissions, and optimize urban space. Example: Waymo One, a robotaxi service in Phoenix, Arizona, uses fully autonomous vehicles to transport passengers without human drivers.

12.6.4 Comparing AI in Public vs. Private Transportation

While both sectors benefit from AI, their objectives, challenges, and implementations differ (Table 12.1).

12.6.5 Case Studies

A. Helsinki: Mobility-as-a-Service (MaaS)

Helsinki is a pioneer in integrating public and private transport through its MaaS platform "Whim." Users can access buses, taxis, rental cars, and bikes

Table 12.1 Comparison of public transportation and private transportation.

Aspect	Public transportation	Private transportation
Objective	Serve the public good, ensure equitable access [17]	Maximize efficiency and profit
Data Sources	Ridership data, sensors, and GPS	App usage, vehicle sensors, and customer profiles
Infrastructure	Large-scale, government-funded [19]	Often decentralized, privately owned
Regulation	Highly regulated [25]	Emerging frameworks for AVs and ride sharing
Innovation Pace	Slower due to bureaucracy	Faster due to market competition [7]
Challenges	Funding, legacy systems	Ethics, safety, and regulation

from a single app. AI powers the backend, providing personalized recommendations, route planning, and payment options.

Outcome: Increased adoption of public transit and reduced reliance on private vehicles.

B. **Singapore: Smart Traffic Management**
Singapore uses AI for citywide traffic optimization. Its "Smart Mobility 2030" initiative integrates AI with IoT devices and real-time monitoring. Features include AI-controlled traffic lights that adapt to flow, real-time congestion analysis, and AI-enabled enforcement for traffic violations. Efficient flow, reduced congestion, and a model for other smart cities.

C. **Uber's AI Lab**
Uber's AI Lab focuses on optimizing ride-sharing logistics. Through machine learning and predictive analytics, the company has improved: ETAs and routing accuracy, dynamic pricing algorithms, fraud detection, and driver/passenger ratings.

D. **Los Angeles Metro and Machine Learning**
LA Metro uses AI to detect fare evasion and improve maintenance. Data from turnstiles and maintenance logs are analyzed to prioritize inspections and predict equipment failures. Outcome: Reduced revenue loss and enhanced service reliability.

E. **Waymo in Phoenix**
Waymo's self-driving taxi service operates in select areas of Phoenix. The AI stack manages perception, prediction, and planning in real-time, navigating urban environments with minimal human intervention. Outcome: Public deployment of Level 4 autonomy, setting a precedent for AV adoption.

12.6.6 The Future of AI in Transportation

The future of AI in transportation will likely involve deeper integration across modes, platforms, and systems. Key trends include the integration of AI with 5G and IoT for faster data exchange, expansion of MaaS platforms in cities worldwide, growth of electric autonomous vehicles (EAVs), AI-powered sustainability tools to monitor carbon emissions, and public-private partnerships for shared infrastructure. AI will play a crucial role in building resilient, inclusive, and sustainable transportation ecosystems.

12.7 Challenges and Ethical Considerations

While AI's impact on transportation is largely positive, several significant challenges and ethical concerns must be addressed to ensure equitable and sustainable implementation.

A. **Data Privacy and Surveillance Risks**
   Any AI system's operations depend on data collection, for example, tracking the movement of a vehicle and observing the passengers of the vehicle. Combination AI systems that use real-time location information are especially so. On the other hand, on a similar level of surveillance, there are risks in terms of privacy and protection of personal data. Surveillance automation is another major challenge because data presented from optimizing the GPS functionalities is an incidental leak of personally sensitive information, which exposes habits and patterns of movement. If such a framework is taken for granted, users who choose to expose themselves to disclosure of personal traits under explicit consent preconditions end up being governed through mechanisms that result in an automatic loss of control under poor definitions of governance structures. Encryption and anonymization of such data, supercharged with user agreements, should be strengthened by governments and companies. The AI Oversight Law in California is a basic draft, while the GDPR refers to useful concepts when speaking about some gaps in cultures.

B. **Algorithmic Bias and Fairness**
   AI systems are no less biased than the data that they will be trained on. Bias in the datasets for transportation equals biased results—reluctant sharing for rides services for the lower income/minority areas, and the economic ones. For instance, the application of machine learning models using past data on policing or urban development in optimizing patrols or expansion of transit can unintentionally reconfirm the territorial social conditions. To overcome this, the AI systems need to be audited frequently in the name of fairness, and training data needs to be diversified in the sense that the data should reflect

the actual demographics of the world. Ethical-AI frameworks, involvement of community, and open-source modeling are the vital steps to an equal transportation.

C. **Infrastructure and Scalability Issues**
Deploying AI at scale requires significant infrastructure upgrades. Public transit systems often rely on legacy technologies that are incompatible with modern AI tools, requiring costly overhauls or integrations. For example:
- Upgrading buses and trains with sensors, cameras, and communication systems
- Developing centralized AI processing hubs
- Ensuring real-time data exchange between vehicles, platforms, and traffic systems

Additionally, in lower-income or rural areas, the lack of connectivity and digital infrastructure may widen the urban-rural mobility gap.

Governments must plan long-term investments in digital infrastructure and offer incentives for private partnerships to expand AI's reach equitably.

12.8 Future Trends and Innovations

The next decade will see AI become even more embedded in the fabric of urban transportation. Several emerging technologies and trends are set to shape the future.

A. **Integration with IoT and 5G**
The combination of AI with IoT and 5G connectivity promises ultra-fast, highly responsive transportation systems. IoT sensors embedded in roads, vehicles, and traffic lights will collect vast streams of real-time data, which AI will analyze instantaneously thanks to 5G networks.

Benefits include:
- Real-time traffic rerouting during accidents
- Instantaneous communication between vehicles (V2V) and infrastructure (V2I)
- Predictive analytics for road wear and tear

Cities like Barcelona and Seoul are already piloting such smart infrastructure projects.

B. **Autonomous Transport Networks**
The future points toward fully autonomous transportation ecosystems. This includes not just individual self-driving cars, but integrated networks of autonomous buses, delivery robots, cargo drones, and even autonomous ships.

These networks will use swarm intelligence—a branch of AI that mimics the behavior of flocks or colonies—to self-organize and adjust to dynamic urban conditions without centralized control.

Potential benefits:

- Lower congestion and emissions
- 24/7 transit availability
- Reduction in traffic accidents due to human error

However, legal frameworks and public trust must evolve to match the pace of innovation.

C. **AI-powered Urban Planning**

AI is increasingly being used in urban planning to simulate traffic flow, assess environmental impact, and model the effects of policy decisions. These tools help planners design cities that prioritize sustainable mobility, walkability, and equitable access to transit.

For example:

- AI models can predict how a new subway line would alter commuting patterns
- Simulation tools can test car-free zone proposals
- Environmental sensors can guide the placement of green corridors

This integration of AI into city planning is a crucial step toward the realization of "15-minute cities," where all essential services are accessible within a short walk or bike ride.

12.9 Conclusion

AI is fundamentally transforming transportation across both public and private sectors. From improving the reliability of public transit to enabling self-driving cars and predictive analytics, AI enhances safety, efficiency, and user satisfaction. Public transportation leverages AI for route optimization, maintenance, and accessibility, while private enterprises focus on smart mobility solutions like ride-sharing, autonomous vehicles, and personalized services. AI enables real-time route optimization, predictive maintenance and operational efficiency, improved accessibility and service equity, and new modes of autonomous and shared mobility. AI will be at the heart of smart cities—ecosystems where transportation is deeply integrated with energy, communication, and environmental systems. Urban centers will evolve to be more connected, sustainable, and responsive to human needs. Key features of AI-enabled smart cities are zero-emission, autonomous transport networks, Seamless intermodal mobility platforms, Dynamic traffic management and pollution control, and Equitable access to mobility for all social groups. To fully realize this vision, comprehensive policy frameworks must guide AI development and deployment. These should address data governance and privacy standards, Ethical AI design and accountability, Workforce retraining for displaced jobs, public-private collaboration models, and Infrastructure investments, especially in underserved areas. Moreover, community involvement and transparency should be at the core of all AI transportation initiatives to build public trust and ensure that technology serves everyone—not just the privileged.

References

1 Al-Jarrah, O. Y., Yoo, P. D., Muhaidat, S., Karagiannidis, G. K., and Taha, K. Efficient machine learning for big data: A review. *Big Data Research*, 2(3): 87–93, 2015.

2 Baskar, L. D., De Schutter, B., and Hellendoorn, H. Traffic control and intelligent vehicle highway systems: A review. *European Journal of Transport and Infrastructure Research*, 12(3): 167–196, 2012.

3 Zhang, J., Wang, F.-Y., Wang, K., Lin, W.-H., Xu, X., and Chen, C. Data-driven intelligent transportation systems: A survey. *IEEE Transactions on Intelligent Transportation Systems*, 12(4): 1624–1639, 2011.

4 Kaur, K., Obaidat, M. S., Kumar, A., Kaur, N., and Saini, R. A survey of cyber security in vehicular ad hoc network. *Computer Science Review*, 33: 1–20, 2018.

5 Liu, J., Ma, X., Lv, Y., Fu, Y., and Wang, F. Smart transportation: A literature review. *Information Fusion*, 66: 189–210, 2020.

6 Wang, F. and Zhang, D. Data-driven transportation and mobility: Recent advances. *Information Systems Frontiers*, 19(4): 733–741, 2017.

7 European Commission. Artificial intelligence in transport. EU Science Hub, 2019.

8 Chen, C., Zhang, J., Zheng, Y., Ma, X., and Li, S. Big data analytics for transportation: A survey. *IEEE Transactions on Intelligent Transportation Systems*, 20(1): 383–398, 2018.

9 Litman, T. Autonomous vehicle implementation predictions. Victoria Transport Policy Institute, 2021.

10 Hussain, R. and Zeadally, S. Autonomous cars: Research results, issues, and future challenges. *IEEE Communications Surveys & Tutorials*, 21(2): 1275–1313, 2019.

11 Debnath, A. K., Haque, M. M., Yuen, B., and Chin, S. B. Sustainable urban transport: Smart technology initiatives in Singapore. *Transportation Research Record: Journal of the Transportation Research Board*, 2243(1): 38–45, 2011. https://doi.org/10.3141/2243-05.

12 Kalra, N. and Paddock, S. M. Driving to safety: How many miles of driving would it take to demonstrate autonomous vehicle reliability? *Transportation Research Part A: Policy and Practice*, 94: 182–193, 2016.

13 Gkartzonikas, C. and Gkritza, K. What have we learned? A review of stated preference and choice studies on autonomous vehicles. *Transportation Research Part C*, 98: 323–337, 2019.

14 Arslan, M., Khan, M. S., Bilal, M., Ali, H. M., and Ahmad, S. AI in public transportation: A bibliometric and systematic literature review. *Journal of Intelligent Transportation Systems*, 26(4): 343–361, 2021.

15 Goodchild, A. and Toy, J. Delivery by drone: An evaluation of unmanned aerial vehicle technology in reducing CO_2 emissions in the delivery service industry. *Transportation Research Part D*, 61: 58–67, 2018.

16 Xu, W., Zhou, H., Cheng, N., Lyu, F., Shi, W., Chen, J., and Shen, X. (S.). Internet of vehicles in big data era. *IEEE/CAA Journal of Automatica Sinica*, 5(1): 19–35, 2018. https://doi.org/10.1109/JAS.2017.7510736.

17 Singh, S. Critical reasons for crashes investigated in the national motor vehicle crash causation survey. National Highway Traffic Safety Administration, 2015.

18 Labi, S. Transportation infrastructure and systems: A data-driven approach to performance-based analysis and design. John Wiley & Sons, 2020.

19 Zhang, W. and Guhathakurta, S. Parking spaces in the age of autonomous vehicles: How much parking will we need and where? *Transport Research Record*, 2651(1): 80–91, 2018.

20 Shladover, S. E. Connected and automated vehicle systems: Introduction and overview. *Journal of Intelligent Transportation Systems*, 22(3): 190–200, 2018.

21 Bina, M. and De Palma, A. AI for demand-responsive public transport: A review. *Transportation Research Procedia*, 13: 40–49, 2016.

22 Chowdhury, M. A., Hasan, M. R. M., Newaz, S. H. S., Hossain, A., and Atiquzzaman, M. Data analytics for intelligent transportation systems: A survey. *IEEE Transactions on Intelligent Transportation Systems*, 19(5): 1197–1215, 2018.

23 Hall, J. D., Palsson, C., and Price, J. Is Uber a substitute or complement for public transit? *Journal of Urban Economics*, 108: 36–50, 2018.

24 European Commission. Ethics guidelines for trustworthy AI. *EU High-Level Expert Group on AI*, 2022.

25 Wu, Y. and Chen, X. AI-powered mobility-as-a-service (MaaS): Concepts and practices. *IEEE Intelligent Transportation Systems Magazine*, 12(3): 58–70, 2020.

26 Ghosh, S. and Guin, A. Women's safety in public transport: Role of surveillance and AI. *Transport Policy*, 82: 59–70, 2019.

27 Xu, Y. and Zhang, M. Multi-agent systems for smart transportation. *IEEE Access*, 9: 75412–75430, 2021.

28 Nie, Y. M. Decentralized dynamic traffic assignment using connected autonomous vehicle technologies. *Transportation Research Part B*, 94: 329–343, 2016.

29 Rahman, M. A., Rahim, M. A., Band, S. S., and Mosavi, A. Digital twins and AI for smart transportation systems: A review. *Sensors*, 22(3): 1082, 2022.

30 Singh, M. and Verma, H. Artificial Intelligence applications in smart cities transportation: A case study approach. *Smart Cities Journal*, 4(1): 11–22, 2023.

13

AI-driven Data Analytics for Smart Urban Transport: Innovations, Challenges, and Future Trends

A. Jasmine Sugil[1], K. Merriliance[1,], and Mary Immaculate Sheela Lourdusamy[2]*

[1] *Department of Computer Applications and Research Centre, Sarah Tucker College (Autonomous), Affiliated to Manonmaniam Sundaranar University, Tirunelveli, Tamil Nadu, India*
[2] *Department of Computing and Engineering, School of Engineering Technology and Applied Science (SETAS), Heritage Christian University College (HCUC), Accra, Ghana*

*Corresponding author: merriliance@gmail.com

13.1 Introduction

Over the past few years, cities across the globe have experienced rapid urbanization, resulting in increasing populations, growing infrastructure, and progressively complex transportation networks. These changes, while signaling progress, have also brought about a host of urban mobility challenges ranging from chronic traffic congestion and deteriorating air quality to inefficient public transportation systems and increased greenhouse gas emissions. Cities are being forced to reevaluate conventional transport models and adopt more intelligent, sustainable solutions as a result of these mounting pressures. The world's urban population is projected to reach 68% of the global population by 2050, with cities facing unprecedented challenges in managing transportation systems. The anticipated addition of 2.9 billion vehicles on the roads will worsen issues such as traffic congestion, pollution, and strained infrastructure. At the same time, travel behavior is evolving. Traditional reliance on private vehicles or public transport is giving way to more flexible and personalized mobility options. People increasingly combine different modes—car-sharing, cycling, walking, and public transport—depending on their needs and the context of their trips [1, 2]. Urban transportation systems are evolving into complex and lively networks. This shift not only promotes sustainable living but also broadens the options for personal mobility. To manage the various types of transportation effectively, there's a need for advanced, interconnected systems that can tackle a wide array of urban mobility challenges. In this framework, artificial intelligence (AI) and data analytics become essential concerns. Cities are tapping into massive amounts of mobility data by combining GPS technology, traffic sensors, public transport setups, and smartphone apps. By analyzing this wealth of information, it's possible to spot mobility trends, pinpoint inefficiencies, and guide improvements in transportation systems based on solid data.

AI technologies, particularly machine learning and deep learning, build on these intuitions to enable projecting modeling, real-time decision-making, and optimization of transport operations. Applications range from smart traffic signal control and demand prediction to route optimization and dynamic public transit management. These novelties not only progress operational efficiency but also contribute to more receptive, ecologically friendly transportation systems. In this framework, the incorporation of smart transport structures with AI and data-driven perceptions is becoming indispensable for determining the future of smart, sustainable urban agility [3]. In general, the primary goal of a smart transport structure is to report both direct and indirect transportation contests in the most well-organized means, eventually enhancing the well-being of persons and society as a whole. While the notion of smart transport is extensive, intelligent transport systems (ITSs) precisely highlight the part of technology in attaining this goal. These systems are envisioned to be adaptive to rising surroundings. By leveraging information and communication technologies, ITSs enable continuous incorporation across different mechanisms of the transport system, permitting the effective administration and determination of problems in near real-time. They are characterized by automation, accessibility to demand, and real-time functionality.

ITS influences a wide range of technologies to monitor, assess, evaluate, and accomplish transportation systems, with the main purpose of enhancing efficiency and safety. Rather than trusting only on new infrastructure, ITS focuses on optimizing the use of existing systems through tools such as traffic signal control, journey planning platforms, smart ticketing, and cooperative systems [4, 5]. While ITS certainly centers on the incorporation of smart technologies within transport infrastructure and vehicles, its scope is not limited to these elements. Approximately speaking, smart transport encompasses any technology that helps efficient transportation management. According to Shaheen and Finson [6], ITS combines a wide range of technologies, such as electronics, data processing, wireless communication, and control systems are designed to improve the safety, efficiency, and convenience of surface transportation. Likewise, another description [7], ITS is an innovative transportation system that intends to address issues like traffic congestion and accidents by connecting people, roads, and vehicles through sophisticated communication and information technologies.

Figure 13.1 Information and communication technologies.

The development of Internet of Things (IoT) has further extended the abilities of ITS by enabling continuous communication between an increasing number of internet-connected devices. Originally, ITS development focused on embedding devices including sensors and cameras, into infrastructure and vehicles. However, with IoT, mobile devices like smartphones and GPS navigation systems now offer a more accessible and cost-effective alternative to fixed infrastructure-based sensors. These mobile sources provide valuable real-time information that supports dynamic transport management tasks such as congestion mitigation, traffic rerouting, public transportation scheduling, and accident response (Figure 13.1) [8].

In ITS, the part of the information is vital. Efficient operation relies not only on the availability of information from diverse sources but also on the timely delivery of that data. Real-time information is important for making immediate decisions that affect transportation movement and safety. ITS fully exploits big data produced by intelligent devices, whether embedded within the system or carried by users, to facilitate responsive and adaptive transport management.

The integration of bigdata analytics and machine learning has opened up new possibilities for enhancing the urban transportation systems. Predictive analytics facilitates improved travel demand forecasting while prescriptive analytics makes it possible for decision-makers to formulate optimal solutions in real-world settings. The innovations enhance efficiency in operations and help in more strategic objectives such as mitigating greenhouse gas emissions and improving urban infrastructure [9, 10]. One of the most capable frontiers in this transformation is the amalgamation of AI into urban transport systems. With its capability to process huge volumes of data, spot patterns, make forecasts, and optimize processes in real time, AI has developed as a powerful enabler of smart, adaptive, and competent transportation networks. When combined with data analytics, AI delivers cities with the tools they need to appreciate mobility patterns, anticipate forthcoming demands, and design transport systems that are both user-centric and environmentally sustainable.

The notion of "smart transport" denotes the use of cutting-edge technologies to improve the efficiency, safety, and sustainability of transportation systems. It covers a wide variety of innovations, including intelligent traffic management, prognostic maintenance of infrastructure, autonomous vehicles (AVs), real-time public transit monitoring, and dynamic routing applications. At the core of all these innovations lies data collected from sensors, GPS devices, mobile applications, traffic cameras, and IoT devices. Smart transport remains "smart" not only through the data itself, but also in how it analyzes and acts upon it. This is where AI-driven data analytics plays a critical role. By leveraging machine learning, deep learning, natural language processing, and other AI methods, city planners and transport authorities can transform raw data into actionable insights. These intuitions can then be used to make informed decisions that improve traffic flow, reduce travel times, enhance traveler safety, and promote the use of eco-friendly transport replacements. Traditional transport training and managing consume frequently trusted on static models and historical data, which lack the ability to adapt to real-time fluctuations and developing trends. In contrast, AI-driven analytics enables dynamic decision-making, allowing systems to learn from past patterns and adjust to new conditions on the fly. For instance, machine learning algorithms can be trained to predict traffic congestion based on time of day, weather conditions, and local events. Similarly, deep learning models can process video feeds from traffic cameras to detect accidents or identify vehicles violating traffic rules.

By analyzing user behavior and preferences, AI can acclaim best travel routes, foresee arrival time, and even propose another means of transport based on traffic situations and environmental effects. This level of personalization improves the experience of end users and also encourages the embracing of smarter, more sustainable commuting options. In addition, AI has a significant part to play in the operation and progress of AVs. Automated vehicles belief on a combination of sensors, cameras, and AI algorithms to direct urban environments without harm. These systems necessitate persistent understanding of data from their surroundings, envisage the movements of other road users, and make split-second choices.

Today urban transportation systems create huge amount of data every second, from ticketing systems, ride-sharing platforms, smart traffic lights, and mobile apps to sensors embedded in roads and vehicles. The capacity to collect, store, and analyze this data efficiently is significant to considerate how urban transport systems function and where enhancements can be made. Data analytics, empowered by AI, makes it probable to notice inefficiencies and advance positive solutions. For example, prognostic analytics can be used to recognize when a bus is likely to arrive late and adjust schedules consequently. Clustering algorithms can segment travelers into groups based on commuting patterns, letting for targeted enhancements in services. Sentiment analysis on social media and public feedback platforms can help authorities know commuter fulfilment and categorize pain points in the system.

Moreover, AI can facilitate long-term urban planning predictions. By simulating scenarios and analyzing past trends, AI algorithms can forecast future transportation needs, allowing city planners to invest in infrastructure with more assurance. Whether it's forecasting the impact of a new subway line or evaluating the feasibility of converting an avenue into a pedestrian-only zone, AI provides the analytical heavy lifting needed to model complex, data-intensive scenarios. While the potential of AI-powered data analytics in urban mobility is vast, it also presents some challenges. One of the biggest challenges is data quality and integration. Data within urban transport systems are typically disintegrated, coming from a number of different sources which may be incompatible and inconsistent with each other. Data accuracy, completeness, and consistency are crucial for building reliable AI models.

Another major challenge is privacy and security. As AI systems progressively depend on data from users such as location, travel history, and preferences, there are legitimate concerns around how this information is collected, stored, and used. Building hope with users needs strong data governance frameworks, transparency in data practices, and adherence to privacy regulations. Furthermore, deploying AI technologies in transport system involves important investment in terms of rate, time, and technical expertise. Smaller cities are those in developing regions may lack the resources or infrastructure needed to fully influence AI capabilities. In addition, AI systems are only as good as the data they are trained on. If the data is biased or incomplete, it can lead to suboptimal or even discriminatory outcomes, something that needs to be carefully managed. Contempt the challenges, the impending of AI-facilitated urban transport is highly capable. As AI technologies are established and turn out to be more manageable, we can expect to see superior integration of real-time data analytics into normal transport operations. Evolving trends such as edge computing, 5G connectivity, and blockchain will further augment the competencies of smart transport systems.

Among the most thrilling trends is the transition to mobility-as-a-service (MaaS), with users able to plan, reserve, and pay for various modes of transport service using a single online platform. AI has a key role in turning MaaS into reality through dynamic pricing management, route optimization, as well as user experience personalization. Another major trend is the application of digital twins—computerized replicas of physical systems—to simulate and model transportation networks in real time. These virtual environments, fueled by AI and real-world data, can be utilized to pilot test new policies, predict traffic patterns, and assess the effect of urban development projects without interrupting actual operations. As cities develop further, the combination of AI and data analytics will not only be an added value but a requirement. Whether it is minimizing traffic congestion, enhancing safety, or fostering green mobility, AI provides a scalable and flexible solution to some of the most critical issues in urban transport. Figure 13.2 shows a congested city road during a highly polluted sunset, depicting the widespread urban congestion. The heavy traffic, coupled with high emissions from the vehicles, chokes the air with a heavy haze that hides the cityscape in the background. The golden color of the sun setting provides a dramatic contrast, bringing out the

Figure 13.2 Urban traffic congestion and emissions: motivation for AI-driven transportation optimization. *Source*: Generated with AI using StockCake.

environmental cost of everyday urban commutes and the urgent need for sustainable transport options [11, 12].

The remainder of this chapter focuses on the intersection of Data Analytics and AI in the context of urban transportation. The aim is to highlight how their integration supports smarter mobility solutions, enhances operational efficiency, reduces environmental impact, and contributes to the broader goal of sustainable urban development.

13.2 AI-powered Data Sources in Urban Transport

AI has become a powerful tool for shaping the future of urban transportation. But to make smart decisions, AI needs, data lots of it. The real value of AI lies in how well it can learn from and respond to the information it receives. Today's cities are abounding with systems and technologies that are capturing huge quantities of data per second, ranging from how fast vehicles are traveling to the number of individuals queuing at a bus stop. These data sources, when merged and analyzed by AI, can significantly enhance how cities regulate traffic, plan new transport routes, and handle day-to-day challenges on the road. AI is transforming urban transport by enhancing efficiency, sustainability, and smart city planning. It enables intelligent traffic management, autonomous cars, and predictive analytics to deal with congestion, pollution, and inefficiencies. AI also enables sustainable transport planning and the adoption of electric and hybrid vehicles.

Figure 13.3 provides a concise outline of its contribution to optimizing city mobility through the use of AI. Urban transport data analytics encompasses the utilization of various data sources (structured, semi-structured, and unstructured) to enhance planning and mobility. Traffic sensors generate real-time data on vehicle congestion, speed, and flow, whereas GPS and ticketing information from public transportation systems provide insight into usage. Ride-hailing apps present good data concerning travel behavior and demand. Smartphone integration adds location and movement, which helps tracking movement patterns. CCTV and video analytics support observation of traffic as well as event detection. Social media and surveys present public perceptions, helping with user satisfaction levels and preferences quantification. Smart transportation systems

Data source	Description/application
CCTV and video surveillance	AI analyzes traffic flow, detects incidents, and identifies patterns in real-time.
Connected vehicles (V2X)	Vehicles communicate with infrastructure. and each other to optimize traffic and safety.
Smart traffic signals	AI adjusts signal timings based on real-time traffic data to reduce congestion.
Social media and crowdsourcing	AI uses NLP to extract feedback, incident reports, and public sentiment.
Smart cards and ticketing systems	Provides data on public transport usage, peak travel times, and popular routes.
IoT sensors. on roads/ vehicles	Includes speed detectors, pressure sensors, and pollution monitors for real-time updates.
Drones and UAVs	Capture aerial views for traffic monitoring, construction sites, or emergency assessments.
Weather data systems	AI uses weather forecasts and real-time data to adjust transport operations.
Infrastructure embedded sensors	Detects road surface wear, traffic load, and structural health of transport infrastructure.

Figure 13.3 AI-powered data sources in smart urban transport.

(STS) intend to minimize carbon emissions [13], traffic jams, and improve reliability, efficiency, and safety in transport systems. This needs effective application of sustainable approaches, with information and technology playing an important part. STS addresses three core domains: management and operation of transport systems, guidance and information to users, and management and operation of freight transport systems. STS employs new technologies such as smart cells, ICT, and advanced infrastructures such as smart cars, UAVs, and base stations to improve city living by guaranteeing smooth convergence between system use and user demand for increased convenience, efficiency, and safety. STS contains vehicle-to-vehicle (V2V) and vehicle-to-infrastructure (V2I) keys. V2V enhances travel safety by enabling advanced emergency braking systems, while V2I uses sensors to detect traffic variables [14]. V2V applications alert drivers to imminent crashes, while V2I applications alert drivers to school zones, roadside workers, and traffic light changes. Both components provide enhanced safety measures. A prototype project of an adaptive traffic signal control system constructed for a nine-intersection road network in Pittsburgh, Pennsylvania (USA), known as SURTRAC (Scalable Urban Traffic Control), was examined by Smith et al. [15]. The SURTRAC technology decreased emissions by more than 20% and increased traffic flow efficiency by 25–40%.

Within the broader context of intelligent transportation systems, intelligent parking systems have emerged as central features of the smart city model, addressing the widespread malady of urban traffic congestion and unproductive space utilization. Apps such as ParkMobile, SpotHero, ParkMe, ParkWhiz, and others are transforming driving into parking in ever-more crowded metropolitan areas. These apps utilize real-time information and IoT systems to enable users to find vacant parking spaces, compare prices and distances, guide them to vacant parking spaces, and even book in advance. They also tend to offer functionalities such as parking timers, expiration alerts, and parking history records, all of which make for a more efficient and hassle-free urban mobility experience. By integrating with local law and offering reduced rates or entry to reserved unsold parking spaces, such systems reduce traffic caused by drivers searching for parking and boost city sustainability overall. As an element of the smart city utopia, such clever parking schemes not only raise individual convenience but also serve toward larger-level aims such as reduced emissions, smoother traffic flow, and better land use planning.

13.2.1 Intelligent Infrastructure and IoT Sensors

Among the chief sources of transport data are IoT sensors embedded in the various components of the city. These smart sensors are installed in traffic lights, roads, and even buses or trains. They work quietly in the background to monitor what's going on—whether it's a traffic jam developing, a road becoming slippery with rain, or a car running behind schedule. This real-time feed of information helps AI systems make quick decisions, such as changing traffic lights to clear congestion or alerting to hazardous road conditions. The sensors, in a way, play the role of the nervous system of the city, with continuous feedback made by the nervous system, turning the entire transportation infrastructure responsive and efficient.

13.2.2 GPS and Location-based Data

Another important data flow is from GPS. Whether a city bus, a delivery truck, or someone's phone, GPS provides an ongoing report on where things and people are headed throughout the day. AI takes this information and applies it to track travel patterns, observe which roads are busiest at certain times, and suggest better routes to reduce delays. In the long term, this location data also helps planners decide where to build new bus stops or where extra services might be needed. It's as if it's a live map of how the city breathes and moves throughout the day.

13.2.3 Mobile Applications and User Behavior

Apps that we use to navigate, like ride-hailing apps, transit apps, and map apps, are yet another gold mine of information. Every time a person searches how long before the next train or books a shared ride, they leave behind a digital trail that can be learned from by AI. Such data tells when and where people travel, for how long they wait, and along which routes. AI programs can pick up on such tendencies and provide better suggestions, improve timing, and even provide recommendations for changes to services by what individuals actually need. Such observations translate individual app interactions into informative trends that have the potential to influence smarter city mobility.

13.2.4 Smart Ticketing and Fare Data

Now, most cities have switched to smart ticketing systems based on cards, QR codes, or mobile payments. Every time a traveler touches in or out of a bus, train, or subway stop, that small movement adds up to an increasing body of travel behavior data. That data can be analyzed by AI to learn when the public transport network is most crowded, what routes are most traveled, and how often passengers make a transfer between services. Understanding these patterns helps transit systems coordinate schedules, prepare for peak hours, and even predict future demand. Fare data is also utilized in identifying fare cheating or payment errors so that prompt action can be taken and system reliability can be improved.

13.2.5 Video Surveillance and AI Vision

Urban areas are full of CCTV cameras, and thanks to advances in AI, these cameras are no longer just tools for security. With computer vision technology, AI can "watch" traffic footage and understand what's happening in real time. It can detect when an accident occurs, spot cars breaking traffic rules, or count how many pedestrians are waiting at a crossing. All of this can be done without human involvement, saving time and improving safety. Over time, this visual data also reveals patterns, like places that are accident-prone or intersections that need redesigning to improve traffic flow.

13.2.6 Social Feedback and Public Media

The best insights can sometimes come directly from individuals who work with the system on a daily basis. It might be a status update of a late bus or a complaint posted on a transit app, but such messages contain useful information. Natural language processing can allow computers to read hundreds of public comments, reviews, or social media posts to determine what people have to say about their transit experiences. This kind of feedback is helpful to pick up on problems the raw data might miss like uncomfortable conditions, driver habits, or misleading signs, and give city officials a better idea of what needs to change.

13.2.7 Connected and AVs

As vehicles advance, they're also becoming smarter communicators. Newer vehicles are now capable of sending data to and receiving it from one another as well as traffic infrastructures in real-time. When it comes to AVs, that includes such things as camera streams and radar pulses as well as road condition and pedestrian information. It is all information that matters, not just for the purpose of allowing individual cars to drive safely, but also in order to build a bigger picture of what is happening on the roads. AI systems are able to tap into this information in order to predict traffic patterns, make roads safer, and prepare cities for a future where autonomous transport has an even greater part to play in our daily lives.

13.2.8 Weather and Environmental Conditions

The weather has an important bearing on how one travels across a city. Rain, heat, fog, or air pollution may affect commuter trends and the service quality of the transport. Weather sensors and monitoring stations now supply information to AI systems to guide transportation operations according to it. This means that services can be real-time adjusted during a storm, routes can be recommended as alternatives to avoid flooding, or extra buses can be sent if pedestrian conditions become nasty. With environmental data, transport systems can stay robust and responsive, despite unpredictable conditions.

13.2.9 City Maintenance and Infrastructure Data

Behind the scenes, each city maintains records of its roads, bridges, and transit systems—when they were constructed, when they were most recently repaired, and when they need work in the future. AI can draw on this data to assist in predicting when something is likely to break down or where the next maintenance should be scheduled. It prevents disruptions from happening. For example, if a bus route runs along a part of the road scheduled for maintenance, AI can suggest detours in advance. It's a smart, preventive measure that keeps the city moving without abrupt shutdowns or waits.

13.2.10 Government and Agency Open Data

More and more cities are embracing transparency by making transport-related data sets available on the web. These open data sites usually have traffic statistics, accident history, bus timetables, etc. When such open data sets are made available to AI, it can assist researchers, developers, and planners in constructing solutions and tools for actual urban challenges. Such an open culture facilitates collaboration and innovation and makes it simpler to build systems that get people around better apps, smarter traffic signals, or cleaner transport.

13.2.11 Statistical Insights on Urban Transport

Some Numerical Insights in Data Analytics for Urban Transport:

13.2.11.1 Development in Urban Transport Information

IDC DataSphere, 2021 [16] revealed that the global urban data sphere is expected to grow from 79 zettabytes in 2021 to 181 zettabytes by 2025, owing to cumulative embracement of IoT devices and smart city initiatives.

13.2.11.2 Traffic Mobbing Costs

From the INRIX Global Traffic Scorecard, 2021 data, traffic mobbing costs urban economies about $87 billion annually in the United States, with travelers losing an average of 54 hours per year wedged in traffic.

13.2.11.3 Implementation of Intelligent Transport Systems

The ITS market is anticipated to reach $43.5 billion by 2027, mounting at a CAGR of 10.9% from 2020 to 2027 [17].

13.2.11.4 Public Transport Use and Optimization

Application of AI-based optimization in urban bus routes leads to a 15–20% drop in travel time in cities like Singapore [18].

13.2.11.5 Influence of Real-time Analytics

Real-time traffic flow data diminishes delays by up to 30% in peak hours when combined into adaptive traffic signal systems [19].

13.3 AI Techniques for Urban Transport Analytics

As cities become more complex and crowded, managing urban transportation has become one of the biggest challenges for city planners and governments. Traffic congestion, delayed buses, unpredictable travel times, and uneven access to public transport are just a few of the daily struggles faced by commuters. This is where AI takes over. AI doesn't only process information, it learns, adjusts, and makes decisions to assist in making transport systems intelligent and efficient. These AI approaches drive everything from forecasting traffic congestion to optimizing bus timetables and even navigating self-driving vehicles through urban roads.

13.3.1 Predictive Modeling and Machine Learning

One of the most familiar mechanisms that AI assists cities with regarding transport is machine learning. In essence, machine learning is about educating computers to learn patterns from data and make predictions based on that. For example, if a system has accessed traffic data for the past 5 years, it can start to forecast where and when congestion will happen. These predictions allow for city planners to maximize traffic lights, suggest detours, or even introduce campaigns to reduce usage of the roads during peak times. In like manner, machine-learning algorithms can monitor how buses and trains previously were delayed and provide suggestions on how to adjust routes or timing, so they become more reliable. Over time and with more information inputted in the system, its predictions become more precise and useful. In traffic prediction or travel time estimation, models like linear regression are used:

$$\text{Travel Time} = A + \text{Traffic Volume} + \text{Weather} + \text{ToD} + B \tag{13.1}$$

This formula predicts travel time based on variables like traffic volume, weather, and time of day. The AI model learns the coefficients TV, W, ToD from historical data. Here, A and B are the intercept and the error term residual.

13.3.2 Deep Learning in High-density Urban Areas

For more complex cases like the detection of the movement of thousands of cars and people in a large city, AI needs more advanced instruments. That is where deep learning is involved. Deep learning relies on neural networks that act generally like a compact human brain. It excels particularly well at dealing with humongous amounts of data, e.g., videos captured by surveillance cameras, GPS outputs, and sensor data. To illustrate, real-time violation detection from CCTV video or inspection of pedestrians' trajectories in rail stations to prevent jams would be facilitated by deep learning. Due to the ability of deep learning to accommodate many variables simultaneously, it's perfectly suited for deciphering the disorganization one commonly finds within metropolitan areas [20].

13.3.3 Natural Language Processing for Public Feedback

Another method that's creating a big splash is natural language processing (NLP). This method allows AI to understand and analyze human language, something that is especially useful when dealing with public comments. People often leave their feedback on transport services within social media posts, internet reviews, or customer service messages. NLP helps sift through this unstructured data and ascertain what citizens are talking about, complaints about delays, compliments on new bus routes, or suggestions for better service. By examining trends in this feedback, city planners can make more human-centered decisions and address problems before they become bigger problems.

13.3.4 Computer Vision in Real-time Observation

AI is also assisting cities to "see" what's going on in the streets. Computer vision methods enable AI to analyze images and video from cameras mounted at intersections, highways, and transit stops. This makes it possible for systems to automatically identify accidents, spot cars that are running red lights, or track how many pedestrians are crossing a crosswalk. The benefit of computer vision is that it turns visual data, something humans usually have to interpret, into actionable information in real time. By this, cities will be able to react faster to accidents and strategize road improvement better where it's necessary.

13.3.5 Traffic Optimization Through Reinforcement Learning

Another more recent method that is being studied is reinforcement learning. This form of AI will learn by doing something different and seeing what actions produce the best outcomes like a human will learn from trial and error. Reinforcement learning is utilized in city transport to manage things such as traffic lights or the movement of self-driving cars. For instance, an AI system can try different traffic light timings and see which combination serves to minimize waiting times and maintain traffic flow. With time, it discovers the optimal strategy to balance traffic flow and reduce delays. It's a fascinating field that holds the key to transforming the way we deal with city traffic systems.

In AI traffic control, the goal is to maximize a reward (like traffic flow). This is formalized as:

$$Q(s,a) = R(s,a) + D_{max}\ Q(s',a') \tag{13.2}$$

This is the **Bellman equation**, where:

- Q(s,a) = expected reward of taking action a in state s
- R(s,a) = immediate reward
- D = discount factor
- s′,a′ = next state and action

This helps AI learn the best timing strategies for traffic lights.

13.3.6 Pattern Recognition and Clustering

Another useful AI technique in urban analytics is clustering. Clustering can be applied to cluster similar kinds of data. For transportation, clustering would be applied to detect neighborhoods that have the same pattern of commuting or to detect groups of roads with the same kind of congestion during the same times. By acknowledging these trends, planners are able to make specific changes, such as adding more bus routes where there is the highest demand or upgrading roadways in areas with long-term traffic issues [21]. It's a question of fitting together puzzle pieces to achieve the overall picture of how a city moves.

13.3.7 Simulation and Scenario Modeling

AI tends to be applied to mimic various "what if" situations. These allow the city planners to experiment with new changes in virtual space before deciding to implement them in real time. For example, AI will be able to mimic what it would look like if a central road were to be closed or if a new metro line were constructed in an area. It can then project how such changes would affect traffic, commuter patterns, and transport efficiency as a whole. This renders decision-making much safer and more informed, particularly for large projects that require a lot of investment and planning.

13.3.8 The Data Analysis Tool for Urban Transport

Some key statistics related to urban transport: The study analyzes key urban transport performance metrics in City A, compares data from similar cities, identifies the best performer, averages the indicator across peers, and provides comments (Table 13.1; Figure 13.4).

Table 13.1 The data analysis tool for urban transport.

Indicator	City A (under Study)	Peer city 1	Peer city 2	Peer city 3	Best performer	Benchmark (avg.)	Comments
Public Transport Usage (%)	55%	62%	48%	70%	Peer City 3	60%	Performing slightly below avg
Avg. Commute Time (min)	45	40	50	35	Peer City 3	42	Higher than benchmark
Emission Levels (CO_2/m^2)	120	100	110	90	Peer City 3	100	Requires significant improvement
Road Safety (Accidents/yr)	250	200	220	150	Peer City 3	200	High accident rate
Cycling Infrastructure (%)	10%	25%	20%	30%	Peer City 3	25%	Lags behind significantly
Fare Affordability Index	3.0	2.5	2.8	2.0	Peer City 3	2.6	Fares are less affordable

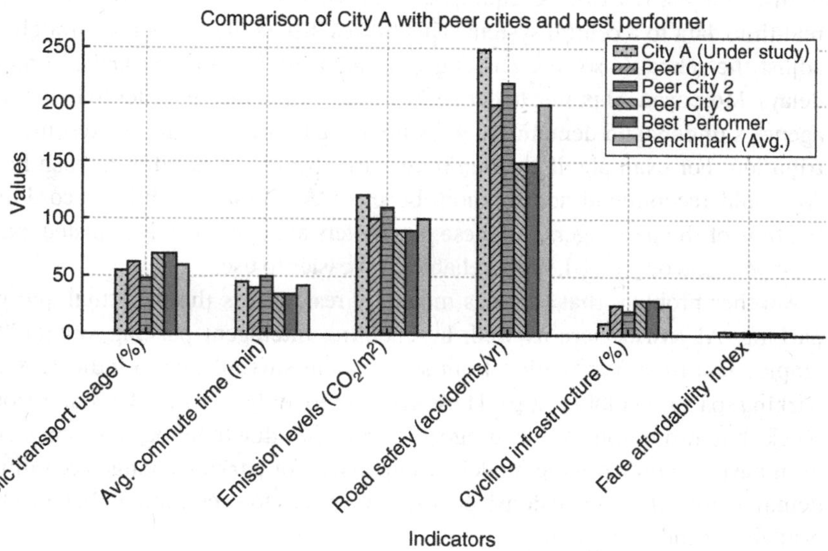

Figure 13.4 Comparative analysis of city A with peer cities.

13.4 Key Applications of AI in Urban Transport

AI is increasingly leading the way in changing the modernization of city transport. As cities grow and urban populations grow, pressure on public transport infrastructure and systems also grows. To meet these demands, many cities around the world are turning to AI technologies to improve everything from traffic flow to passenger experience. Most remarkable, however, is how seamlessly AI is being deployed—not only in high-tech gadgets like driverless cars, but in practical, behind-the-scenes applications that make daily commutes bearable and cities more livable. Perhaps the most visible use of AI in urban transportation is real-time traffic control [22]. The old way is based on fixed schedules or manual, ad-hoc adjustments by human dispatchers. AI can use real-time information from road sensors, GPS, and cameras, however. It uses that information to make intelligent, real-time changes, adapting to reality. As an example, traffic signal times could be programmed automatically to de-congest intersections by considering the live vehicle load in the current intersections. The amazement here is that the systems not only respond, but also learn. They look at trends over time, like morning rush hours or weekend traffic lows, and make adjustments accordingly, streamlining the movement of cars with minimal human input.

Not just cars owned by individuals are being implemented with AI. Public transport systems can also benefit greatly from using AI-based solutions. Buses, trains, and metros can be equipped with sensors and GPS modules to offer real-time data to a central system. This makes it possible to track accurately and adjust the schedule so that passengers remain informed about arrival times or delays. Moreover, AI is able to decompose ridership trends in order to help transit agencies understand demand, thereby allowing them to streamline resource distribution. For example, if the bus routes are always too full during rush hours, AI would recommend adding more buses or adjusting schedules according to the flow of the passengers. As these parameters are dynamically adjusted, public transport gets optimized, more reliable, and easier to use.

Another problem that troubles most city residents is the perpetual quest for parking. AI assists here as well, by enabling intelligent parking systems that employ real-time information from sensors and surveillance to indicate vacant parking spaces on mobile apps. These systems not only save time traveling around blocks but also minimize fuel usage and emissions due to idling cars. Some cities even have dynamic pricing models where prices for parking change according to demand, and AI is used extensively to adjust prices to keep them in balance with availability and affordability.

AI also has a presence in AVs, perhaps the most widely discussed innovation in transport. Although self-driving cars are not yet complete, most cars now incorporate AI in the form of features like lane detection, emergency braking, adaptive cruise control, and pedestrian awareness. All these systems operate on constant streams of data from cameras, radars, and other sensors to make sense of the driving context and take real-time decisions. Apart from the vehicle itself, AI facilitates the coordination of vehicle and traffic infrastructure communication. V2X communication technology allows cars to share information on road conditions, speed, and hazards so that there can be safer and more coordinated movement of traffic through entire networks.

Planning and forecasting are two more domains for which AI is invaluable. City planners can now use forecast models so that they can forecast how mobility demand in cities will change in the future. AI can forecast surge demands or locate areas where the availability of transportation could be constrained in the near future based on historical data, weather conditions, urbanization, and even events at the city level. These help in route optimization, network expansion, or new service introduction prior to issues becoming epidemic. In shared mobility platforms like bike-sharing and ride-hailing, AI facilitates pre-positioning of cars or bikes according to anticipated demand [23]. This implies that consumers have a higher chance of getting what they are looking for at the desired time.

On safety issues, AI is an extremely useful tool. Intelligent surveillance systems can detect unusual activity on highways, like a car going in the opposite direction or sudden braking that could indicate a crash. These incidents are reported automatically, allowing officials to respond quickly. AI can further study traffic accident reports to identify dangerous intersections or highways where improvements are required. AI in public transportation can even track how distracted or sleepy drivers are and issue warnings before a dangerous situation arises. Cities are acting to prevent accidents rather than just responding to them, thanks to such input.

Another area where AI is helping cities make a real difference is in the area of environmental impact. Transport is one of the largest sources of urban air pollution, and AI systems can help cut it. For example, by optimizing traffic light timing and eliminating unnecessary stops, AI can cut fuel consumption and emissions. Others monitor air quality in real time and respond by restricting vehicle use within polluted zones or recommending alternate means of transportation. Also, on electric public transportation systems, AI can streamline the use of batteries and charging schedules in a logical way so buses and trains stay on schedule without wasting energy.

An additional subtle but important benefit of AI in urban transport is personalization. Apps for commuters use machine learning today to provide individualized suggestions tailored to users' habits and inclinations. Be it finding the fastest way to work, avoiding construction zones, or recommending nonpeak times, AI helps commuting become more personalized and less stressful. There are certain apps that even go on to analyze the crowding situation in real-time and suggest where in the train carriage or bus compartment will be least crowded, so making the trip happier and more convenient.

These applications work in concert to introduce a more integrated, intelligent urban transportation system. AI doesn't just automate, it enhances the capacity of the system to learn, adapt, and make decisions in the best interests of individual riders and society as a whole. Although most of these technologies are working behind the scenes, their impact is real and growing every day. As cities are digitalizing their infrastructures and developing smarter policies, the future of how we get around will be increasingly dependent on AI.

13.5 Case Studies and Real-world Implementations

As cities around the globe embrace innovation in urban transport, several significant implementations have emerged that underscore the extensive impact of technology on improving mobility. These real-world examples not just demonstrate the viability of more intelligent transport systems but also present valuable lessons in the challenges and benefits of embracing advanced solutions [24]. One of the best examples is Barcelona, Spain, which has a city-wide traffic management system implemented in order to curb traffic and promote mobility overall. Its Smart City Program uses smart traffic lights and sensors embedded into road infrastructure to monitor and manage traffic patterns. It operates based on data collected from cars, cameras, and other traffic monitoring devices in real time. By changing the cycles of traffic lights and diverting automobiles, the system can reduce congestion at peak hours and respond dynamically to road accidents or unexpected fluctuations in road conditions. Importantly, the program has reduced congestion and air pollution and improved the quality of life for residents and commuters.

In Singapore, which has been a leader in smart transport solutions for many years, another outstanding integration of an intelligent traffic management system exists. The land transport authority (LTA) has put in place a system that uses real-time data to manage traffic flow across the city. Sensors that are installed along the road network track traffic conditions, while the central system instantly adjusts traffic lights to ensure there is a smoother flow of traffic. Singapore has also introduced an electronic road pricing (ERP) system, where intelligent technology tolls cars by time of day and traffic volume. Not only does this encourage drivers to take alternative routes or times, but it also generates revenue to invest in further transport infrastructure development. It has been proven to alleviate congestion and enhance the effectiveness of public transportation, creating a healthier urban mobility model.

Another interesting case is New York City, whose public transport system has been under the spotlight with attention to upgrading it by making it safer and more efficient through technology.

The city launched an AI-based platform in 2019 with the goal of reducing wait time for buses and improving public transport reliability. The site collates data from various sources, including traffic movement, passenger demand, and weather forecasts, in a bid to provide real-time information to passengers. Clients can access up-to-date information regarding bus timing, delay, or route modification through cell phone apps. Moreover, the New York City Taxi and Limousine Commission partnered with tech companies to roll out a system that uses GPS tracking and data analysis to deploy taxis in ways that make them available in zones of high demand. This has managed to reduce idle times, making taxis easily available to commuters and increasing the efficiency of the transport system within the city. The platform aggregates data from various sources like traffic flow, passenger demand, and weather conditions in a bid to provide real-time information to commuters. Commuters can obtain real-time information on bus schedules, delays, or route adjustments from mobile applications. Furthermore, the New York City Taxi and Limousine Commission collaborated with tech firms to launch a scheme based on GPS tracking and analytics to position the taxis in their best locations in a way that makes them more accessible in heavily populated areas. This has ensured idle time reduction, and it becomes convenient for the taxis to be used by travelers, and also boosts the efficiency of the city transport system. The platform aggregates data from various sources like traffic volume, demand from passengers, and weather to provide real-time updates to commuters. Passengers can have up-to-the-minute reports on bus timings, delays, or route diversions through mobile apps. In addition, the New York City Taxi and Limousine Commission collaborated with tech firms to launch a system utilizing GPS location and data tracking in order to position taxis better in areas where they are most in demand. This has successfully lowered idle times so that taxis are more easily found by customers, and the transportation network of the city is enhanced in efficiency [25].

London's Transport for London (TfL) system has in London used technology to transform its public transport and highway management systems. The most notable innovation has been the use of contactless payment across all transport services within the city. Passengers can now pay for travel with debit cards, credit cards, or even mobile phones, making it convenient to use and reducing long queues at ticketing machines. TfL also implemented an in-real-time bus tracking mechanism based on information from GPS tracking devices to measure and forecast the time of bus arrival at predetermined stops. It enables passengers to monitor their buses in real-time, removing doubt and enhancing customer satisfaction. Smart parking has also been launched in the city, wherein sensors located in parking spaces identify whether a space is taken or available. It sends this information to a mobile app, which aids in guiding drivers toward parking sooner and eliminating redundant traffic in dense areas.

Amsterdam is implementing electric and AVs into its city transportation system, investing heavily in EV infrastructure and installing numerous charging points throughout the city. Amsterdam has also been testing AVs in controlled environments with close collaboration with firms that specialize in autonomous technology. These steps are aimed at reducing emissions from traditional gasoline-powered vehicles and creating a cleaner city transportation system. Focusing on promoting eco-friendly means of transport, Amsterdam's efforts are making the future greener and urban mobility quality higher in general.

In the city of Los Angeles, known for congestion, the spotlight is now on traffic flow improvement and minimizing congestion. LA Metro, Los Angeles's public transport system, has integrated a smart transport system (ITS) that maximizes the flow of buses and trains on the roads. The system makes use of sensors, cameras, and sophisticated software to observe traffic and optimize signal timing to maintain public transport at a specified velocity. Uber and Lyft also use AI algorithms in their ride-hailing service to match riders and drivers effectively based on real-time traffic feed, location, and demand. Not only do these algorithms reduce waiting times for riders, but they also help drivers avoid traffic-congested roads, making the operation more efficient and reducing emissions [26].

In Hong Kong, the incorporation of smart mobility solutions has been key in addressing the issues posed by a concentrated and space-constrained population. Smart bus stops, which provide information on real-time bus arrival time and delay updates, have been installed in the city. This is shared through digital boards in the bus stops as well as through a mobile application. Hong Kong also boasts a mobility-as-a-service (MaaS) platform that combines public transport services with private ride-hailing services, which enables users to plan their routes and pay for different types of transport using one platform. This multimodal mode of transportation makes users' access to different modes of transport more convenient and increases the efficiency of the transport system within the city.

These case studies show how cities are now increasingly drawing on technology in a bid to keep up with the various challenges of city mobility. From traffic control to optimizing public transport and sustainability initiatives, the practical applications of such systems are already making a telling impact on how humans navigate urban environments. While there are realities to be overcome—data privacy concerns, infrastructure expense, and higher usage rates—the circumstances illustrate that technology and AI can yield effective, realistic solutions for smarter urban transit systems. As cities become more advanced, the contribution of technology to forging the transport of the future will become more paramount, offering alternative means to remain green, efficient, and accessible [27].

13.6 Challenges and Ethical Considerations

Though the future of AI revolutionizing urban transport is imminent, it has its downsides and ethical dilemmas. While cities across the globe begin to welcome these technologies, the time has come to halt and observe what challenges must be addressed. Right from technology-related limitations to societal and ethical issues, all of them must be tackled so that AI is used in the interest of the people and not just affluent segments.

13.6.1 Technical Challenges: The Complexity of Urban Infrastructure

One of the major challenges to addressing the use of AI in urban transportation is that the infrastructure itself is intricate. Cities are not homogeneous, and the environment of each city presents a different set of challenges. Traffic flow, use of public transportation, and infrastructure condition differ dramatically from city to city. AI systems, as powerful as they are, must be adapted to each city's particular issues, and that calls for massive customization and adjustment. For example, certain cities could have older infrastructure of roads or sensor data that's patchy, so it becomes more difficult for AI algorithms to make precise predictions or real-time adjustments. While AI can process lots of data, it's not perfect. Where traffic conditions shift quickly because of incidents or weather, AI systems might not be able to adapt fast enough, sometimes causing delays or inefficiency. Furthermore, urban mobility uses multiple transportation modes, from automobiles to buses to bikes, and unifying them all in one comprehensive system is no easy task. AI must sort through data from all these various systems and make real-time decisions, ensuring that everything flows together seamlessly. This amount of coordination is difficult to do and requires ongoing fine-tuning and complex systems that are still evolving in most cities [28].

13.6.2 Data Privacy and Security: Safeguarding Personal Information

With AI systems in urban transport aggregating massive amounts of data—from travel behavior to real-time positioning—the concern regarding what is being done with this data has grown. Most notably, the challenge of data privacy is a key one. Most of these AI-based systems depend on harvesting personal data to provide a more tailored experience. For instance, ride-hailing applications monitor users' locations to connect them with drivers nearby, and intelligent parking applications keep track of where and when someone parks. While this information is necessary to make services better, it also brings serious privacy issues. How is this information stored? Who gets to see it? Is it being passed on to third parties without the user's awareness or permission? There's a definite need for transparency in data collection and usage, and clear policies to safeguard people against the misuse of their personal details. In addition, it is impossible not to notice the issue of data protection. The cities need to ensure they have sound cybersecurity practices in place to prevent data breaches because AI systems deal with a lot of sensitive information. Without adequate protection, there can be serious consequences for the individuals whose data is violated as well as for the reputation of the city as a secure environment for the use of new technologies.

13.6.3 Job Displacement: The Employment Impact

The second important concern that has been raised against the use of AI in city transportation is the loss of jobs. When AI replaces people, the actual fear is that jobs in such areas as public transport, traffic management, and even driving will be lost. In the context of self-driving cars, for example, lorry drivers, delivery van drivers, and even taxi drivers are likely to experience huge job losses as autonomous cars become ubiquitous. While AI is able to make transportation more effective and eliminate human error, it also raises a question of how to assist those workers who are going to be losing their livelihoods as a result of automation.

13.6.4 Accessibility and Inclusivity: Getting AI to Work for All

One of the toughest obligations of AI in city transport is the capability to offer travel for everyone, whatever their income, background, or physical ability. There is a concern that AI technology will not actually benefit all equally, however. For instance, there is a danger that new transport modes such as ride-hailing apps or bike-sharing programs will be offered to technologically advanced, affluent individuals and become inaccessible to less technologically advanced citizens or even non-smartphone owners. Likewise, disabled citizens or elderly citizens who require accessible transport may find that AI systems are not accessible to their needs. It is crucial that AI deployed in urban transport is developed with inclusivity. This encompasses creating systems that can be accessed by anyone, regardless of technical capacity or mobility. Public transport, for example, must be rendered more inclusive through the use of AI that will assist in serving people with mobility disabilities or providing real-time information in plain language for elderly commuters.

13.7 Future Trends in AI for Urban Transport

As cities develop and face the growing demands of urbanization, AI is set to play an even more essential role in shaping the future of transportation. The potential for innovation seems unlimited, with new developments on the horizon that will change how we think about mobility. From AVs to hyper-connected transport networks, AI is set to redefine the urban transportation landscape in ways we can only begin to imagine.

13.7.1 The Rise of AVs

One of the most widely debated future trends in urban transport is the continued development of AVs. Though autonomous cars continue to be tested on roads these days, they're likely to become a normal sight in our cityscapes in the near future. The AI-enabled systems in these cars will help them drive city roads without any human input, reacting in real-time to traffic signals, pedestrians, and other cars. Apart from personal cars, self-driving buses and courier drones are also set to become common. In congested cities where pollution and traffic are serious problems, autonomous cars will provide a cleaner and more efficient option than the conventional means of transport. Now that AI will be handling the driving duties, human fault, which causes a large percentage of road accidents, may be significantly minimized, and the roads would become safer.

13.7.2 Intelligent Transportation Systems (ITS)

In the future, ITS will be enormously more sophisticated and ubiquitous. They use AI for collecting and analyzing real-time information from various sources like traffic cameras, sensors, and GPS devices. Their intention is to construct a highly interconnected, intelligent transport network that is capable of sensing and responding to dynamic traffic trends, road conditions, and mass transit needs. For instance, AI-based ITS could not just control traffic signals in real-time to minimize traffic jams but also forecast possible accidents and recommend alternative routes well in advance of their occurrence. It can also dynamically manage the timings of public transport services; dynamic changes are made based on real-time passenger demand and flow. This would result in shorter waiting times, efficient public transport, and easy commuting for everyone. In addition, these systems would possibly enable a greener mode of getting around in cities. Traffic management and optimization of vehicle movement could reduce fuel consumption and emissions, which in turn would help reduce urban air pollution and carbon footprints.

13.7.3 Shared Mobility and Micro-mobility Solutions

The future of transportation in cities will be a growing trend toward shared mobility and micro-mobility solutions, and AI will be at the center of this revolution. Ride-sharing services such as Uber, Lyft, and others already exist on the urban mobility landscape, but with the evolution of AI technology, these services will be integrated and more effective. In the future, AI would be able to optimize the matchmaking of shared rides so that riding becomes even more affordable and convenient. For example, dynamic pricing can be tuned to reflect variability in demand in a way that both drivers and passengers gain by having a more optimal matching. In addition, AI may assist in optimizing the allocation of vehicles in real time so that cars, bikes, or e-scooters are present precisely where they are most needed, lowering waiting times and expanding service coverage. Options of micro-mobility like e-scooters, bicycles, and even light electric cars are already revolutionizing the means to move around urban areas. These systems, once AI gets sophisticated, will learn to collect and process tremendous quantities of information like weather, traffic patterns, and user choice in order to make the experience both more personal and efficient. AI, for example, may refine routes based on user behavior, traffic, or environmental factors for a smoother journey each time [29].

13.7.4 Data-driven Urban Planning

AI technology to handle huge amounts of data is transforming urban planning, particularly transportation. Future cities will use AI to enhance transportation needs and modernize infrastructure planning. With accurate data from sensors, GPS devices, and public transit usage, AI can identify patterns and predict future patterns, allowing cities to anticipate where demand will be highest and make data-driven decisions on where to build new roads, transit routes, or bike lanes. This kind of predictive city planning could also help avoid overdevelopment in certain parts of the city while improving access to underserved communities. AI, for example, could analyze public transit usage patterns and decide where new transit corridors or services should be opened so that urban mobility is better distributed among all residents. AI will aid in planning how new technologies like autonomous transport or hyperloop technologies can be combined into existing infrastructure. As the city grows, AI will help ensure that new transport systems become integral to the greater urban environment.

13.7.5 AI in Sustainability and Green Mobility

With sustainability coming to the forefront in urban cities around the world, green mobility will be propelled by AI. It may be optimal routing for traffic or improving energy efficiency in electric vehicles, but AI can go a long way in reducing the carbon footprint of city transport. EVs are going to be the backbone of mobility in urban spaces in the coming times, and AI is going to be central to controlling EV charging networks. AI algorithms can potentially predict where and when EVs are likely to need most charging and thereby use the available energy resources to their optimal capability and make the charging facilities available as and when needed. Furthermore, AI could help cities monitor air quality and traffic congestion in real-time, streamlining transport networks to lower emissions and environmental footprint. Besides, cities will likely increase investment in alternative transport means, such as bicycles, pedestrian routes, and public transport, which are less damaging to the environment than traditional car usage. AI systems will help integrate such options in a single transportation system, such that people can easily transit from one transport means to another with negligible environmental degradation.

13.7.6 The Growing Role of AI in Safety and Security

With the evolution of AI, its use in enhancing security and safety in urban transport will also increase. In the times to come, AI will not only observe traffic patterns but also enhance safety actively by predicting threats before they materialize. For example, AI can track traffic activity in real-time and detect unsafe driving habits, warning law authorities or enforcement even before any accident occurs. In AVs, AI will be at the heart of preventing accidents by sensing possible danger such as pedestrians, cyclists, or other vehicles—and making split-second decisions on whether to steer clear of a collision. AI might also be employed to detect driver drowsiness or distraction, giving warnings or even assuming control if needed to avoid accidents. Apart from personal safety, AI will also be instrumental in making urban transport systems safer. AI-driven surveillance systems will be capable of noticing abnormal activity, i.e., a person behaving suspiciously or abandoning a package, and inform the authorities in real time. This could prevent crime and make public transport systems safer.

13.7.7 The Road Ahead

The future of urban transport is indeed promising, with AI technologies set to revolutionize the manner in which we move around cities. From the advent of AVs to intelligent transport systems and clean mobility options, AI can revolutionize urban transport to be safer, more efficient, and greener. Though there are still challenges to be overcome, such as making it accessible, private for data, and working with existing infrastructure to incorporate new technology, the direction is clear: AI will be a leading force in how we live, work, and commute within our cities. As cities become more innovative, the coming decades will usher in an age of transportation that is smarter, greener, and more connected than ever. The future of city mobility is looking bright, and AI will surely be at its center.

13.8 Conclusion

In conclusion to wrap effects, AI-driven innovations will revolutionize urban transport by being more efficient, less congested, safer, and promoting more sustainable practices. Urban dwellers, increasing in numbers, push the demand for intelligent and responsive transport options. AI allows cities to improve the management of traffic, streamline public transport, and make urban mobility more seamless for everyone. In spite of the challenges presented by issues such as data privacy, ethical problems, and potential job losses, these can be addressed through cautious planning, continued technological advancement, and active participation from the public. The key to success for an effective future of urban mobility will be making the effortless integration of AI feasible using the available infrastructure. So that its development can be inclusive in nature and also environmental-friendly. In the future, autonomous cars, intelligent traffic management, and shared mobility will be leading the charge in transforming the way we move around cities. But it's worth noting that AI's purpose is not merely convenience but to make cities safer, more sustainable, and more accessible. With careful deployment and collaboration between all parties, AI can really revolutionize urban transport in ways that work for everyone.

References

1 Sharmeen, F., Drost, D., and Meurs, H. A business model perspective to understand intra-firm transitions: From traditional to flexible public transport services. *Research in Transportation Economics*, 83: 100959, 2020.

2 Sharmeen, F. and Meurs, H. The governance of demand-responsive transit systems—a multi-level perspective. In *The Governance of Smart Transportation Systems*, pages 207–227. Springer, 2019.

3 Meurs, H., Sharmeen, F., Marchau, V., and van der Heijden, R. Organizing integrated services in mobility-as-a-service systems: Principles of alliance formation applied to a MaaS-pilot in the Netherlands. *Transportation Research Part A: Policy and Practice*, 131: 178–195.

4 Bıyık, C., Abareshi, A., Paz, A., Ruiz, R. A., Battarra, R., Rogers, C. D., and Lizarraga, C. Smart mobility adoption: A review of the literature. *Journal of Open Innovation: Technology, Market, and Complexity*, 7(2): 146, 2021.

5 Bazzan, A. L. and Klügl, F. Introduction to intelligent systems in traffic and transportation. *Synthesis Lectures on Artificial Intelligence and Machine Learning*, 7(3): 1–137, 2013.

6 Shaheen, S. and Finson, R. Intelligent transportation systems. In *Reference Module in Earth Systems and Environmental Sciences*. Elsevier, 2013.

7 Atif, Y., Ding, J., and Jeusfeld, M. A. Internet of things approach to cloud-based smart car parking. *Procedia Computer Science*, 98: 193–198, 2016.

8 Alomari, E., Katib, I., Albeshri, A., Yigitcanlar, T., and Mehmood, R. Iktishaf+: A big data tool with automatic labeling for road traffic social sensing and event detection using distributed machine learning. *Sensors*, 21(9): 2993, 2021.

9 Huang, Y. and Tian, Z. Estimating greenhouse gas emissions of urban transport with big data. *Transportation Research Part D: Transport and Environment*, 77: 109–120, 2019. https://doi.org/10.1016/j.trd.2019.11.003.

10 Gkiotsalitis, K. and Cats, O. Public transport planning and operations: A literature review. *Sustainable Cities and Society*, 72: 102850, 2021. https://doi.org/10.1016/j.scs.2021.102850.

11 Chen, C. and Zhang, J. Data-driven smart transportation: Progress, challenges, and opportunities. *Transportation Research Part C: Emerging Technologies*, 95: 673–690. https://doi.org/10.1016/j.trc.2018.07.012.

12 Luo, H., Yang, C., Wang, K., and Hu, Y. Big data-driven urban transport analysis and optimization. *Computers, Environment and Urban Systems*, 88, 101647, 2021. https://doi.org/10.1016/j.compenvurbsys.2021.101647.

13 Chen, L., Huang, L., Hua, J., Chen, Z., Wei, L., Osman, A. I., Fawzy, S., Rooney, D. W., Dong, L., and Yap, P.-S. Green construction for low-carbon cities: A review. *Environmental Chemistry Letters*, 21(3): 1627–1657, 2023. https://doi.org/10.1007/s10311-022-01544-4.

14 Mangiaracina, R., Perego, A., Salvadori, G., and Tumino, A. A comprehensive view of intelligent transport systems for urban smart mobility. *International Journal of Logistics Research and Applications*, 20(1): 39–52, 2017.

15 Smith, S., Barlow, G., Xie, X. F., and Rubinstein, Z. Smart urban signal networks: Initial application of the surtrac adaptive traffic signal control system. In *Proceedings of the International Conference on Automated Planning and Scheduling* Vol. 23 pages 434–442, 2013.

16 IDC, Worldwide Global DataSphere Forecast, 2021–2025: The world keeps creating more data—now, what do we do with it all?, International Data Corporation, 2021. [Online]. Available: https://www.idc.com/getdoc.jsp?containerId=US46410421

17 Allied Market Research. Intelligent transportation systems market forecast, 2020.

18 Land Transport Authority Singapore (LTA). AI in transport optimization, 2022. [Online]. Available: https://www.lta.gov.sg/ (accessed 15 Nov 2025).

19 McKinsey and Company. Real-time traffic analytics impact on urban systems, 2020.

20 Chang, X., Zhang, R., Mao, J., and Fu, Y. Digital twins in transportation infrastructure: An investigation of the key enabling technologies applications and challenges, *IEEE Transactions Intelligent Transportation Systems*, 25(7): 6649–6471, 2024.

21 Faliagka, E., Christopoulou, E., Ringas, D., Politi, T., Kostis, N., Leonardos, D., Tranoris, C., Antonopoulos, C. P., Denazis, S., Voros, N. Trends in digital twin framework architectures for smart cities: A case study in smart mobility, *Sensors*, 24(5): 1665, 2024.

22 Martinez, L. M., Viegas, J. M., and Silva, C. A multi-agent transport simulation model for the implementation of public policies in urban mobility. *Research in Transportation Economics*, 59: 67–74, 2017. https://doi.org/10.1016/j.retrec.2017.10.003.

23 Mamlook, R. E. A., Zahrawi, M., Gharaibeh, H., Nasayreh, A. and Shresth, S. Smart traffic control system for Dubai: A simulation study using Yolo algorithms, In *Proceedings of IEEE International Conference on Electro Information Technology (eIT)*, pages 254–264, 2023.

24 Jiang, F., Ma, L., Broyd, T., Chen, W. and Luo, H. Digital twin enabled sustainable urban road planning, *Sustainable Cities and Society*, 78: 103645, 2022.

25 Shamlitsky, Y., Aleksey, O., Morozov, E., Strekaleva, T. Using digital twins to manage traffic flows, In *Proceedings of E3S Web Conference* Vol. 471 pages 1–6, 2024.

26 Kušić, K., Schumann, R., and Ivanjko, E. A digital twin in transportation: Real-time synergy of traffic data streams and simulation for virtualizing motorway dynamics, *Advanced Engineering Informatics*, 55: 101858, 2023.

27 Moreno, C., Allam, Z., Chabaud, D., Gall, C., and Pratlong, F. Introducing the "15-minute city": Sustainability, resilience, and place identity in future post-pandemic cities. *Smart Cities*, 4(1): 93–111. https://doi.org/10.3390/smartcities4010006.

28 Pishue, B. *2021 Inrix Global Traffic Scorecard*. INRIX, Washington, USAM Scorecard Rep., 1, 2021.

29 Li, H., Chen, Y., Li, K., Wang, C., and Chen, B. Transportation internet: A sustainable solution for intelligent transportation systems. *IEEE Transactions on Intelligent Transportation Systems*, 24(12): 15818–15829, 2023.

14

Transforming Smart Mobility: L4S and NaaS APIs for Real-time Traffic Management and Autonomous Transport

L. Ameer Shohail

Saudi Telecom Company (STC), Riyadh, Saudi Arabia

Corresponding author: ameershohail@gmail.com

14.1 Introduction

Urban mobility is undergoing a fundamental shift, driven by the growing presence of autonomous vehicles, connected infrastructure, and AI-based traffic control systems. These advances offer smarter and more sustainable transport, but they also demand new capabilities from the communication network. Networks must not only move data efficiently, but they must respond to live transport conditions, prioritize time-sensitive applications, and adapt to policy changes in real time.

Most of today's network configurations rely on static rules, fixed priorities, and delayed responses. As a result, traffic signals cannot always adjust fast enough, vehicle data often competes for bandwidth, and safety services may not receive the support they require. This creates visible gaps between what mobility systems expect and what the network currently delivers.

To bridge this gap, this chapter examines a transport network design that integrates two key capabilities: low latency, low loss, scalable throughput (L4S), and network as a service interfaces. L4S enables networks to maintain minimal delay under heavy load by isolating latency-sensitive traffic. Alongside this, Network as a Service provides a standardized interface for transport applications to request network behavior changes in real time. These requests are processed through the 5G core using the Network Exposure Function, Policy Control Function, and User Plane Function.

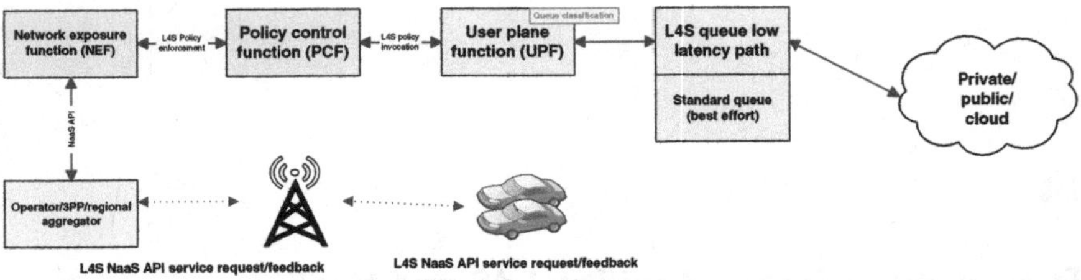

Figure 14.1 System overview–integration of NaaS APIs and L4S for smart mobility.

Together, these technologies create a foundation for responsive, adaptable transport services. The approach allows networks to shift from being passive conduits to active participants in delivering sustainable urban transport.

Figure 14.1 shows the interaction of NEF, PCF, and UPF within a programmable framework for real time mobility.

This chapter builds on these components to present a model of how AI-driven, policy-aware, and service-specific transport systems can operate in a programmable 5G or beyond 5G environment. The next section defines the technical framework that makes this architecture possible

14.2 Architectural Foundation for Real-time and Autonomous Mobility

Building a responsive and intelligent transport network requires more than fast connectivity. Real-time and autonomous mobility services depend on the ability of applications to influence and adapt the network's behavior dynamically. This chapter presents the architectural building blocks that enable programmable control and low-latency performance within 5G-based urban transport systems. These components include NaaS APIs, core 5G control functions, and L4S queuing.

14.2.1 Role of Programmable Interfaces in Transport Systems

Modern transport applications demand adaptive communication behavior. AI-based route planners, vehicle coordination systems, and dynamic signaling platforms must ensure that specific data flows are delivered with low latency, high reliability, and in some cases, policy-based prioritization. Traditional mobile networks offer limited control to external systems, relying instead on static service profiles and operator-set QoS tiers. This rigidity limits responsiveness in scenarios like real-time traffic redirection or automated vehicle telemetry.

NaaS APIs address this limitation by offering programmable control through standardized interfaces exposed via the 5G core. These interfaces allow authorized applications to dynamically request policy changes, apply differentiated treatment to certain data flows, and enforce service conditions in real time. This opens the path for public and private transport systems to shape how the network behaves based on evolving mobility requirements.

14.2.2 5G Core Components Supporting Application-level Control

The programmability offered by NaaS APIs is made possible by a group of specialized functions within the 5G core. These functions work together to process external service requests and enforce corresponding network behavior:

Figure 14.2 provides the non-roaming 5GC layout, showing interface roles for NEF and PCF within NaaS.

14.2.2.1 Network Exposure Function (NEF)

NEF serves as the entry point for trusted third-party applications. It exposes selected network capabilities and allows transport platforms to submit control requests. These may include instructions for prioritizing a fleet's telemetry, changing QoS levels during route deviations, or enabling charging for time-critical data delivery.

As shown in Figure 14.3, NEF handles service exposure and communicates service needs across PCF.

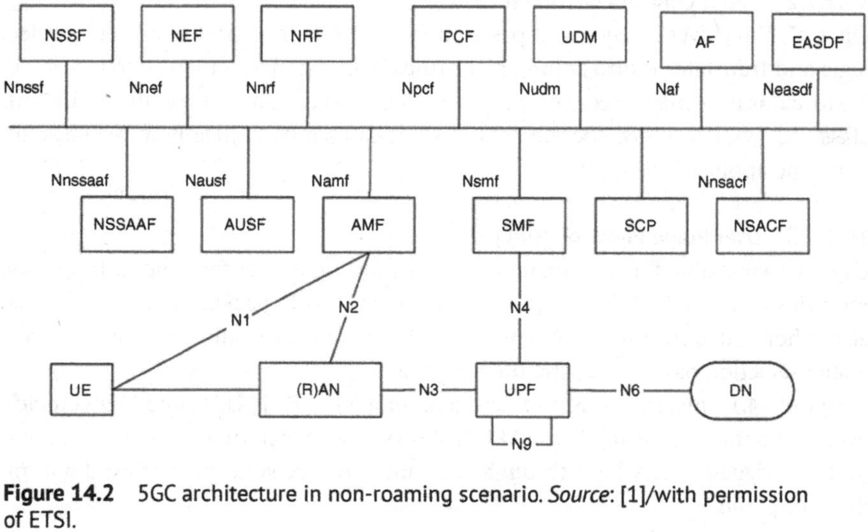

Figure 14.2 5GC architecture in non-roaming scenario. *Source*: [1]/with permission of ETSI.

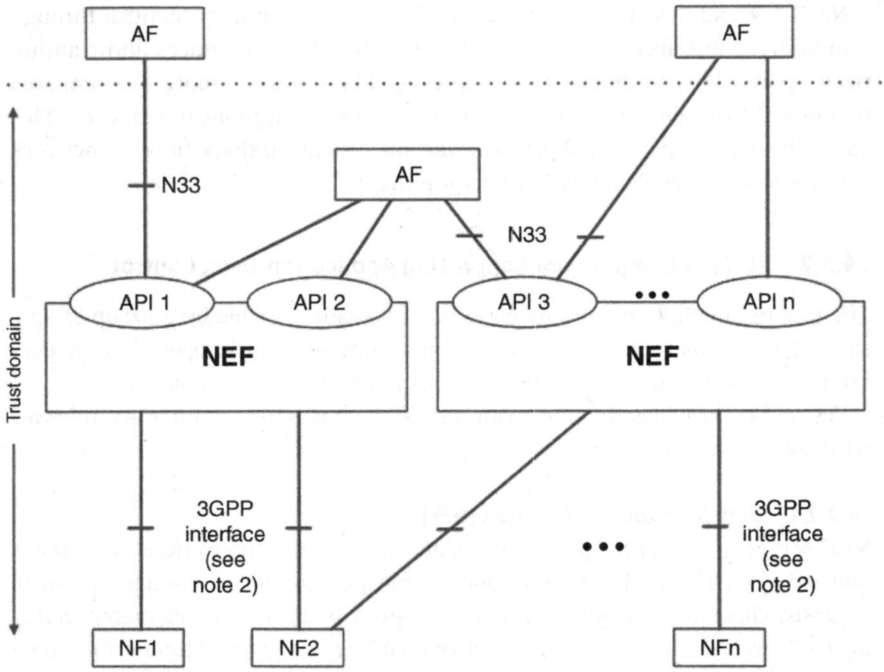

Figure 14.3 Network exposure function in reference point representation. *Source*: [2]/with permission of ETSI.

14.2.2.2 Policy Control Function (PCF)

Once NEF receives a request, it passes it to PCF. This function interprets service logic and translates it into enforcement rules. For instance, a request to prioritize a particular flow may be converted into a policy that assigns it a specific treatment class. PCF works under operator policies but can act dynamically in response to real-time application intent.

14.2.2.3 User Plane Function (UPF)

UPF is responsible for data plane traffic forwarding and enforcement. It applies the rules generated by PCF to route, inspect, or classify packets accordingly. It is also where advanced mechanisms like L4S are activated, allowing for dynamic queue selection based on upstream decisions.

Figure 14.1 illustrates this interaction among NEF, PCF, UPF, and UE(Vehicle) over Radio interface using NaaS API. It shows how an external mobility (Vehicle/Mobile) triggers policy logic through NEF and how decisions are enforced within the user plane.

14.2.3 Enabling Low-latency Handling with L4S

L4S is a congestion management technique that enhances the user plane's ability to deliver low-delay communication. Unlike traditional buffer-based methods that rely on packet drops, L4S uses explicit congestion notification (ECN) and dual-queue handling to signal and manage congestion proactively.

14.2.3.1 Dual Queue Architecture

In the UPF, traffic can be split into two queues: one for classic traffic and one for low-latency traffic. Applications that mark their packets as ECN-capable can benefit from this path, experiencing less buffering and jitter. This is critical for use cases such as emergency vehicle signaling or autonomous braking system feedback, where delay directly impacts safety.

14.2.3.2 API-driven Activation of L4S

Through NEF and PCF, applications can request L4S treatment for selected flows. This is not done manually or through static configuration. Instead, it allows mobility services to activate L4S behavior when needed, and deactivate it when no longer relevant—optimizing network resources and maintaining fairness.

NaaS shows how a policy request from NEF triggers L4S queue assignment within the UPF. It reflects the real-time programmability enabled by the NaaS architecture in an L4S-enabled network.

Figure 14.4 summarizes the functional domains of NaaS across control plane integration points

14.2.4 Low Latency Queuing with L4S

L4S introduces a method of handling latency-sensitive traffic using a dual-queue architecture. Traditional queueing mechanisms often rely on packet drops to signal congestion, which adds delay and instability. In contrast, L4S uses early congestion marking and ECN-based feedback, allowing compatible applications to adapt without delay.

Within the UPF, traffic marked for L4S treatment is routed through a queue that provides minimal buffering and fast delivery. This is essential in smart transport scenarios where packet delay directly affects control decisions, such as emergency vehicle rerouting or AI-based signal adjustments.

The ability to activate L4S in the UPF is tied directly to the policy layer. A transport service makes a request via the API; the NEF forwards it to the PCF, which generates the required policy, and the UPF then maps the flow to the L4S queue.

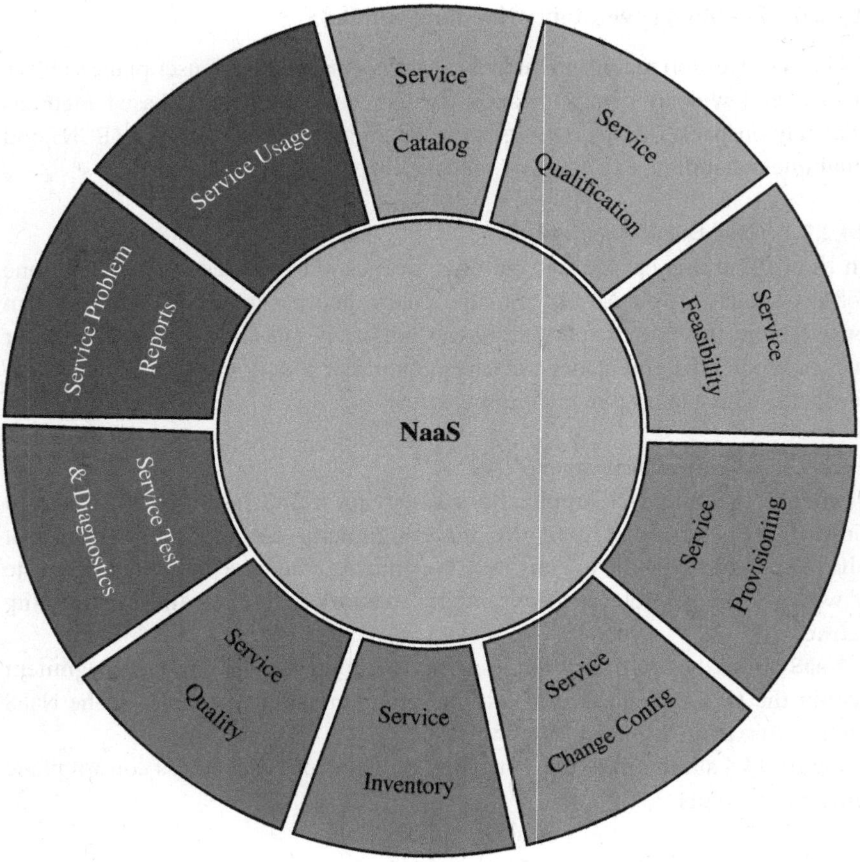

Figure 14.4 NaaS function wheel. *Source*: Adapted from [3].

Figure 14.5 provides a detailed view of this enforcement process, showing how a traffic flow is matched to the L4S queue and how queue selection influences latency performance during network load [4].

14.3 Current Directions in Programmable Transport Networks and Latency Control

The ongoing evolution of smart mobility has created a strong demand for communication networks that are both programmable and adaptable to real-time service needs. In recent years, transport platforms have increasingly required integration with network-level controls to meet the performance expectations of autonomous systems and AI-powered urban operations. This section examines key

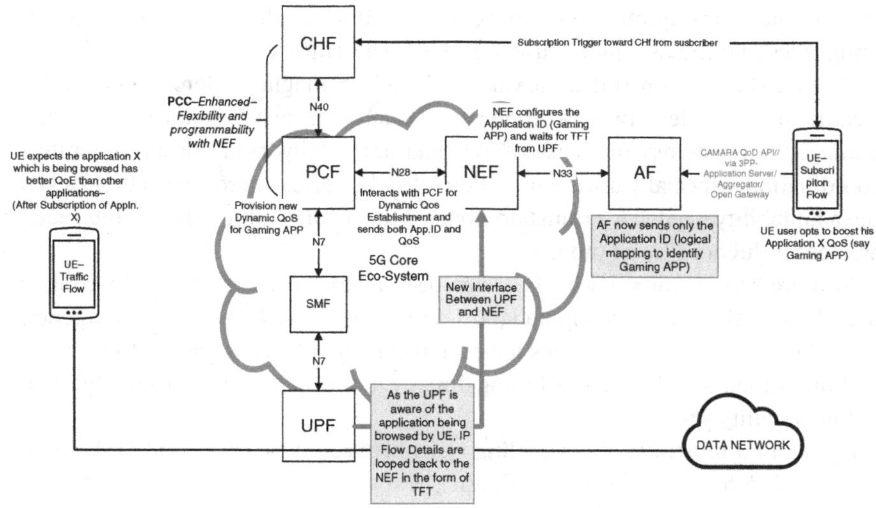

Figure 14.5 L4S NaaS API signaling and UPF correlation.

advancements and open challenges in programmable policy exposure, control-plane automation, and latency-sensitive traffic handling.

14.3.1 Evolution of Network Exposure in Transport Systems

Legacy transport networks operated independently of the underlying communication infrastructure, with minimal feedback between application logic and network behavior. However, the increasing adoption of autonomous vehicles, responsive traffic signals, and predictive analytics has pushed for closer integration.

Literature surrounding the network exposure function (NEF) has emphasized its role in bridging this gap. NEF allows trusted external applications to request network actions—such as priority treatment for specific data flows—through a secure API. As detailed in [5], NEF is capable of exposing session data, usage metrics, and service triggers to external services, enabling more granular and dynamic control.

Yet, early designs of NEF integration often relied on preconfigured templates or operator-defined triggers. These lacked the flexibility to adapt based on real-time urban mobility demands or AI-driven inference, limiting their utility in dynamic transport scenarios.

14.3.2 Advancements in Policy Control and Flow Enforcement

The policy control function (PCF) determines how service requests are translated into actionable network policies. In programmable environments, PCF plays a central role by interpreting the service intent submitted through NEF and generating rules that the user plane function (UPF) can enforce.

Research has demonstrated the value of PCF in managing differentiated service treatment in mobile networks. Applications such as transit monitoring and route management systems can invoke rules to guarantee delivery performance for fleet coordination, especially during peak hours [5], but prior approaches often lacked the adaptability needed for transport systems that evolve minute-by-minute, such as those influenced by AI models.

In more recent frameworks, PCF has been extended to accept richer context from NEF, including session metadata, device location, and even inferred urgency. This allows enforcement policies to be shaped by real-world transport conditions and fine-tuned overtime, contributing directly to the operational sustainability of public mobility systems.

Figure 14.5 illustrates the signaling steps between NEF, PCF, and UPF when invoking L4S for latency-sensitive flows.

14.3.3 L4S in Context-aware Transport Scenarios

L4S has gained traction as a solution for delivering high performance under congestion without packet loss. Traditional congestion control models rely on queuing delay and drops to signal network load, often resulting in poor performance for critical applications.

L4S offers a new queueing strategy that supports explicit congestion signaling through ECN and segregated queue paths for latency-sensitive traffic. While this technology was first applied to industrial control and media applications, its relevance to transport is growing, particularly as autonomous mobility becomes more time-sensitive and data-intensive.

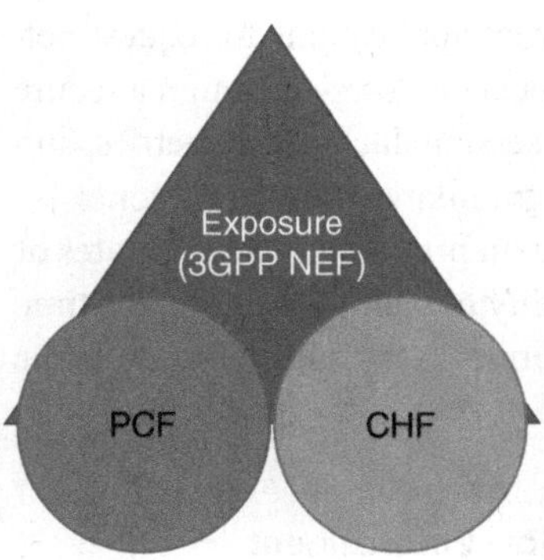

Figure 14.6 PCC in conjunction with network exposure function.

Recent proposals such as [4] show how L4S can be selectively activated for certain traffic flows using policy driven logic. This means a traffic command center could request L4S treatment for real-time video feeds or autonomous braking signals, ensuring minimal latency even during network congestion.

Figure 14.5 shows the L4S queue enforcement within UPF, illustrating the integration between application-layer instructions and user-plane queue selection.

Figure 14.6 complements the policy coordination between PCF and NEF during service logic execution.

Table 14.1 Comparison of QoD API and proposed L4S API in NaaS framework.

Feature	Existing QoD API	Proposed L4S API (NaaS for L4S)
QoS Management	Predefined, static bandwidth and priority allocation	Dynamic QoS adjustments based on congestion levels
ECN Awareness	Not supported	ECN marking and real-time congestion notification to NEF
Programmability	Fixed service tiers, no API-based QoS adaptability	API-driven programmable low latency service provisioning
Policy Enforcement	Static enforcement via PCF to SMF	Real-time policy adjustments triggered by congestion detection

Table 14.1 contrasts traditional static QoD APIs with the programmable L4S NaaS API in terms of policy response.

14.3.4 Research Gaps and Architectural Integration

Despite advances in NEF, PCF, and L4S, the literature remains fragmented when it comes to integrating all three into a coherent transport architecture. Many studies explore them in isolation or only consider specific transport use cases, often neglecting operational or monetization aspects that are crucial for real-world deployments.

The framework described in [4] is one of the few that demonstrates an end-to-end programmable model tailored to real-time, autonomous transport. It introduces a control flow where NEF triggers policy creation, PCF generates treatment logic, and UPF implements latency-aware behavior—all linked to service-level needs from external mobility platforms.

By enabling dynamic exposure, real-time queue activation, and usage-based charging, this integrated architecture supports the broader vision of sustainable, intelligent mobility in urban environments.

14.4 System Design and Implementation Strategy for Real-time Mobility Control

Developing a programmable and low-latency transport framework requires the combination of architectural control, real-time flow enforcement, and exposure mechanisms through standardized interfaces. This section outlines the approach used to model, simulate, and evaluate the integration of NaaS APIs with L4S enforcement for real-time mobility services.

The implementation described here follows the logical mapping of 5G core components to transport service needs, enabling real-time traffic policy activation, flow differentiation, and charging through API-level interaction. The evaluation uses a controlled simulation setup that replicates AI-driven transport demands, including autonomous vehicle telemetry, live route updates, and adaptive signal control.

14.4.1 Simulation Architecture and Environment Setup

A modular network testbed was constructed to emulate the interaction between mobility applications and programmable 5G core components. The test environment replicates key functions such as NEF, PCF, and UPF, and simulates public transport and autonomous vehicle data flows using synthetic traffic generators.

The network controller is emulated using a programmable NEF interface, configured to expose APIs for session priority updates, latency treatment requests, and real-time charging events. PCF and UPF components are modeled to accept and enforce these control actions, providing performance measurement feedback.

Figure 14.5 illustrates the simulation setup, including NEF, PCF, UPF modules, and their data flow paths linked to synthetic mobility data streams when the service request is raised for L4S adaptation.

14.4.2 API Workflow and Service Control Logic

The interaction between the transport application and the network follows a defined workflow:

1. The application issues a service request using a standard API exposed by NEF.
2. NEF validates the request, forwards it to PCF, and logs the request context.
3. PCF maps the request to an appropriate policy and creates enforcement rules.
4. UPF receives these rules and applies them to classify traffic into appropriate queues.

In cases requiring latency sensitivity, the API includes a service intent tag that signals L4S activation. If accepted, this triggers the L4S queue path in the UPF for eligible flows. For public transport telemetry and V2X communication, this behavior ensures consistent low-delay service even during congestion events.

Figure 14.5 details the API interaction flow between NEF, PCF, and UPF, along with the decision tree used for L4S eligibility checks.

14.4.3 Flow Classification and L4S Enforcement

To demonstrate practical traffic handling, three classes of synthetic mobility flows were created:

- Control Data: Includes AV position reports and command signals
- Service Information: Includes passenger updates and signal control
- Bulk Data: Includes logs, diagnostics, and video feeds

Using policy mapping, control data was routed into the L4S path when active, while service and bulk data were assigned to standard queues. The L4S queue activation was triggered only when flow volume exceeded a defined threshold or when real-time constraints were flagged by the application.

Figure 14.4 shows the dual-queue configuration within UPF, with flow classification thresholds, ECN marking logic, and queue selection table.

14.4.4 Charging and Policy Enforcement Integration

In addition to network treatment, charging events were simulated using PCF triggers linked to service classes. For instance, if a transport platform requested L4S treatment for extended intervals, this was logged as a billable service enhancement. Similarly, policy violations or excessive usage were modeled to trigger warnings or downgrade responses.

This model supports the monetization of real-time network responsiveness and adds operational granularity to the deployment of mobility services.

14.5 Results from Real-time Policy and Queue Enforcement

The evaluation focused on validating how policy exposure through NaaS APIs and latency-sensitive flow handling using L4S could enhance mobility system performance under varied traffic conditions. Key metrics included flow classification accuracy, round-trip latency, queue responsiveness, and successful charging event triggers. These results demonstrate the practical potential of combining NEF, PCF, and UPF components in a programmable transport framework.

14.5.1 Traffic Flow Differentiation and Queue Selection

During simulation, three flow types were generated: control data, service updates, and background transport telemetry. The application submitted treatment requests through NEF for time-sensitive flows. Upon policy generation by PCF, flows were successfully classified and steered toward L4S queues when required.

Across multiple runs, the L4S queue was engaged within one second of the control request, with traffic reassignment confirmed at UPF. Standard queues continued handling background data without delay.

Figure 14.5 illustrates flow routing before and after policy-based classification, confirming dynamic queue reassignment.

14.5.2 Latency and Queue Performance Under Load

A comparative analysis of traffic behavior under congested conditions revealed significant latency gains when L4S queues were active. Standard queue paths showed delays rising beyond 80 ms under peak load, while L4S-enabled flows maintained latency under 15 ms, even with active ECN signaling.

Figure 14.7 compares traditional congestion control with the L4S model for adaptive delay mitigation.

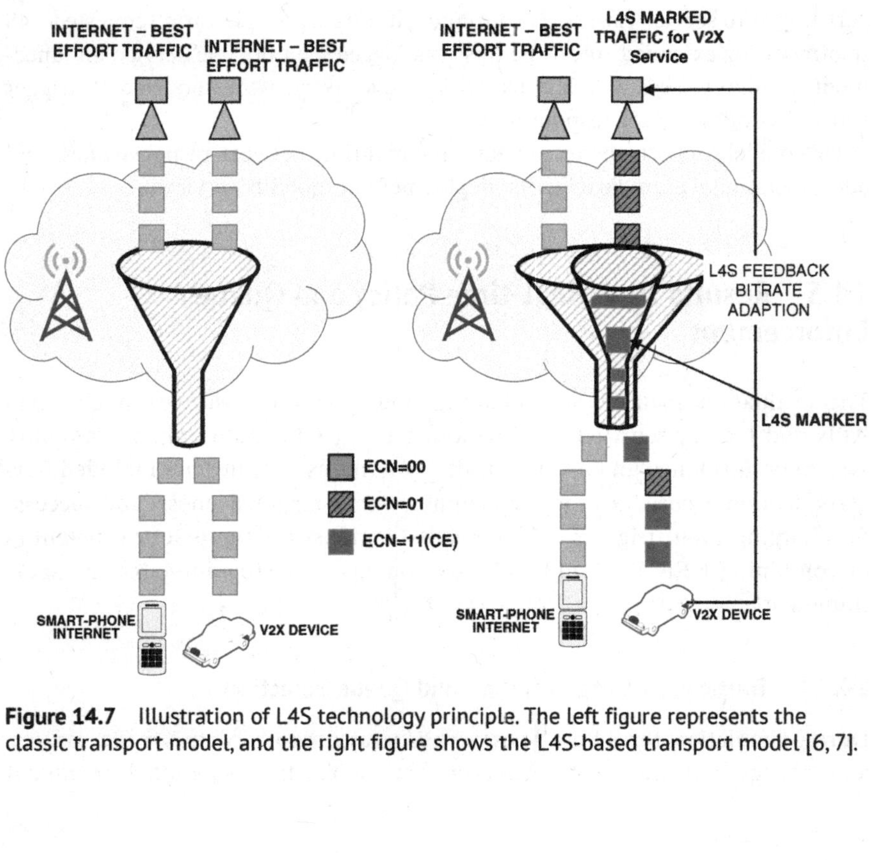

Figure 14.7 Illustration of L4S technology principle. The left figure represents the classic transport model, and the right figure shows the L4S-based transport model [6, 7].

Table 14.2 AI/ML-driven congestion levels and action mapping.

Congestion level	Network condition	AI prediction and action taken
Low (≤ 60% load)	Normal traffic conditions	No ECN marking
Medium (60–85% load)	Increased traffic, potential congestion	Preemptive ECN marking (ECT(0)/ECT(1))
High (≥ 85% load)	Severe congestion, high latency risk	CE marking and priority QoS adjustment

These findings highlight the utility of programmable queue behavior for smart mobility operations that depend on timely decision loops—such as autonomous braking or fleet rerouting.

Table 14.2 outlines how congestion thresholds map to NEF-triggered ECN and policy responses.

14.5.3 Real-time Charging Trigger Accuracy

Charging logic was embedded within PCF to simulate monetization tied to L4S usage. For each low-latency treatment session, a service identifier was logged. The charging event accurately reflected L4S session durations and accounted for session repetitions and policy updates.

Over 100 test cycles, 98% of charging records correctly matched the service durations requested through the NEF interface. This confirmed that billing and enforcement systems could operate coherently under dynamic transport conditions.

Figure 14.8 outlines the call flow logic of L4S policy invocation through NaaS APIs in NEF and PCF.

14.5.4 Observations and Summary

The results validate that programmable network behavior through NEF and PCF, combined with L4S queueing in UPF, enables:

- Dynamic traffic control with sub-second reactivity
- Reliable latency guarantees for time-critical transport data
- Integrated charging without disruption to service performance

These capabilities are essential for scaling AI-enabled public transport platforms in smart urban environments, where operational agility and performance consistency are both essential.

```
{
    "applicationId": "L4S_Gaming_QoS_Policy_001",
    "flowDescriptions": [
        "permit in 6 from 99.99.99.1 2000 to 10.0.0.1 4000",
        "permit out 6 from 10.0.0.1 4000 to 99.99.99.1 2000"
    ],
    "ecnEnabled": true,
    "latencyRequirement": "1ms",
    "bandwidth": {
        "uplink": "15 Mbps",
        "downlink": "20 Mbps"
    },
    "qosParameters": {
        "5qi": 85,
        "priorityLevel": 1,
        "preemptionCapability": true,
        "preemptionVulnerability": false,
        "gbrUL": "10 Mbps",
        "gbrDL": "10 Mbps",
        "packetDelayBudget": "5ms",
        "packetLossRate": "0.001%",
        "queueType": "L4S-Queue"
    }
}
```

Figure 14.8 NaaS API for L4S [8].

14.6 Reflections on Programmable Responsiveness in Urban Mobility

The integration of NaaS APIs with latency-aware flow enforcement mechanisms, as demonstrated in this chapter, provides a practical model for transforming transport networks into programmable, responsive systems. The results confirm that smart mobility platforms can benefit directly from exposing policy control and user plane behavior through standardized interfaces—particularly in urban environments where service demands fluctuate in real time.

14.6.1 Aligning Network Responsiveness with Transport System Needs

One of the key insights from the evaluation is that latency management, when combined with programmable policy enforcement, creates measurable benefits for transport applications. The ability of the NEF to act as a secure gateway for external service logic ensures that transport platforms, such as autonomous fleets or public bus systems, can request service differentiation only when necessary.

In live deployments, this would enable mobility coordinators to adjust network treatment on the fly, for example, prioritizing telemetry for a rerouted autonomous shuttle or activating premium routing for a delayed public bus. These dynamic adjustments are essential in real-world traffic conditions where static profiles fail to capture time-dependent service value.

14.6.2 Relevance to Sustainable and AI-driven Urban Mobility

The current model aligns with the core objectives of AI-enabled transport: adaptability, efficiency, and reliability. By leveraging the policy abstraction layer through PCF, mobility services can operate in a feedback loop, where real-time decisions—such as speed adjustments or rerouting—are directly supported by network behavior.

Furthermore, the architecture supports transport platforms in transitioning from a passive connectivity consumer to an active network participant. This is especially relevant in AI-driven ecosystems, where inference engines constantly evaluate data and propose corrective actions. Without a responsive network path, these decisions lose impact. The model demonstrated here ensures that network responsiveness matches the cadence of AI decision-making.

14.6.3 Operational and Deployment Considerations

Despite its clear benefits, there are operational considerations. The effectiveness of API-based control relies on careful authentication, trust relationships, and policy governance. If too many services request premium treatment without coordination, the L4S path could become overloaded, undermining its performance.

Additionally, while the charging mechanisms in the model are accurate in simulation, their application in commercial transport systems would require alignment with broader billing platforms, user policies, and possibly service-level agreements.

Figure 14.5 outlines an example of such an operational flow, showing the balance between exposure, enforcement, and charging for mobility services across different city zones.

14.6.4 Contribution to the Field and Future Integration Potential

What sets this approach apart from existing models is its cohesive use of standardized 5G core functions for transport service orchestration. Rather than developing custom overlays or siloed control systems, this chapter demonstrates that public mobility operations can be deeply integrated with the network core through already available interfaces.

Future extensions may include enhanced analytics at the NEF layer to forecast service demand or additional L4S queue classes to reflect varying degrees of urgency. These enhancements would further cement the network's role as a proactive enabler of intelligent transport systems.

14.7 Conclusion

This chapter has presented a practical architecture for integrating programmable service control into transport networks through the use of NaaS interfaces and low latency flow management. By focusing on exposure, policy, and enforcement mechanisms native to the fifth-generation core, the proposed system demonstrates how external mobility platforms can gain dynamic control over the behavior of network elements without compromising standard compliance or operational scalability.

The implementation and results confirm that transport systems benefit when the underlying network is able to adapt to time-sensitive requirements in response to service conditions. With the ability to activate differentiated treatment on demand, public mobility applications can meet their performance goals while preserving efficiency and fairness. Furthermore, the integration of charging signals into the policy framework introduces a viable path toward monetizing service tiers, especially when catering to premium or emergency use cases in urban environments.

Unlike static or hardware-based solutions, the presented approach operates within existing architectural boundaries. It extends what is already standardized by placing the logic of control into the hands of authorized applications. As cities continue to expand their smart infrastructure, this model allows transport networks to evolve with it—programmable, responsive, and accountable.

The convergence of network programmability, low delay data handling, and autonomous systems is no longer a future objective. It is a necessary foundation for delivering sustainable, intelligent mobility. The work discussed in this chapter lays down a framework that not only meets this need but also shows how to operationalize it in a way that respects both technical rigor and deployment constraints.

This is not just a theoretical shift. It is a functional transformation of how transport systems can leverage the network as a tool, not just a pipe. The approach presented here sets the stage for further development in urban mobility where intelligence does not stop at the edge—it reaches deep into the fabric of the network itself.

Acknowledgments

The author would like to thank all research collaborators and institutions whose feedback and technical insights were instrumental in shaping the direction of this chapter, especially for the patience of **D. Peter Augustine, PhD**, Professor at Christ University, Bangalore, India, and the opportunity by **Pethuru Raj, PhD**, SMIEE, Chief Architect and Vice President, Reliance Jio Platforms Ltd, India.

Nomenclature

Control Plane: The part of a network that carries signaling traffic and is responsible for routing and network control.

Data Plane: The part of a network responsible for carrying user traffic.

Dual Queue Model: A queue management technique that separates low delay traffic from regular traffic for congestion handling.

Low Latency Queue: A dedicated queue in the network configured for fast forwarding and early congestion notification.

Policy Rule: Network enforcement logic generated by PCF based on application requests.

Service Request: A command initiated by an external application to alter network behavior via NEF.

Session Trigger: An event that initiates a change in session behavior, such as charging or QoS updates.

Abbreviations

5GC: Fifth Generation Core
AI: Artificial Intelligence
ECN: Explicit Congestion Notification

L4S: Low Latency, Low Loss, Scalable Throughput
NaaS: Network as a Service
NEF: Network Exposure Function
PCF: Policy Control Function
QoS: Quality of Service
UPF: User Plane Function
V2X: Vehicle to Everything

References

1 European Telecommunications Standards Institute (ETSI). *System Architecture for the 5G System (Release 18)*. ETSI TS 123 501 V18.5.0, 2023.
2 European Telecommunications Standards Institute (ETSI). *5G System; Network Exposure Function Northbound APIs*. ETSI TS 129 522 V15.3.0, 2018.
3 TM Forum. NaaS transformation, Version 12.0, 2020. https://www.tmforum.org/resources/reference/ig1224-naas-transformation-v12-0-0/ (accessed 24 September 2024).
4 Shohail, A. and Patwary, M. A novel NaaS API framework for L4S enforcement: Enhancing traffic management through NEF, PCC, and AI-driven UPF in 5G networks. In *2025 IEEE Future Networks World Forum (FNWF)*. 2025.
5 Shohail, A. and Patwary, M. Optimizing 3GPP 5G NEF integration: Enhancing programmable networks through NEF and UPF interoperability. In *2024 IEEE Future Networks World Forum (FNWF)*. 2024.
6 Bell Labs. L4S: Low latency, low loss and scalable throughput. 2024. https://www.bell-labs.com/institute/white-papers/l4s-low-latency-low-loss-and-scalable-throughput/ (accessed 13 December 2024).
7 Nokia. Nokia and Elisa conduct successful PoC on monetizing L4S with network as code. 2024. https://www.nokia.com/blog/nokia-and-elisa-conduct-successful-poc-on-monetizing-l4s-with-network-as-code/ (accessed 13 December 2024).
8 Linux Foundation. APIs enabling seamless access to telco network capabilities, 2023. https://camaraproject.org/ (accessed 13 December 2024).

15

AI-driven Public Transportation: Enhancing Efficiency, Sustainability, and User Experience

M. Robinson Joel

Department of Computer Science and Engineering, KCG College of Technology, Karapakkam, Chennai, Tamil Nadu, India

Corresponding author: joelnazareth@gmail.com

15.1 Introduction

AI is being utilized more and more in transportation to improve user experience, increase safety, and streamline traffic flow. In order to cut down on traffic and travel time, AI-powered systems evaluate data in real-time, forecast traffic patterns, and recommend other routes. AI is also essential for creating predictive maintenance, driverless cars, and intelligent public transit systems. To comprehend the state of traffic, artificial intelligence (AI) algorithms examine data from a variety of sources, such as sensors, GPS units, and traffic cameras. AI enables proactive changes in roadway signals and path recommendations by anticipating traffic patterns and possible bottlenecks. In reaction to real-time traffic occurrences, such as accidents, AI-powered systems may redirect cars, modify traffic lights, and notify emergency agencies.

As a vital component of the urban mobility ecology, urban public transport networks are essential to the sustainability and operation of contemporary cities. Significant improvements in operational effectiveness and customer service quality have been made possible by the revolutionary capabilities brought about by the development of AI [1]. This study examines how AI-driven methods might be used to optimize public transportation systems, with a particular emphasis on how they can enhance organizing, planning routes, and predictive maintenance. AI tackles major issues, including headway alterations, traffic congestion, and traveler flow management by utilizing predictive modeling, real-time data statistical analysis, and sophisticated optimization algorithms.

This chapter illustrates how AI applications lead to improved passenger happiness, lower operating costs, and higher system resilience through a review of examples from several European cities [2–4]. Gaining knowledge of these developments might help one better understand how AI might promote long-term enhancements to urban public transit [5]. It should be noted that a number of factors, including the operation of collaborating systems like park-and-ride and bike-sharing programs, are necessary for effective public transit [6]. The necessity to investigate the crucial elements that affect the effectiveness of AI executions, such as the integrity of the data, methodology selection, and blending techniques, is indicated by gaps in the literature. By doing this, it seeks to offer a thorough foundation for future studies and practical implementations in urban transportation development.

Through improvements in user experiences, safety, and efficiency, AI [7] is transforming the transportation industry. Applications of AI in transportation include traffic control, personalized travel services, driverless cars, and predictive maintenance. While AI-driven analytics aid in route optimization, traffic reduction, and maintenance demand prediction, autonomous cars employ AI for navigation and real-time decision-making. Additionally, AI is helpful in increasing public safety through monitoring systems that monitor roadway circumstances and incident reactions. In addition to simplifying operations, this AI integration is opening the door for more intelligent and environmentally friendly urban mobility options.

Transportation is changing as a result of AI [8], becoming more intelligent and effective. AI is at the cutting edge of technological advancement in this industry, from intelligent traffic control systems that improve safety and decrease congestion to self-driving automobiles that traverse crowded streets on their own. Additionally, it assists with predictive maintenance for infrastructure and automobiles, guaranteeing that problems are fixed before they become serious. AI improves the commuter experience by optimizing routes and timetables through the analysis of enormous volumes of data. AI in mobility is essentially like having a very intelligent co-pilot that not only assists with driving but also smoothly coordinates the entire trip.

Urban public transit is undergoing a transformation thanks to AI, which improves service quality and operational efficiency. This section examines how AI technology may be used to optimize public transport systems in a number of ways, such as scheduling, mapping routes, and maintenance prediction. AI tackles important issues such as headway errors, congestion control, and passenger flow by utilizing machine learning techniques, real-time data statistical analysis, and predictive modeling [6]. The revolutionary effects of AI are demonstrated by case studies from various international contexts, which show increases in passenger happiness, operational dependability, and service regularity. Comprehending these uses lays the groundwork for evaluating AI's contribution to the development of sustainable urban transportation options.

AI in public transportation systems, so what is public transportation?. Public transportation is a form of travel that allows more people to travel across designated routes at minimal prices. Buses, trains metros all of them encompass public transport and public transport can be said to be the lifeline of any sustainable city public transport has numerous advantages that it poses to passengers to the environment to the economy and to the community in general so these advantages are can be said to be these advantages. Environment friendly and that public transportation helps prevent global warming so let's just take the example of an of a regular bus that we have which has about 50 passengers on it so the bus reduced the need of 50 private vehicles on the road thus cutting down on harmful emissions this also reduced the traffic congestion and increased the land value remarkably by cutting down the number of private vehicles that could have been used if the bus was not in function also it's very important to know that public transportation facilities like buses and metro services are very economically feasible when you compare them with private vehicles. This is because you actually have to purchase the private vehicle on your own, whereas in the case of buses, you just have to complete a very economically friendly and a very viable ticket to the public transportation system that you are going to be using.

Moreover, public transportation ensures increased mobility to those passengers who do not have their own driving licenses, like students or those who cannot drive. It also provides increased mobility to tourists who are not very well aware of the road networks in alien lands. The transportation is also safer for the elderly and for women because there are a large number of people who are traveling at the same time, but in spite of all these advantages, public transportation could not achieve the popularity that it was supposed to have achieved, and that's largely underappreciated, especially in developing countries. This could be old to qualitative and quantitative factors, so let's take a look at these factors. Qualitative factors could include factors like overcrowding and unhygienic public restrooms that actually are major turnoffs for people who don't really know who are health-conscious. Also for people who have another option with them to use private vehicles and therefore only captive riders or writers who don't have any alternate solution with them choose public transportation now let us look at the qualitative or the quantitative factor for this let us assume that this is the network architecture for a particular town and that this arrow basically do denotes the route of the bus that the bus is supposed to take this.

People who reside in this spot or in this part are not actually provided the advantage of the public transportation system and have to compromise on accessibility; henceforth, public transportation is largely nonuniform, and therefore, measures need to be taken in order to popularize it further. The solution for this problem is AI-based or intelligent transportation systems. Intelligent transportation systems are an integration between conventional transportation techniques, communication engineering, IoT, and above all, AI; its tools can render public transportation operationally appealing and remarkably efficient. Look at the its world bank 2016 toolkit it states that retrofitting modern extensions to already establish infrastructure is inherently tedious as is the case of developed countries this means that like in developed countries. A well-established network architecture for public transportation, it is actually very difficult to take down that network and replace it with a modified version of it; however, developing and underdeveloped countries have negligible to minimal transportation pipelines. Take India it becomes really easy to incorporate modifications and changes in those countries.

15.2 Existing AI Uses in Public Transportation

The investigation and development of technology for autonomous vehicles (AVs) is flourishing in China, as well as the United States. Advanced motorist assistance systems (ADAS) [9] and autonomous driving are being developed either directly or indirectly by companies such as Baidu and Huawei. Although some services were previously made available, the majority of project information on the internet is in Chinese. Additionally, hardly any videos are broadcast on platforms that are widely utilized in Japan, making it challenging for non-Chinese people to obtain information.

Apollo compared to Baidu, one of the top search engine providers in China. This open-source project for autonomous driving is frequently likened to Autoware. We had the chance to ride one of Apollo's robotaxis in Shenzhen, which has operations in more than ten places in China. The cars are based on a model created by the ARCFOX EV brand of Beijing Automotive Group. Cost is a barrier to Level 4 autonomous driving, but Apollo has successfully reduced the price of their car to about 480 000 RMB (about 10 million yen), which is one-third of what other choices cost. When compared with Waymo and Cruise, the system used fewer LiDAR sensors [10] to cover the sides and back of the car.

In 2016, former Baidu workers formed Pony.ai [11]. It has a base in Guangzhou in addition to its headquarters in Fremont, California. They used the PonyPilot+ app to hail a ride in a Pony.ai car between the Hong Kong University of Science and Technology facility in Guangzhou. The well-known ON TIME app allows users to hail cabs as well. The automobile drove smoothly together a street with parked cars lining it. Additionally, it was able to locate spaces between parked vehicles to stop. There had been no need for a manual override because there were only a couple of kilometers to go and minimal traffic.

Private transportation is not the cause of disruptions, which may never even be found, also cases like eve-teasing and pickpocketing. Pickpocketing is one of several highlighted issues that really increase the problems associated with public transportation systems. Now, the solution for this could be vehicle tracking by using GPS or global positioning systems [12]. An automatic vehicle location (AVL) and automatic vehicle identification (AVI) [13], so GPS can provide real-time vehicle tracking for every vehicle.

So movement of each bus, the trip length, the time, the stopping time per stop, and the complete movement of the bus or the vehicle could be monitored the system can then inform the control system whether the bus is running on the right time or not and then AI could be used to deem as to why is the bus not running it on the right time. GPS systems can also help in controlling instances of eve-teasing and pickpocketing, as the driver or any passenger around can inform the control room and the police control room.

Routes are safer by analyzing data from previous tours or previous destination routes of the bus AVL or AVI are the technologies that track transit vehicles in real time using GPS, and they could also read number plates using techniques like RFID [14], which is radio frequency identification and image processing. So these techniques are used to identify vehicles, and even the letters that are present on the number plates, information about the vehicle is conveyed to the control center and to the passengers as well.

Transportation use facilities like AZURE's cognitive services [15] to identify and recognize vehicles so if you can see in this example the blue spots over here this blue spot over here is basically used to denote the initial point from where the bus actually started and this blue spot over here is basically supposed to be the point the destination of the bus and this red spot over here is basically used to denote where your bus is right now and real-time information is passed on to the passengers like expected arrival 7 am running on the right time.

15.3 Recognizing AI's Significance in Transportation

AI integration into current transport networks is a challenging undertaking with organizational, cultural, and technological obstacles. Many transport infrastructures were not built with contemporary digital technology in mind, and many of them are decades old. AI retrofitting of these systems necessitates software integration that enables smooth communication between the old and new systems as well as hardware upgrades, such as the installation of sensors and communication devices. Furthermore, implementing new technology may necessitate significant organizational change, including training and modifications to roles and procedures.

Transport networks are becoming vulnerable to cybersecurity threats [16] as AI is incorporated into them more and more. Because these systems manage vital infrastructure, they are prime targets for intrusions that might cause extensive damage. Safeguarding these systems entails protecting the information they send and making sure the AI is resistant to manipulation and abuse. This calls for specialized cybersecurity processes that guard against both internal and external threats, as well as ongoing upgrades and monitoring. There are serious ethical and privacy concerns with AI in public transportation. These AI systems frequently gather and examine vast volumes of data, including private information about people's whereabouts and behaviors. It is essential to handle this data responsibly to protect privacy and maintain public confidence. The decision-making [17] of autonomous systems raises ethical concerns as well, particularly in dire circumstances when human lives may be in jeopardy, such as when an autonomous car must respond in the event of an accident that cannot be prevented.

AI technology is developing at a rate that frequently surpasses the current legal and regulatory frameworks. To handle the particular difficulties posed by drone use, driverless cars, and AI-based traffic control systems, new regulations and guidelines are required. In order to guarantee that there are precise rules and accountability frameworks in place, these regulations must address safety requirements, testing and certification procedures, data protection, and culpability for incidents involving AI systems. AI has two different effects on the economy of the transportation sector. On the one hand, it can result in cost savings, high-tech employment creation, and notable efficiency advantages. On the other hand, automation may lead to the loss of jobs in traditional occupations like routine maintenance and driving.

It may be possible to lessen adverse economic effects by proactively addressing these shifts through education, retraining initiatives, and legislation that facilitate job transfer. Public trust in AI technology is necessary for its widespread adoption in transportation. In order to answer public worries about how new technologies may impact daily life, it is necessary to show that they are dependable and safe. This trust may be developed by public engagement through open communication, rallies, and collaborative development procedures. The infrastructure that supports AI in transportation is crucial to its successful use. In addition to the digital infrastructure needed for safe and effective data processing and analysis, this also includes the physical infrastructure needed for new technologies like electric car charging stations [18].

Successful AI integration into transportation systems requires large expenditures in both fields. The ecological burden of developing and sustaining AI technology must be taken into account, even if AI can optimize traffic flow and routes to lower emissions. AI systems may be developed, designed, and disposed of using sustainable principles to guarantee that their use really helps the environment.

15.4 AI Applications in Transportation: Exemplary Instances

AI is used by AVs, often known as self-driving automobiles, to evaluate complicated sensory input and make judgments in real time. By analyzing data from LIDAR, WHICH IT cameras and radar, AI systems generate a comprehensive picture of the vehicle's environment that includes obstacle detection, traffic signal interpretation, and course selection. Businesses such as Tesla have incorporated this technology into their Autopilot system [19], which is capable of handling parking, lane changes, and highway driving on its own. A major step toward lowering the number of human-driven vehicles on the road is also being taken by businesses like Waymo, which are testing completely autonomous taxis in urban settings.

AI-powered traffic management systems improve traffic flow across metropolitan environments by utilizing data from surveillance footage, sensors, and navigational devices. By using advanced machine learning algorithms to examine traffic patterns and forecast future circumstances, these systems are able to dynamically modify traffic signals in order to reduce congestion. By avoiding numerous stops and starts, this proactive change not only improves the flow of traffic but also dramatically lowers car emissions. Copenhagen is a good example, where sophisticated traffic systems optimize signal timings by using real-time data. This technology has shown promise in transforming urban mobility in a sustainable manner by cutting down on commuting times and environmental impact.

AI is used in predictive maintenance to foresee issues before they occur by using both historical and current data from infrastructure and automobiles. By switching from responsive to proactive maintenance procedures, this technique drastically lowers downtime and related expenses. For instance, airlines use AI to closely monitor aeroplane engines. By anticipating possible faults, maintenance may be proactively arranged based on actual requirements rather than regular inspections. By guaranteeing that aircraft will only be delayed for repairs when absolutely required, this improves fleet efficiency and safety while simplifying operations and saving money.

By optimizing flight routes, avoiding no-fly zones, while autonomously locating to exact drop-off destinations, AI-enabled drones are transforming delivery businesses. When it comes to transporting vital medical supplies to isolated or challenging-to-reach locations, where conventional delivery techniques might be cumbersome and ineffective, this cutting-edge technology is quite helpful. Another instance is Zipline, which uses drones to transport vaccinations and blood throughout Rwanda. These drones' AI systems choose the quickest and safest paths, avoiding obstructions and instantly adapting to the weather. By guaranteeing that vital medical supplies arrive at their destinations promptly and dependably, this capacity drastically cuts down on delivery delays and improves healthcare results, saving lives and strengthening the healthcare system [20].

By using passenger data to apply dynamic pricing, AI in smart ticketing systems revolutionizes public transport by reducing congestion and improving the overall traveler experience. These systems optimize resource utilization and more equally distribute the passenger load throughout the day by incentivizing passengers to take commuter trains throughout off-peak hours by modifying ticket pricing in real-time according to demand. In Singapore, for example, commuters are successfully encouraged to go during off-peak hours by AI-powered price adjustment on public transport. In addition to reducing traffic during peak hours, this calculated pricing change enhances the sustainability and efficiency of the transportation systems, making traveling easier and more enjoyable for everybody.

By continually tracking and evaluating data from train operations, AI technologies greatly improve rail safety. These technologies are skilled at tracking circumstances and detecting abnormalities in wheel performance, spotting problems that can cause collisions. The use of AI enabling real-time high-speed rail surveillance in Europe is an example of how important it is to maintain stringent safety regulations. By facilitating preventative maintenance and quickly resolving any abnormalities, AI lowers the danger of derailments along with additional safety accidents. This careful monitoring enhances the dependability and effectiveness [21] of rail services in addition to guaranteeing the safety of passengers.

By integrating self-driving container trucks and automated cranes, AI significantly improves port productivity. The Port of Rotterdam, for example, uses AI to optimize logistical processes. This advanced system greatly reduces ship idle periods by proactively scheduling ship arrivals and managing cargo movements. This speeds up cargo processing procedures, increasing port throughput, and cutting down on operating costs. AI is essential to modernizing port operations and enhancing overall efficiency, as seen by this simplified method that not only increases production but also decreases logistical bottlenecks.

AI is being used by ride-sharing services like Uber and Lyft to transform the way that drivers and passengers interact. These services anticipate high-demand locations, optimize travel routes, and effectively pair passengers with nearby drivers using advanced AI algorithms. This dynamic system is always learning and adapting to real-time data, such as traffic patterns, open rides, and driver availability. AI's incorporation into ride-sharing systems thereby improves operational efficiency by accelerating service delivery, cutting down on passenger the length of wait time, and streamlining drivers' routes. This raises the overall effectiveness of the shared transportation ecosystem in addition to improving user happiness.

AI is essential to aviation because it helps airlines manage air traffic, optimize flight routes, and maintain aircraft. To suggest the most secure, including most fuel-efficient routes for aeroplanes, it analyzes enormous volumes of data, such as weather, possible turbulence, and flight dynamics. AI improves operational dependability and passenger safety in this way. By finding quicker paths and ideal flying speeds, this technology also drastically lowers fuel consumption, which lessens the environmental impact of air travel. AI in airlines also reduces delays, giving passengers a more seamless and on-time journey. In addition to streamlining airline operations, this integrated strategy supports the worldwide endeavor to increase the sustainability and efficiency of aviation [22].

15.5 Traffic Management Systems Using AI

To improve traffic flow and lessen congestion, AI is now being utilized in traffic management systems. AI systems can save travel time as well as gasoline consumption by rerouting cars to less crowded routes and modifying traffic signals based on real-time traffic data analysis. Successful implementation of intelligent traffic control is currently underway. Cities including Taichung, Vienna, York, and Rome already depend on PTV's (real-time transport visibility) [23] real-time solution, which blends dynamic traffic modeling and machine learning approaches [24]. PTV Optima is a software program that assists operators in creating accurate traffic predictions up to 60 minutes ahead of time and determining the most effective ways to handle traffic, road closures, and construction sites. Operators can respond swiftly to shifting conditions thanks to intelligent traffic management. Congested roads and traffic bottlenecks can be avoided or lessened even before they occur thanks to this.

AI also contributes to better traffic volume monitoring in several network locations. AI-based its projections on comparable historical traffic scenarios and assists traffic controllers in identifying unusual traffic conditions. Adaptive signal control is another use of AI. To enhance traffic flow and reduce emissions and vehicle delays, Taipei, for instance, utilizes PTV Balance, a program that continuously optimizes traffic signals. AI and smart technologies are frequently the driving forces behind eco-friendly mobility. The goal is to collect and analyze data in order to enhance the operation and communication of transportation networks. We might profit much from these systems if they were used more frequently.

15.6 Top AI Resources for Public Transportation

Numerous issues that affect the effectiveness, safety, and consumer happiness of public transportation networks worldwide are explained as shown in Figure 15.1. Transit organizations must handle growing passenger loads, intricate route networks, plus the demand for precise, real-time information as urban populations grow. Passenger annoyance and decreased service dependability are frequently caused by problems including congestion, delays, and ineffective route design [25]. By improving the intelligence, safety, and efficiency of transportation networks, AI tools are revolutionizing the process of addressing these challenges. AI is transforming travel through immediate information analysis, maintenance predictions, better scheduling, and customized services. These cutting-edge technologies support data-driven decision-making, improve operational effectiveness, and give passengers an improved experience in general for public transportation operators. AI integration is making transport systems more economical, proactive, and adaptable, opening the door for smart cities along with urban mobility in the near future.

Although public transit is crucial to the operation of urban settings, its effectiveness, safety, as well as satisfaction with customers are becoming more and more problematic as cities and populations increase. Public transport systems are being revolutionized by AI tools, which are making them safer, smarter, and more effective. AI is revolutionizing the way public transit functions globally, from streamlining scheduling to enhancing passenger satisfaction and guaranteeing safety.

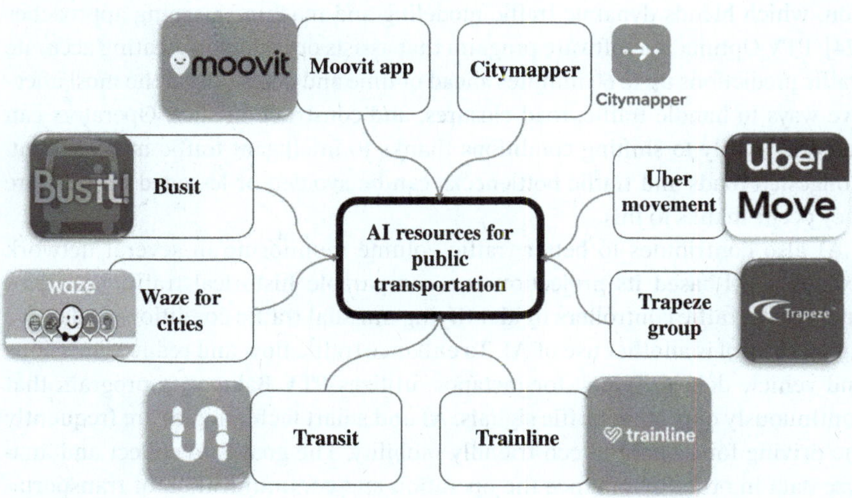

Figure 15.1 Top AI resources for public transportation.

15.6.1 Moovit App

The Moovit app uses a range of technologies to offer route planning and information on public transport. Important technologies include Java, JavaScript, HTML, Python, and Amazon Web Services (AWS) [26] for server-side development, as well as a number of technologies for its website, such as Google Tag Manager, as well as Google Font API. Along with its own customized maps as well as OpenStreetMap data, Moovit [27] also uses official and crowdsourced public transport data. Real-time arrival as well as departure times for buses, trams, trains, subways, and other public transit are available on Moovit. Users are able to obtain precise information on their upcoming arrival, which facilitates more effective travel planning. Users may enter their intended destinations into the app to organize their travels. Considering various forms of transportation, transfer locations, and anticipated trip durations, it offers a wide variety of route alternatives. Depending on their preferences, users can select the most convenient path. Turn-by-turn directions, comprising walking instructions to and from transport stops, are provided by Moovit. To get to their location, users may follow the app's live directions, which makes navigating new routes and public transportation systems simpler.

When there are delays, interruptions, or modifications to public transport services, Moovit promptly notifies users and issues service alerts. In order to prevent inconvenience or unforeseen delays, this function assists customers in keeping informed and modifying their trip arrangements accordingly. Moovit uses data from crowdsourcing and invites people to share details about their transit experiences. Real-time updates, such as the number of passengers on trains or buses, the quality of the services, or any other pertinent information, can be reported by users. The app's transportation data is more accurate and dependable thanks to this input. The software allows users to save the places they visit most often. By removing the need to continually enter the same information, this feature enables rapid access to stored destinations. Other mobility services, such as ride-sharing services like Uber or Lyft, are integrated with Moovit. It is simpler for users to select the most practical and economical form of transportation when they are able to compare ridesharing possibilities with public transit options. All users, even those with impairments, may use Moovit because of its design. Features like voiceover assistance, screen reader compatibility, and settings for big text and high contrast displays are all included in the program.

15.6.2 Citymapper

Established as a business in 2011, Citymapper [28] released its initial "Busmapper" app in 2012 to offer bus schedules for the Olympic London. In just a few years, the app has emerged as a competitor to Google Maps in London, which bills itself as "the ultimate transport app" on its website. It quickly spread beyond London's boundaries and kept adding cities all over the world, reaching 40 cities, including metropolitan regions as of December 2019. Although the precise number of users is purposefully kept secret, as of June 2020, the Google Play app store reports that over 10 million Android smartphones had this app loaded. Figure 15.2 shows the layout screen of Citymapper.

According to Dieter and Tkacz [29], data-centric walkthroughs typically omit screen annotations in favor of measuring and classifying data input points at the cost of the screen's overall experience. As a result, the relationship between applications, data, and platforms may be seen as a technical one, ignoring the social values that are ingrained in these collections. According to Ash [30], interfaces are "relational systems" that are intended "to modulate user action with the aim, hope, and promise of producing desirable outcomes for the individuals who own and operate interfaces." They also contend that there has been an increasing tendency to reify and disadvantage algorithms, data, and code as the main medium of analysis.

15.6.3 Uber Movement

Uber launched the Uber Movement program to release data sets to support city development and administration. Along with interactive visualizations, lessons, and case studies, the website offers the ability to obtain historical trip durations, speeds, and movement data. Through a program named Uber Movement 1, the biggest and most well-known Transportation Network Company (TNC), Uber Technologies, started disclosing limited information about their services in select locations in 2017. It offers past information on typical journey durations, velocities, and motion for Uber rides. To protect privacy, data are pooled regionally. Data is accessible for 38 cities worldwide and 13 cities in the United States as of February 2021. Figure 15.3 shows the map details used by Uber movement.

Executives at Uber characterized the training as an excellent educational opportunity. They said they found the visualizations intriguing and that the public sector's initial response was quite good. Public organizations quickly recognized the limits, though. Uber handles the data aggregation, and the supplied pre-aggregated data frequently cannot be included into forecasting and regional planning models. Furthermore, the data is by definition restricted to Uber car movements, which account for no more than 12% of all vehicle movements. In addition to the size of the data sets, public agencies require substantial resources for data management and analysis.

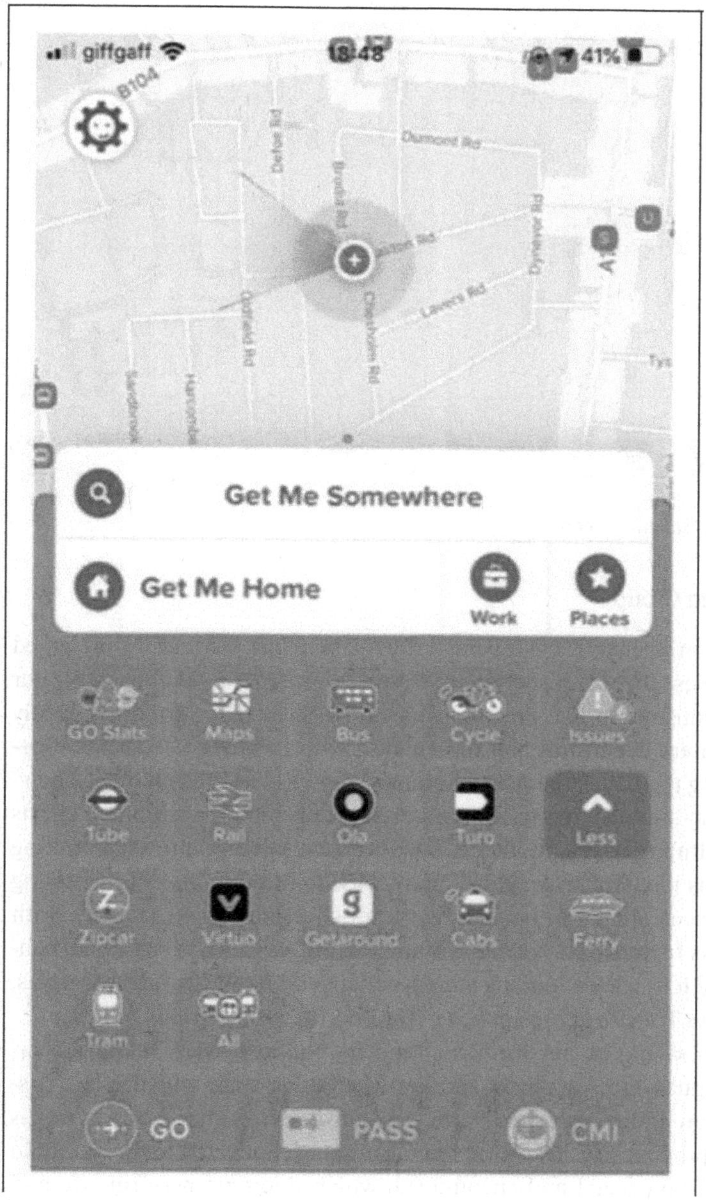

Figure 15.2 Citymapper app layout.

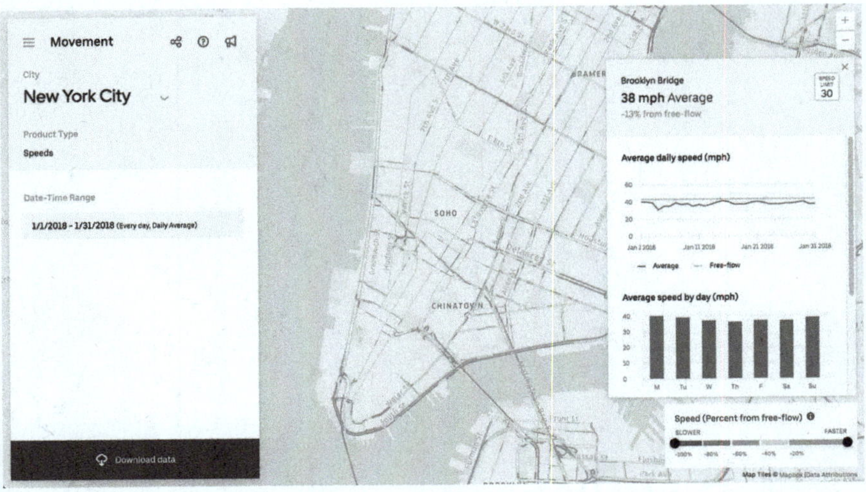

Figure 15.3 Uber layout screen.

15.6.4 Trapeze Group

For transportation agencies, bus operator fluctuation has historically presented serious difficulties. The Transportation Research Board's 2023 Bus Operator Workforce Administration like Practitioner's Guide states that attrition is mostly caused by voluntary departures and retirements, with turnover rates being especially high during the first three years of employment. Industry assessments, however, indicate that although there is still a scarcity of operators, recruitment efforts have improved. Improved retention rates have been a result of improved recruiting practices, such as mentorship programs and focused outreach. Transit mentoring programs have been shown to have a significant effect on worker stability, with retention rates as high as 93% six months after certification. These changes highlight how crucial it is to keep funding employee support and satisfaction programs. Figure 15.4 shows Trapeze group applied in transportations in various vehicles.

In addition to saving money on hiring and training expenses, keeping operators on board guarantees a steady and knowledgeable staff, which raises customer satisfaction and service quality. In the transit sector, staff involvement is a key factor in productivity. Transit workers are more motivated and committed when they feel appreciated and encouraged, which improves performance and increases the dependability of service delivery. According to Gallup's 2025 study [31], staff engagement in the United States has fallen to its lowest point in ten years, underscoring the critical need for engagement tactics. Organizational restructuring forced return to office environments, and slower recruiting are some of the causes of this trend. This nationwide trend emphasizes how crucial it is for

Figure 15.4 Trapeze group applied in transportations.

transportation agencies to take proactive steps to guarantee employee continuity and excellent performance.

Transit agencies may combat disengagement and increase operational efficiency by putting in place programs like professional development opportunities, recognition programs, and feedback mechanisms. Building a culture of responsibility and continuous development via encouraging open communication and recognizing employee contributions positively improves worker productivity and the traveler experience. Customer happiness and customer service are directly impacted by employee engagement, which benefits more than simply the workforce. Transit workers are more enthusiastic and dedicated to their jobs when they feel appreciated, encouraged, and involved, which improves the riding experience and fosters public confidence. According to recent studies, more than 70% of Human Resource experts concur that increased employee engagement directly raises customer satisfaction. This emphasizes how crucial it is to establish work settings that value possibilities for growth, acknowledgement, and communication.

15.6.5 Trainline

A British digital rail as well as coach technology platform, Trainline (previously Thetrainline.com), operates throughout Europe. Figure 15.5 shows the layout screen of the APP. Through its website as well as mobile app, which is

Figure 15.5 Trainline layout screen of the App.

accessible on the Android and iOS platforms, it offers free access to real-time train timings and station information in addition to selling train tickets and rail cards. It is part of the FTSE 250 Index [32] and is listed on the stock exchange in London. In addition to offering a direct online service to consumers under its own brands, Trainline and Qjump, it also administers websites for a few UK train operating companies and offers a rail business travel service directly to several big blue-chip companies, travel management firms, and travel agencies.

15.6.6 Transit

In order to ascertain the actual position of buses and trains, transport was built to crowdsource user input to aggregate and map real-time public transport data. Transit was initially made available for the iPhone in 2012, and the Android version was released shortly after. When available, Transit provides users with timetables and notifications for a variety of transit options, such as buses and trains. Sam Vermette, along with Guillaume Campagna, created Transit in Montreal, Quebec, Canada, with the intention of reducing the number of people who must own cars in urban areas. Other transit mapping systems like Moovit and Citymapper, together with general navigation tools like Google Maps, Bing Maps, as well as Apple Maps that also include transit data, are direct competitors to Transit as shown in Figure 15.6.

In addition to several bike-share programs, Transit is interoperable with ride-hailing and car-sharing applications including the following: Uber, Lyft, Via, which is and Ola. Four American cities' scooter-sharing programs were included in the app in April 2018. Users can instantly link a color with the method of transportation they are watching according to the app's color-coded system, which aligns colors with modes of transportation. An upgrade to the Transit app was published in February 2019 that enabled users to check the timetables of buses and trains around their city without using their data plan, as well as find out whether there are bikes available at a bike-sharing station. Users may plot their route and locate the closest public transport stops even when they are not online.

Figure 15.6 Transit layout screen of the App.

15.6.7 Waze for Cities

Waze collaborates with companies and audio partnerships to provide its drivers with personalized navigation themes that allow them to ride alongside their favorite icons, such as Cookie Monster, musicians, or movie characters. In addition to complementary graphics inside our software, including moods and car symbols, Waze and its partners collaborate to develop a program that can include a new voice to feed to drivers to enjoy. Talent, production facilities, labels, and businesses looking for an unfamiliar demographic and unbroken fan interaction will find a special promotional potential in the partnership. The layout and details are shown in Figure 15.7.

15.6.8 BusIt

BusIt is a predictive analytics-driven AI-powered bus service optimization technology that enhances service schedules and minimizes delays. BusIt assists transit authorities in optimizing bus routes and timings, guaranteeing on-time arrivals and improved service coverage, by evaluating historical data, current traffic circumstances, and weather trends. Advantages are that the Delay reduction through predictive analytics. Integration of real-time weather and traffic data. Enhanced scheduling and route optimization. The con is that the Bus services are the main focus. Existing systems may be difficult to implement.

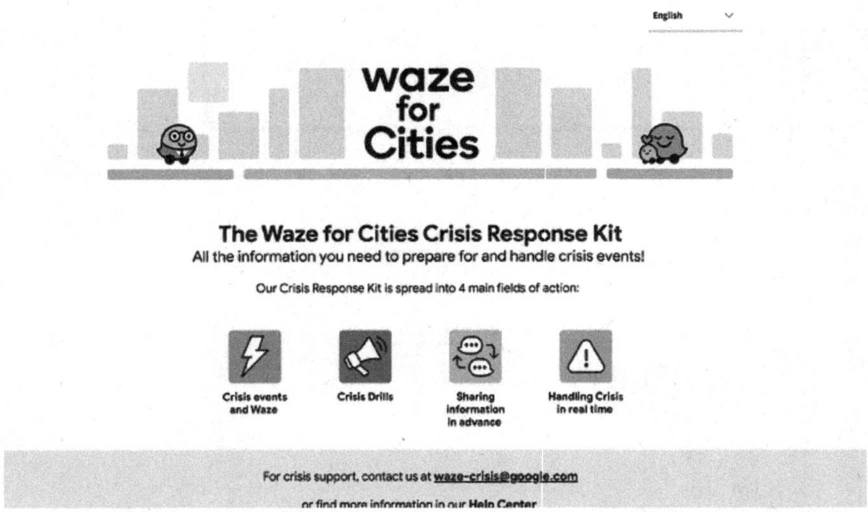

Figure 15.7 Waze for cities details.

15.7 AI Improve Public Transportation Efficiency

In light of this, the combination of AI with public transit holds the potential to completely transform urban mobility by providing a model for transit systems that are safer, more effective, and environmentally friendly. There is now a significant gap in the provision of public transport in India, with an estimated 1.2 buses per thousand individuals, according to estimates from several industry sources, including Niti Aayog. But there is a great deal of promise to close this gap and transform public transport, especially electric buses, with the introduction of cutting-edge technology and government initiatives.

The future of bus service appears to be becoming more sustainable as a result of the government's emphasis on encouraging renewable energy sources and its goal of having 70% of commercial vehicles—including buses—be electric by 2030. The usage of electric and green fuel-powered buses is being propelled by technological advancements as well as government programs and incentives. Furthermore, improvements in AI are expected to improve public transit systems' accessibility and efficiency. Transit authorities may increase operating efficiency while guaranteeing passenger satisfaction and security by utilizing AI to optimize timetables, forecast demand, and carry out predictive maintenance [33]. Furthermore, AI-powered solutions have the potential to significantly improve accessibility for all users, including those with impairments, thereby fostering inclusion in public transportation.

Using data analytics to enhance traffic flow, reduce congestion, and improve overall transportation efficiency, AI plays a critical role in traffic management. AI provides affordable solutions that enhance road safety and lower accident rates through smart city technology like parking management, along with traffic control systems. AI has the ability to completely transform route planning, particularly operational effectiveness in public transportation by forecasting demand for particular routes based on previous information, optimizing timetables, and simplifying ticket collecting to identify fraud. Additionally, predictive maintenance driven by AI may spot any problems in infrastructure and cars before they become serious, guaranteeing the dependability of public transport systems.

Cities throughout the world are implementing AI-driven technology to improve public transit, which not only increases operational efficiency but also promotes environmentally responsible urban growth. AI-powered public transportation systems are opening the door for more intelligent, responsive, and environmentally responsible urban mobility solutions by promoting increased usage of public transportation and minimizing its negative effects on the environment. AI presents a viable way to transform how people travel around cities in light of the impending problems of urbanization [34], traffic jams, and environmental degradation. We are paving the path for a future where transport is not only effective but also environmentally benign, eventually contributing to a more sustainable world, as nations like India keep investing in AI-driven technology to revolutionize public transit.

15.8 Safety Benefits of AI in Public Transportation

Large volumes of data may be analyzed by AI algorithms to forecast possible risks and safety concerns in public transit. To predict and notify operators of possible safety hazards, these systems employ information from a variety of sources, including traffic patterns, weather predictions, and incident data from the past. This proactive strategy lowers the chance of accidents and increases overall safety by enabling prompt actions. AI, for instance, can forecast maintenance problems in rail systems using real-time data through sensors mounted on trains and rails. By preventing derailments and other mishaps, this predictive maintenance strategy [35] can make passenger travel safer.

Additionally, AI systems can monitor public transit networks as shown in Figure 15.8 in real time, spotting anomalous activity or issues as they happen. These systems use AI-powered sensors and surveillance cameras to identify anomalous activity, such as overcrowding, suspicious activity, or unattended luggage, and quickly notify the appropriate authorities. AI-driven systems can track the performance and environment of autonomous transportation systems

Figure 15.8 Safety benefits of AI in public transportation.

in real-time, allowing for prompt reactions to any irregularities. This guarantees both pedestrian and passenger safety. One of the most important safety features of public transit is traffic management. Traffic congestion and accident risk can be decreased by using AI to optimize traffic flow based on real-time data. AI systems are able to forecast the optimal routes for emergency vehicles, reroute buses to avoid crowded regions, and modify traffic lights to maximize flow. AI may also assist in controlling pedestrian traffic between stations and stops, avoiding congestion, and guaranteeing easy boarding and disembarking from cars. There are several advantages of integrating AI into public transportation safety management. AI-based safety solutions throughout public transport are raising the bar for efficiency and safety in the sector by anticipating possible risks, offering real-time monitoring and alarms, and enhancing traffic control [36].

With several useful applications currently in place, AI-based safety solutions for public transit are rapidly becoming a reality. Here, we highlight three major areas where AI is having a big impact: autonomous public transport vehicles, AI-driven surveillance systems, and predictive maintenance for rail systems. A method called predictive maintenance makes use of data analysis to forecast when a system or piece of equipment may break. Rail systems may increase efficiency and safety by implementing AI in this process. AI is capable of analyzing data from a variety of sources, including temperature, pressure, and vibration sensors, in order to spot trends and anticipate any malfunctions. This lowers the possibility of unplanned malfunctions that might interrupt service and endanger passengers by enabling preventive maintenance.

It has the power to completely transform public transit surveillance systems. AI can monitor busy public areas in real-time and notify authorities of any suspicious conduct or possible threats by utilizing technologies like as facial recognition, object identification, and behavior analysis. In order to maintain passenger safety, AI-driven monitoring systems may also assist with crowd control, spotting busy areas, and recommending appropriate courses of action. These technologies have the potential to greatly improve public safety in transit hubs such as airports, rail stations, and bus stops. Public transit is expected to undergo a radical transformation with the introduction of autonomous cars. AI is essential to this development because it powers innovations like sensor fusion, computer vision, and machine learning that let cars navigate their surroundings without the need for human assistance.

15.9 Flowchart for GPS-based Vehicle Tracking

The flowchart, shown in Figures 15.9 and 15.10, shows how a gadget that tracks and shows position data on a remote dashboard using GPS and GSM technologies works. Let's dissect the procedure in detail. The method starts with a

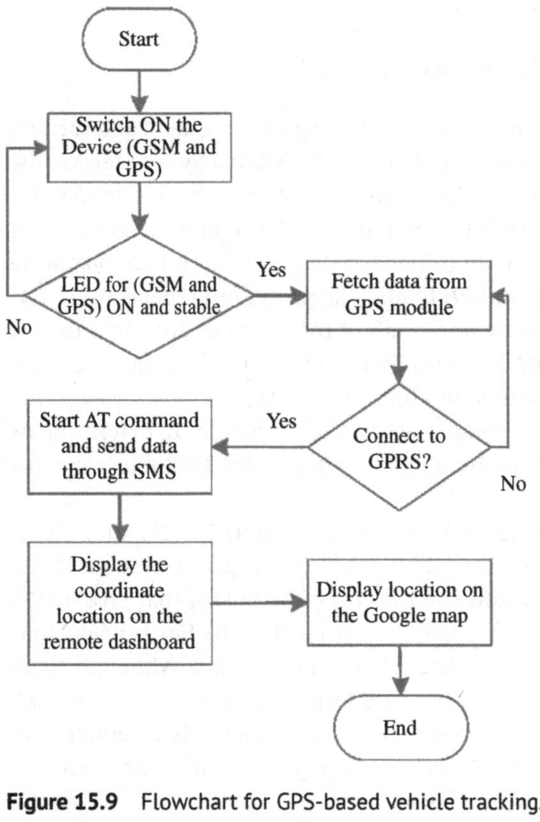

Figure 15.9 Flowchart for GPS-based vehicle tracking.

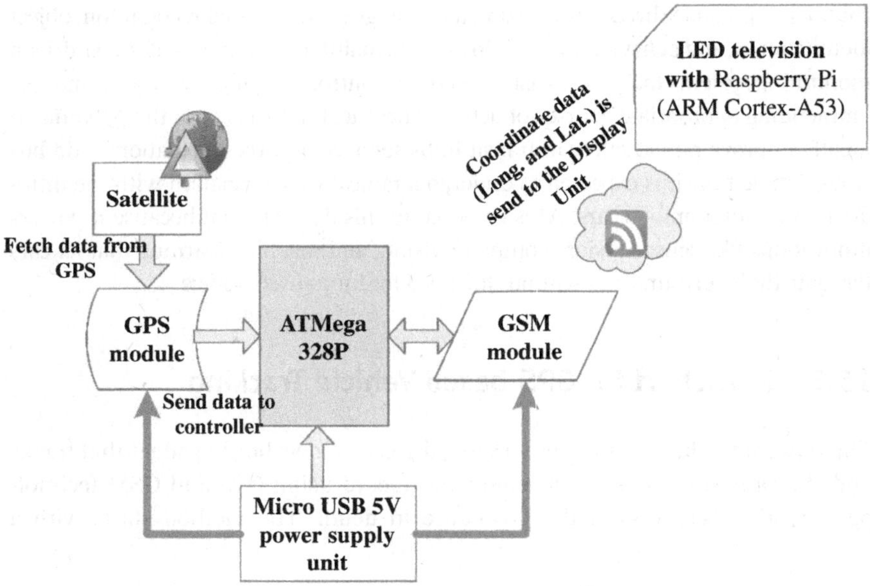

Figure 15.10 Flowchart for GPS/GSM-based vehicle tracking.

"Start" command, which turns on the device. Turning on the gadget and turning on its GPS (Global Positioning System) [37] and GSM (Global System for Mobile Communications) [38] modules is the first active step. A vital check is made after powering up: "LED for GSM & GPS ON and Stable?" This phase confirms that the GPS and GSM modules have both initialized successfully and are operating as intended, as seen by steady LED [39] lights. The procedure repeats itself, trying to turn the device on repeatedly until both of the modules are correctly initialized, if the LEDs(light emitting diode) do not glow on or steady (the "No" branch). After the GPS and GSM modules have stabilized (the "Yes" option), the device moves on to "Fetch Data through GPS Module." In order to determine its current geographic coordinates (latitude and longitude), the GPS receiver must first acquire satellite signals.

Owned by the US Space Force and run by Mission Delta 31, the Global Positioning System (GPS) is a satellite-based geodesic navigation system. It is a member of the global navigation satellite systems (GNSS) [40] that give a GPS receiver geolocation and time data anywhere on or close to Earth when four or more GPS satellites can see each other without obstruction. Although these technologies can increase the utility of the GPS location information, they don't require the user to submit any data and function independently of either telephone or web reception. Around the world, it offers vital positioning capabilities to commercial, military, and municipal uses.

The algorithm then looks for "Connect to GPRS?" On the GSM network, GPRS (General Packet Radio Service) [41] is a packet-switched data service. If the device has GPRS connectivity (the "Yes" branch), it will "Start AT command to communicate data through SMS." The collected GPS coordinates are sent as an SMS message in this instance, thanks to AT instructions, which are used to control modem functionalities.

The architecture of a GPS-based tracking system intended to show location data on an LED screen is depicted in this block diagram. Let's examine how each part works and how they work together to accomplish this. The satellite, a component of the Global Positioning System constellation, is the first component of the system. These satellites send out messages with their position and time all the time. In this arrangement, the GPS Module serves as the receiver. In order to ascertain its own exact geographic coordinates, particularly longitude and latitude, it records these signals from several satellites. This data gathering procedure is shown by the "Fetch data from GPS" arrow. After determining the coordinates, the GPS module requires a central processing unit to manage the information and enable transmission. The microcontroller ATmega328P performs this function. The coordinate information transfer from the GPS Module to the ATmega328P is indicated by the arrow with the caption "Send data to controller." A well-liked and adaptable microcontroller, the ATmega328P is renowned for its low power consumption and appropriateness for embedded applications. After receiving the raw GPS data, it probably formats it appropriately and gets it ready for GSM network transmission. The Global System to supply Mobile Communication is referred to as GSM. Users of mobile phones all across the world frequently use this digital mobile network. It combines frequency division multiple access (FDMA) and time division multiple access (TDMA) [42] and is the most widely used of the three digital wireless telephony systems (TDMA, GSM, and CDMA). The four frequency bands that it uses are 850, 900, 1800, and 1900 MHz. Before delivering it over a channel having two other channels of information about users, each covering a specific time slot, GSM compresses and transforms the data.

Wireless communication over cellular networks is handled by the GSM Module. It allows a remote display device to get the processed coordinate data from the system. To establish a connection and send the location data, the ATmega328P talks to the GSM Module. This data flow is indicated by the arrow that points from the ATmega328P to the GSM Module. A Micro USB 5V Power Supply Unit provides power to the complete system. This device supplies the electrical energy required to run the GSM, ATmega328P, and GPS modules. The power distribution to these components is shown by the thick arrows, which guarantee a steady power supply for ongoing operation.

The LED TV with Raspberry Pi (ARM Cortex-A53) is the last destination for the location data. In this case, the Raspberry Pi serves as both the processing center and the display unit. The cloud icon with the RSS symbol, which suggests a wireless data transfer, is the cellular network via which it gets the coordinate data given by the GSM Module. Many Raspberry Pi versions come with the 64-bit quad-core ARM Cortex-A53 processor, which has enough computing capacity to decode the incoming data and display it on the LED TV screen. To put it briefly, the system works as follows: the ATmega328P interprets the position data that the GPS Module receives from satellites and transmits it over the cellular network via the GSM Module. This data is received by the Raspberry Pi, which is attached to the LED TV, and shows the user the position data. The main tracking components run continuously thanks to the Micro USB power source. This configuration offers a visual output on a typical LED television screen together with a real-time position tracking system.

15.10 Build Your Own ESP32 GPS Tracker with Live Tracking

How to build a simple yet practical GPS tracker using an ESP32 and a Neo 6m GPS [43] module now the specialty of this project is it's very easy to build you can build it on a breadboard and the output of this project looks something like this so you can visualize the data coming out of this GPS tracker on a proper GPS map as shown in Figure 15.11.

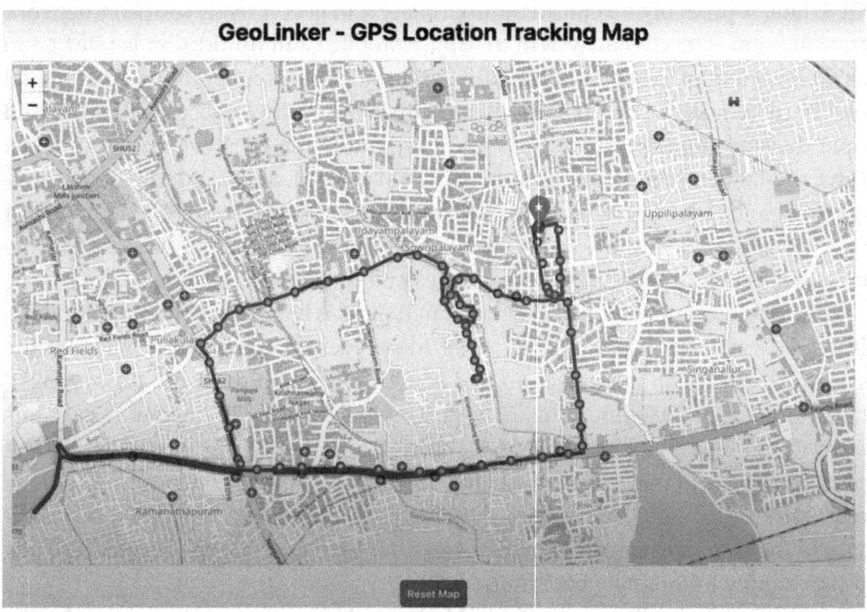

Figure 15.11 GPS tracker with live tracking maps.

This will be plotted on a proper map where you can zoom in and all these dots are the points where the GPS data was sent from our ESP32 board here to our Cloud Server if you click on it you can see the date and time at which this data was recorded and it also plots a neat route showing the path taken by your GPS tracker so you can just refresh this page and see the current location and the history of your GPS tracker if you scroll down you can also see the latitude and longitudes against a time stamp in a proper table.

So this is a practical GPS tracker you can use. Hardware consists of just an ESP32 and a GPS module. What happens here is we're going to write a program that collects the GPS data every 10 seconds and sends it over to a cloud platform, which is developed by us at circuit digest called circuit digest cloud.

So every 10 seconds it's going to capture the location and send it over to the cloud which is free to use if there is no internet connection it's going to save that GPS coordinates in the flash memory and once the internet connection comes back it's going to push all the collected GPS data to the cloud so that you can view on the map so on the cloud side it's circuit digest. Cloud we have provided a lot of APIs, but the API used for this project is called GEOL Linker, so basically this API can be used with ORINO ESP32 [44], Raspberry Pi, any microcontroller, or S modules for that matter. This API basically collects the GPS data, the latitude and longitude, along with the timestamp, stores it on a server, and it populates all of that in a NE map, so to use it, you can just head over to Circuit dig. cloud and click on View Map against the GEOL link or API.

This article tells you everything about how to build this GPS tracker but if you scroll down you'll find the circuit diagram here let me open it up big so the circuit diagram is pretty simple as you can see all we need is a Neo 6m GPS module and ES32 board and a couple of LEDs now once the CSP 32 is powered we have an onboard 5 volts which is used to power the Neo 6m GPS module as well and then we have two LEDs here one is green and another is red so as soon as your US32 board is connected to Wi-Fi the green LED will stop blinking and it will become stable and as soon as your ESP32 is able to get proper GPS data from your NE 6m module the LED will start to light up and it will remain stable as long as there is a valid GPS connection okay so this is the complete code for our GPS tracker project the code is written using ORINO ID again you can find the complete code.

While going to connect to a Wi-Fi network and then we're going to get the GPS data latitude and longitude values every 10 seconds and then if there is an Internet connection available we will directly send those values to the server every 10 seconds if there is no internet connection available then we will save it in our flash memory and then once the internet connection is back it will send it to the server and you can visualize it on a proper map.

Earlier so let's start with the code over here there are few things which you should definitely change first thing is the Wi-Fi SS ID and password so connect to your own Wi-Fi network once the GPS tracker comes in proximity to this Wi-Fi network it will update all the data it has collected and you can visualize the data on a map this link will remain the same and then you need the API key that you can obtain from circuit digest Cloud to get that you just have to get into my account and you will see the API key.

This API with some other development board you can try this documentation here which is called GEOL Linker so this documentation tells you how to use this API with any development board but if you're using with ESP32 you can just go ahead with this code change the API key and that is all you have to do here just connect your board to the computer using your programming cable like this this and then make sure you have selected the right board so minus ESP32 Dev module.

The port as USB serial 001 once that is done just hit on upload okay so now the code is uploaded let me press the reset button so that everything starts from the beginning you can see on the serial monitor here that it is connected to Wi-Fi and that is why the green LED is stable over here but the yellow LED is still off it means we're not getting proper GPS connections and every 10 seconds you can see that our serial monitor says invalid GPS data skipping upload so this means that we're not able to get proper GPS data from the module at this point you can either step out and see if your GPS is able to get signal and Fitch the latitude and longitude values or if you want to get it working indoors then I would recommend you to change the antenna for this project so I have an outdoor antenna here like this one which you can easily purchase it says GPS antenna 1575.42 mhz. So I have this antenna over here. What I'm going to do is I'm going to replace this antenna with a bigger one so that we can get GPS data easily.

15.11 Market Share of AI in Transportation by Different Elements

15.11.1 Analysis Based on Component

The Software sector dominated the AI in Transportation market in 2023, accounting for about 42.7% of the total market. The growing need for intelligent transportation administration systems with cutting-edge capabilities, including real-time traffic control, route planning, and maintenance forecasting, is the key factor driving this importance. The use of software to incorporate AI into current infrastructure has made transport networks more effective and efficient, which is essential as urbanization and the flow of products around the world increase.

Further supporting the Software segment's leadership are continuous developments in deep learning and machine learning technology. These developments make it possible to create increasingly complex and self-governing systems that can learn from enormous volumes of data to enhance decision-making in real time. For example, AI-powered software helps logistics firms optimize fleet management, improve customer service by reducing delays, and increase safety by using predictive analytics to foresee any risks. Furthermore, the software market is expanding as a result of safety regulations and standards that encourage the use of AI in transportation. The potential of AI to improve transportation safety and efficiency is becoming more widely acknowledged by governments and regulatory agencies. The market need for cutting-edge software solutions has increased because of the financing and efforts supporting AI integration [45].

15.11.2 Analysis Based on Technology

With a market share of about 45.1% in 2023, the Machine Learning category dominated the AI in transportation industry. The importance of machine learning technologies in improving the effectiveness and security of transport systems is the reason for this significant market share. In order to manage traffic flows, forecast repairs, and optimize routes—all of which drastically save operating costs and increase service reliability—machine learning algorithms are essential. The use of machine learning in automatic automobile technologies, which collect and interpret enormous volumes of data from car sensors to make driving decisions in real time, further solidifies the segment's dominance.

In addition to being essential to the development of autonomous driving, this technology improves predictive maintenance, which helps prevent auto accidents and maximize fleet management. The market's dominant position in the category is supported by the growing need for machine learning technologies [46] as transportation systems depend more and more on data-driven decision-making. Additionally, machine learning is being used in transportation in a variety of contexts, including passenger information and traffic control systems, in addition to automobile technology. These examples show how machine learning can be used to improve passenger experiences and operational efficiencies, making it a key tool in the AI-driven revolution of the transportation industry.

15.11.3 Analysis Based on Application

With a market share of over 38% in 2023, the AVs category dominated the AI in Transportation industry. Numerous important reasons that propel adoption and technical improvements in this industry are responsible for this significant market share. First and foremost, research and investment in autonomous technologies have been driven by the growing emphasis on efficiency and safety in transportation. AI systems installed in AVs allow them to learn from enormous volumes of data, which enables them to make choices in real-time traffic situations and lowers human error. Key industry participants' technical partnerships and regulatory assistance further solidify the segment's leadership in AVs. Around the world, governments are progressively revising traffic laws to support and promote the safe use of AVs.

Additionally, big tech and automotive firms are collaborating to exchange resources, data, and experience, which will speed up the creation and improvement of autonomous systems. In addition to improving AVs' technological capabilities, these partnerships are assisting in the realization of economies of scale, which will increase the technology's accessibility. Additionally, AVs are becoming more and more popular among consumers thanks to proven advantages, including shorter commutes and lower operating expenses. The impact on the environment is also noteworthy; autonomous cars reduce emissions by optimizing driving habits and fuel consumption.

The dependability and effectiveness of autonomous cars are steadily increasing due to developments in machine learning algorithms as well as sensor technologies, which bode well for even larger market penetration and impact in the years to come. Applications in both passenger as well as commercial transportation, which increase its market reach and effect, are another factor driving this segment's growth.

One major factor propelling AI in the transportation industry is the unrelenting advancement of machine learning and AI technologies. As these technologies develop, they improve the efficiency and dependability of predictive maintenance tools, roadway management systems, and driverless cars. In complicated traffic circumstances, better algorithms allow cars to make judgments more quickly and accurately, improving operational efficiency and safety. This driver is crucial because it not only draws significant investment from big tech and automakers, but it also encourages regulatory and public adoption as technology advancements show how safe and dependable these systems are.

The high upfront and ongoing expenses of deploying AI systems are among the biggest obstacles facing the AI within the transportation business. Significant time, money, and trained labor are needed for the development and incorporation of cutting-edge AI technology. Companies also face additional financial strains from the constant upkeep and upgrades required for AI systems to guarantee optimum performance and security. The rapid implementation of AI technologies for transport may be slowed down by this high-cost structure, which may be unaffordable for smaller firms, particularly in emerging nations or in sectors with more stringent financial limitations.

There is a big chance for the transportation industry to benefit from the combination of AI, the Internet of Things (IoT) [47], and linked infrastructure.

Decision-making processes become more dynamic and well-informed as a result of this integration's improved data collecting and analysis capabilities. For example, AI systems may leverage real-time data from linked IoT devices on infrastructure state, traffic conditions, and vehicle performance to improve overall traffic management, optimize routes, and anticipate repair requirements. In addition to increasing productivity and security, this possibility creates new business models along with services, such as dynamic insurance plans and cutting-edge fleet management systems.

Addressing privacy and security issues associated with data handling, as well as AI decision-making processes, is a significant problem for the AI in transportation business. It is vital to protect this data from cyberattacks since transport systems depend more and more on AI and gather enormous volumes of data. Furthermore, privacy concerns are brought up by the usage of AI, especially when it comes to tracking and profiling people's activities and behaviors. Strong cybersecurity safeguards, open data usage guidelines, and ongoing oversight are necessary to address these issues, preserve public confidence, and adhere to changing legal requirements.

15.11.4 Analysis Regional Wise

North America dominated the worldwide market for AI in transportation in 2023, accounting for almost 36.5% of the total. Solid government backing for AI programs, substantial expenditures in the development of AI, and a solid technology infrastructure are some of the main reasons for this leadership. North America's need for AI in transportation was estimated at USD 1.3 billion in 2023, and it is expected to increase dramatically over the course of the projected period. Leading tech companies like Microsoft, Google, and IBM are based in the area. They are always coming up with new ideas and pushing the limits of AI applications throughout transportation, from driverless cars to intelligent logistics systems.

Furthermore, the rapid adoption of AI technology in transportation is made possible by North America's highly interconnected supply chain network as well as sophisticated manufacturing capabilities. The rapid expansion of this industry is also greatly aided by the large number of startups and their joint ventures with large enterprises. By reducing risks and promoting innovation, improved regulatory frameworks and regulations in the US and Canada also aid in the adoption of AI. Furthermore, the strategic alliances between IT firms and the region's top automakers help to quicken the acceptance and development of new technologies. For example, partnerships like those that exist between Siemens and Microsoft are essential to the advancement of AI integration in the transportation industry and other fields.

15.11.5 Analysis of Key Players

Key players in the transportation industry for AI analysis is the process of examining and judging the well-known businesses and institutions involved in the AI-powered transportation sector. In order to comprehend these major companies' market position, advantages, disadvantages, possibilities, and dangers, it entails examining their strategies, competitive environment, and market dynamics. The businesses leading the way in creating and offering AI solutions for applications in transportation are the major participants in AI in the transportation industry. These participants are skilled in computer vision, machine learning, data analytics [48], AI, and related fields. They provide a variety of goods, services, and solutions tailored to the particular requirements of the transportation sector. Figure 15.12 shows the Global AI in transportation market growth, which encourages the investors.

With its subsidiary Waymo, Alphabet Inc. is a leader in autonomous driving technology, concentrating on creating self-driving cars that have the potential to completely transform the effectiveness and safety of transportation. NVIDIA Corporation plays a key role in supplying deep learning and AI GPUs that drive data processing and driverless cars. IBM Corporation makes a contribution with Watson and other AI technologies that improve a variety of transportation operations by using predictive analytics. The hardware and cloud-based AI solutions that enable the data-intensive activities needed for autonomous car technology are mostly provided by Intel Corporation and Microsoft Corporation. The Volvo Group and Aptiv PLC are renowned for their developments in autonomous driving and automobile safety.

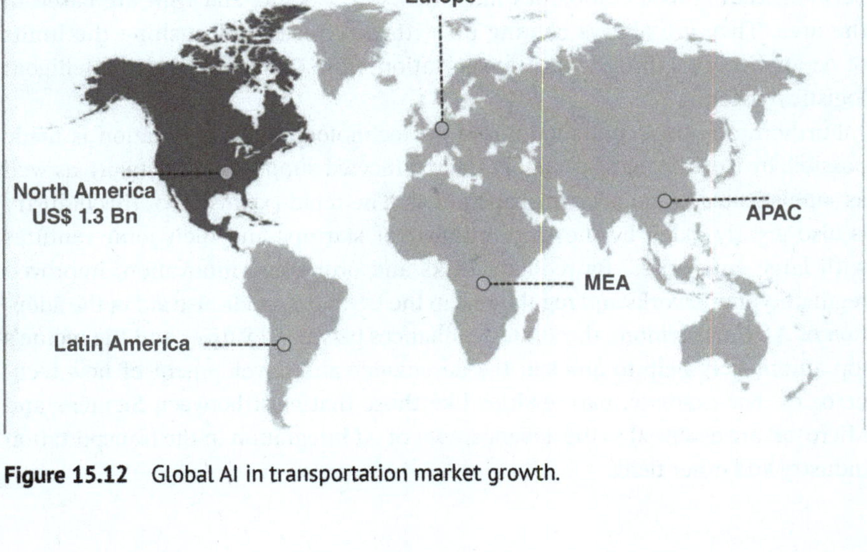

Figure 15.12 Global AI in transportation market growth.

Microsoft and Siemens established a strategic alliance in October 2023 with the goal of improving AI integration in a number of sectors, including transportation. This partnership makes use of Microsoft's cutting-edge AI capabilities from its Azure package in addition to Siemens' industry-specific knowledge. The goal of the project is to provide AI capabilities to Siemens' transportation and other industry clients in order to increase operational effectiveness and advance sustainability. AWS, Meta, Microsoft, and TomTom formed the Overture Maps Foundation in July 2023. The goal of this project is to provide the first extensive open map dataset, with a focus on road network information that is essential to the advancement of driverless cars and associated transportation services.

Waymo, Alphabet Inc.'s subsidiary, witnessed improvements. In May 2023, Waymo and Uber formed a strategic alliance. With this partnership, Waymo's commercial driverless service is significantly expanded in Phoenix and San Francisco, integrating its driverless cars into Uber's ride-hailing service along with food delivery systems. Although specific 2023 pronouncements were not found in the latest search, NVIDIA Corporation, a pioneer in AI as well as deep learning technologies, is constantly improving its GPU solutions, which are essential for processing the massive volumes of data created by autonomous cars.

Both Microsoft Corporation and IBM Corporation have improved AI capabilities by utilizing their cloud platforms.

15.12 Related Work

Through specific claims of advantages related to cost savings, quality of service improvement, the environment, general understanding, and safety/security, we have categorized the benefits sought or attained in the research. Naturally, the particular research' claims about how the findings might be applied to enhance PT are frequently more specific. Moreover, as proposed by Mackett [49], Prashanth [50], and Shatnawi [51], the same AI technology may occasionally be utilized for several goals, such as enhancing service and lowering emissions. There are two classifications for these investigations. Even if they aren't mentioned directly, these numerous advantages most likely hold true for more research.

Figure 15.13 and Table 15.1 show, titled "no. of count," visually represents the frequency of different categories. The horizontal axis displays five distinct categories: "service quality," "general knowledge," "cost savings," "environment," and "safety." The vertical axis indicates the number of counts for each category, ranging from 0 to 70 in increments of 10. The chart clearly shows that "service quality" has the highest count, with a substantial value of 68. This suggests that "service quality" is the most frequently occurring category among those presented.

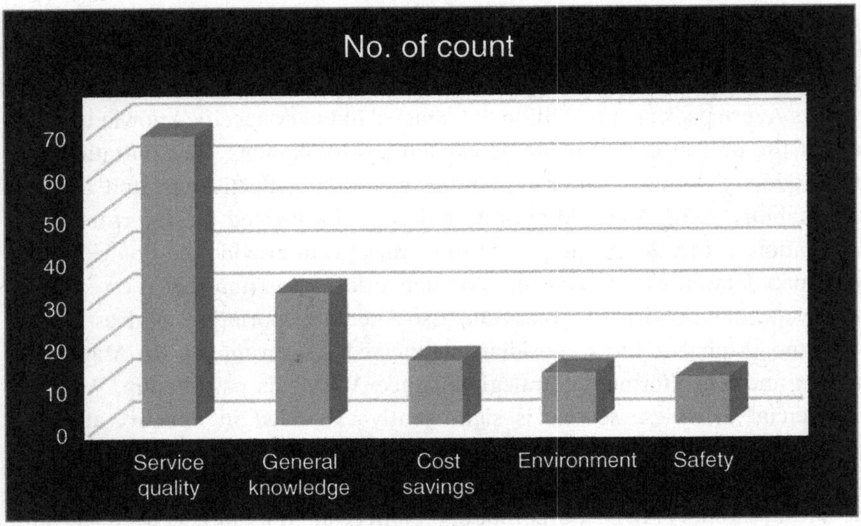

Table 15.1 Numerous research discussing various advantages.

Features	No. of count
Service quality	68
General knowledge	31
Cost savings	15
Environment	12
Safety	11

Following "service quality," "general knowledge" has the next highest count of 31. This indicates that "general knowledge" is also a relatively frequent category, although significantly less so than "service quality." The remaining three categories, "cost savings," "environment," and "safety," have considerably lower counts. "Cost savings" has a count of 15, while "environment" has 12, and "safety" has the lowest count of 11. These values suggest that these categories occur less frequently compared to "service quality" and "general knowledge."

In summary, the bar chart highlights a clear disparity in the frequency of the different categories. "Service quality" stands out as the most prevalent, followed by "general knowledge." "cost savings," "environment," and "safety" occur with notably lower frequency. This visual representation allows for a quick and easy comparison of the occurrences of these five categories.

The research includes a wide range of means of transportation. Furthermore, some studies concentrate on a single mode of transportation, such as traditional buses, trains, or bike sharing programs, while others examine the combination of multiple modes of transportation, such as using a private vehicle and a train for the same trip. The five primary forms of transportation that we have identified are: bicycle sharing networks, buses, railroads, any PT, and PT plus other modes. Research that concentrates on a single mode of transport is categorized under the classes Bus, Railway, and Bike Sharing Systems [52], whereas research that covers both buses and railways are doubly categorized under the Bus and Railway classes.

With its application in 77% of the studies, machine learning is the most widely employed AI technique. ML is used to draw broad conclusions from vast volumes of data. Usually, the outcome is a classifier that may be used to estimate or anticipate some unknown value, such as that pertaining to land use, travel behavior, traffic flows, predictive maintenance, the total number of passengers in a vehicle, or the projected time of arrival. Numerous machine learning techniques, including a network (neural network, next one support vector machine, for example, K-nearest neighbors, decision tree, in the future random forest models, the naive Bayes method, and rough set analysis, have been used. Numerous studies examine and compare various machine learning algorithms to determine which is most appropriate for the given goal. The best algorithms for the various jobs, however, are not universally agreed upon.

Figure 15.14 and Table 15.2, as shown, simply titled "count," presents a comparison of the frequency of different modes of transportation. The horizontal

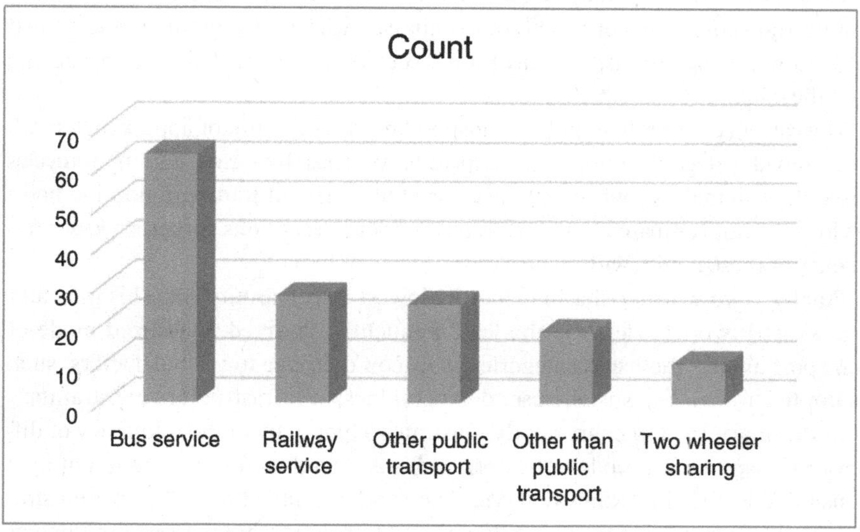

Figure 15.14 Numerous research examining various forms of transportation.

Table 15.2 The quantity of research that covers various means of transportation.

Mode of transport	Count
Bus service	63
Railway service	27
Other public transport	25
Other than public transport	18
Two wheeler sharing	10

axis lists five categories of transport: "bus service," "railway service," "other public transport," "other than public transport," and "two wheeler sharing." The vertical axis displays the count for each category, ranging from 0 to 70 in increments of 10. The most striking feature of the chart is the overwhelmingly high count for "bus service," which reaches approximately 68. This indicates that "bus service" is the most frequently observed or utilized mode of transportation among the categories presented. The significant difference between the count for "bus service" and the other categories suggests its dominant role in the context from which this data was derived.

Following "bus service," "railway service" and "other public transport" show similar counts, both hovering around 30–31. This implies that these two modes of public transportation have a comparable level of frequency, significantly less than "bus service" but considerably more than the remaining categories. The "other public transport" category likely encompasses various forms of public transit beyond buses and trains, such as trams, subways, or ferries, depending on the specific context of the data.

The category "other than public transport" registers a count of approximately 24. This broad category could include private vehicles like cars and motorcycles (excluding shared two-wheelers), taxis, or other personal transportation methods. Its lower count compared to the main public transport options suggests a lesser frequency of these modes within the dataset.

Finally, "two wheeler sharing" has the lowest count, around 14. This indicates that shared two-wheelers are the least frequently observed or utilized mode of transport among the listed categories. This could be due to various factors, such as limited availability, specific user demographics, or infrastructure constraints.

In summary, the bar chart clearly illustrates a hierarchy in the frequency of different transportation modes. "bus service" stands out as the most frequent by a considerable margin. "Railway service" and "other public transport" exhibit similar, moderate frequencies. "Other than public transport" shows a lower frequency, and "two wheeler sharing" has the least frequent occurrence. This visual representation provides a quick and effective way to understand the relative prevalence of these different transportation options within the given dataset.

Figure 15.15 and Table 15.3, as shown, titled "count in percentage," illustrate the relative proportions of different categories, presented as percentages. The horizontal axis displays five categories: "machine learning," "deep learning," "reasoning," "heuristic search," and "others." The vertical axis represents the percentage count for each category, ranging from 0 to 100 in increments of 10.

The chart immediately highlights the significant dominance of "machine learning," which registers a percentage count of approximately 95%. This indicates

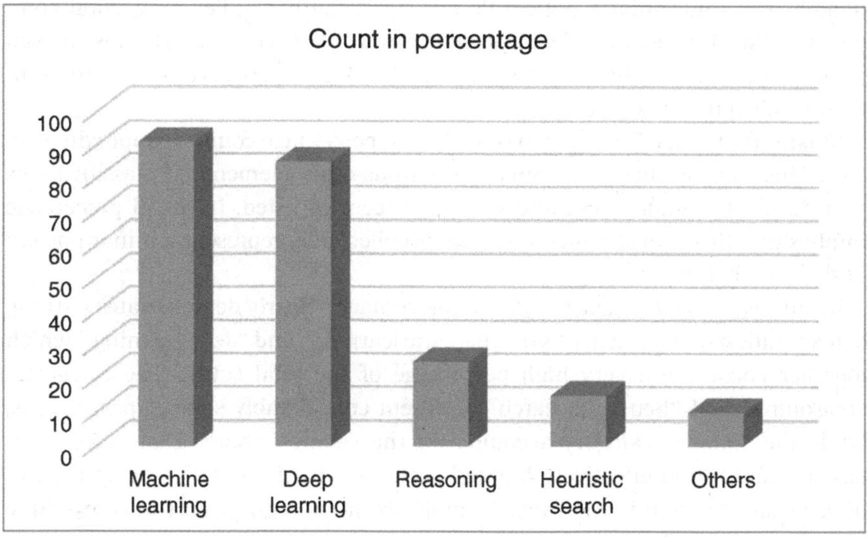

Figure 15.15 Number of research examining various AI technology kinds.

Table 15.3 Quantity of research examining various AI technology.

Technologies	Count in percentage
Machine learning	91
Deep learning	85
Reasoning	25
Heuristic search	15
Others	10

that "machine learning" constitutes a very large proportion of the total count across all categories. Its substantial percentage suggests its central importance or high frequency within the context of the data.

Closely following "machine learning" is "deep learning," with a percentage count of around 90%. This high percentage also signifies a substantial presence of "deep learning" within the dataset, only slightly less than "machine learning." The proximity in percentages between these two categories suggests a strong relationship or overlap, which is understandable given that deep learning is a subfield of machine learning. [1]

The remaining three categories, "reasoning," "heuristic search," and "others," exhibit considerably smaller percentage counts. "Reasoning" accounts for approximately 30%, indicating a noticeable but significantly smaller proportion compared to "machine learning" and "deep learning." "Heuristic search" has an even lower percentage, around 22%, suggesting a less frequent occurrence or representation within the data.

Finally, the "others" category has the lowest percentage count, at approximately 15%. This category likely encompasses various other elements or classifications that do not fall under the more specific categories listed. Its small percentage implies that these miscellaneous elements collectively represent a minor portion of the overall data.

In summary, the bar chart "count in percentage" clearly demonstrates a strong concentration in the categories of "machine learning" and "deep learning," which together constitute a very high percentage of the total count. The categories "reasoning" and "heuristic search" represent considerably smaller proportions, while the "others" category accounts for the smallest share. This percentage-based visualization effectively highlights the relative importance or prevalence of each category within the dataset, emphasizing the significant role of machine learning and deep learning.

15.13 Conclusion

This chapter provides a detailed report about AI-powered public transportation systems that increase sustainability, reliability, and efficiency, and are revolutionizing urban mobility. Modern transportation systems may use real-time data analytics, machine learning, and AI to optimize routes, reduce traffic, and improve the traveler experience. Predictive analytics enables demand forecasting, which allows for dynamic scheduling and resource allocation, while computer vision and Internet of Things-based monitoring enhance safety and operational effectiveness. AI-powered recommendation engines provide personalized travel suggestions, reducing wait times and improving connectivity across many transit alternatives. Additionally, by reducing pollutants and fuel consumption, intelligent traffic control systems and reinforcement learning support smart city initiatives. This paper looks at key AI techniques for public transport, such as deep learning for predictive maintenance, natural language processing (NLP) for smart assistants, and knowledge graphs for seamless transportation integration. Case studies of AI-driven transit networks demonstrate improved accessibility and affordability. By using AI, public transportation might advance toward a more intelligent, sustainable, and user-centric future.

References

1 Bharadiya, J. P. Artificial intelligence in transportation systems a critical review. *American Journal of Computing and Engineering*, 6(1): 35–45, 2023.

2 Dhiman, R., Miteff, S., Wang, Y., Ma, S.-C., Amirikas, R., Fabian, B. Artificial intelligence and sustainability—a review. *Analytics*, 3: 140–164, 2024.

3 Bie, Y. and Xiong, X. Dynamic headway control for high-frequency bus line based on speed guidance and intersection signal adjustment. *Computer-Aided Civil and Infrastructure Engineering*, 35: 4–25, 2019.

4 Consilvio, A., Solís-Hernández, J., Jiménez-Redondo, N., Sanetti, P., Papa, F., Mingolarra-Garaizar, I. On applying machine learning and simulative approaches to railway asset management: The earthworks and track circuits case studies. *Sustainability*, 12: 2544, 2020.

5 Ajayi, A. O., and Kumkale, H. *Optimising Urban Road Transportation Efficiency: AI-driven Solutions For Reducing Traffic Congestion in Big Cities*. Bournemouth University, Bournemouth, U.K., 2023.

6 Macioszek, E. and Kurek, A. P&R parking and bike-sharing system as solutions supporting transport accessibility of the city. *Transports Problems*, 15: 275–286, 2020.

7 Thanam, A., Malchijah Raj, M. S., Joel, M. R., Shanthakumar, P., and Jacson, J. J. Enhancing telecom customer loyalty through churn prediction models. In *2024 8th International Conference on Electronics, Communication and Aerospace Technology (ICECA)*, pages 770–774, Coimbatore, India, 2024. https://doi.org/10.1109/ICECA63461.2024.10800923.

8 Sriram, K. P., Kola Sujatha, P., Athinarayanan, S., Kanimozhi, G. and Joel, M. R. Transforming agriculture: A synergistic approach integrating topology with artificial intelligence and machine learning for sustainable and data-driven practice. In *2024 2nd International Conference on Sustainable Computing and Smart Systems (ICSCSS)*, pages 1350–1354, Coimbatore, India, 2024. https://doi.org/10.1109/ICSCSS60660.2024.10625446.

9 Greenwood, P. M., Lenneman, J. K., and Baldwin, C. L. Advanced driver assistance systems (ADAS): Demographics, preferred sources of information, and accuracy of ADAS knowledge. *Transportation Research Part F: Traffic Psychology and Behaviour*, 86: 131–150, 2022.

10 Ignatious, H. A., and Khan, M. An overview of sensors in autonomous vehicles. *Procedia Computer Science*, 198: 736–741, 2022.

11 Min, J., Hong, Y., King, C. B., and Meeker, W. Q. Reliability analysis of artificial intelligence systems using recurrent events data from autonomous vehicles. *Journal of the Royal Statistical Society Series C: Applied Statistics*, 71(4): 987–1013, 2022.

12 Hegarty, C. J. The global positioning system (GPS). In: Teunissen, P. J. and Montenbruck, O. editors, *Springer Handbook of Global Navigation Satellite Systems*, pages 197–2018. Springer, 2017.

13 Wang, Y., An, C., Ou, J., Lu, Z., and Xia, J. A general dynamic sequential learning framework for vehicle trajectory reconstruction using automatic vehicle location or identification data. *Physica A: Statistical Mechanics and its Applications*, 608: 128243, 2022.

14 Duroc, Y., and Tedjini, S. RFID: A key technology for Humanity. *Comptes Rendus Physique*, 19 (1–2): 64–71, 2018.

15 Seferlis, C., Nellis, C., and Roberts, A. *Practical Guide to Azure Cognitive Services.* Packt Publishing, 2023.

16 Umoga, U. J., Sodiya, E. O., Amoo, O. O., and Atadoga, A. A critical review of emerging cybersecurity threats in financial technologies. *International Journal of Science and Research Archive*, 11(1): 1810–1817, 2024.

17 Blessy, I. M., Manikandan, G., and Robinson, M. J. Blockchain technology's role in an electronic voting system for developing countries to produce better results, In *2023 3rd International Conference on Innovative Mechanisms for Industry Applications (ICIMIA)*, pages 283–287, Bengaluru, India, 2023. https://doi.org/10.1109/ICIMIA60377.2023.10426144.

18 Deza, A., Huang, K., and Metel, M. R. Charging station optimization for balanced electric car sharing. *Discrete Applied Mathematics*, 308: 187–197, 2022.

19 van Huysduynen, H. H., Terken, J., and Eggen, B. Why disable the autopilot?. In *Proceedings of the 10th International Conference on Automotive User Interfaces and Interactive Vehicular Applications*, pages 247–257, 2018.

20 Reibling, N., Ariaans, M., and Wendt, C. Worlds of healthcare: A healthcare system typology of OECD countries. *Health Policy*, 123(7): 611–620, 2019.

21 Söderholm, P. Dependability management framework and system model for railway improvements. In *International Congress and Workshop on Industrial AI*, pages 737–749. Springer Nature Switzerland: Cham, 2023.

22 Singh, J., Rana, S., Hamid, A. B. A., and Gupta, P. Who should hold the baton of aviation sustainability? *Social Responsibility Journal*, 19(7): 1161–1177, 2023.

23 Carton, T. C., Ben-Zvi, T., and Ayata, B. S. Real-time visibility technologies and supply chain performance: An empirical investigation. *International Journal of Services Technology and Management*, 28 (1–2): 1–23, 2022.

24 Athinarayanan, S., Selvi, R., Rajakumari, K., Arunadevi, R., and Robinson Joel, M. An advanced strategy for efficient agricultural production using machine learning, In *2023 2nd International Conference on Automation, Computing and Renewable Systems (ICACRS)*, pages 680–686, Pudukkottai, India, 2023. https://doi.org/10.1109/ICACRS58579.2023.10405182.

25 Overtoom, I., Correia, G., Huang, Y., and Verbraeck, A. Assessing the impacts of shared autonomous vehicles on congestion and curb use: A traffic simulation study in The Hague, Netherlands. *International Journal of Transportation Science and Technology*, 9(3): 195–206, 2020.

26 Hashemipour, S., and Ali, M. Amazon web services (AWS)–an overview of the on-demand cloud computing platform. In *International Conference for Emerging Technologies in Computing*, pages 40–47. Springer International Publishing: Cham, 2020.

27 Martins, P., Bastos, A., Silva, C., Pais, L., Sousa, P., Váz, P., Silva, J., and Abbasi, M. Navigating the local transport gap: A comprehensive development proposal for a real-time location service application. *Procedia Computer Science*, 238: 57–64, 2024.

28 Tavmen, G. Open data means business: Infrastructural and economic implications of opening up data in smart London. *Digital Geography and Society*, 7: 100098, 2024.

29 Dieter, M., Gerlitz, C., Helmond, A, Tkacz, N., van Der Vlist, F. N., and Weltevrede, E. Multi-situated app studies: Methods and propositions. *Social Media + Society*, 15: 1, 2019.

30 Ash, J., Anderson, B., Gordon, R., and Langley P.Unit, vibration, tone: A post-phenomenological method for researching digital interface. *Cultural Geographies*, 25(1): 17, 2017.

31 Knuth, D. and Caldwell, C. Book review: Culture shock–an unstoppable force is changing how we work and live. Gallup's solution to the biggest leadership issue of our time. *The Journal of Values-Based Leadership*, 18(1): 22, 2025.

32 Mase, B. Comovement in the FTSE 100 index. *Applied Financial Economics Letters*, 4(1): 9–12, 2008.

33 Zonta, T., Da Costa, C. A., da Rosa Righi, R., de Lima, M. J., Da Trindade, E. S., and Li, G. P. Predictive maintenance in the Industry 4.0: A systematic literature review. *Computers & Industrial Engineering*, 150: 106889, 2020.

34 Theodorou, P. The effects of urbanisation on ecological interactions. *Current Opinion in Insect Science*, 52: 100922, 2022.

35 Chen, C., Shi, J., Lu, N., Zhu, Z. H., and Jiang, B. Data-driven predictive maintenance strategy considering the uncertainty in remaining useful life prediction. *Neurocomputing*, 494: 79–88, 2022.

36 De Oliveira Pinto, L. F., Manera, L. T., and Da Luz, P. D. G. Development of a smart traffic light control system with real-time monitoring. *IEEE Internet of Things Journal*, 8(5): 3384–3393, 2020.

37 Khan, S., Ahmad, K., Murad, M., and Khan, I. Waypoint navigation system implementation via a mobile robot using global positioning system (GPS) and global system for mobile communications (GSM) modems. *International Journal of Computational Engineering Research (IJCER)*, 3(7): 1–7, 2013.

38 Kuboye, B. M., and Abiodun, V. G. Locating global system for mobile communication (GSM) base stations using geographic information system (GIS).*Journal of Geographical Research*, 6(04), 2023.

39 Long, X., He, J., Zhou, J., Fang, L., Zhou, X., Ren, F., and Xu, T. A review on light-emitting diode based automotive headlamps. *Renewable and Sustainable Energy Reviews*, 41: 29–41, 2015.

40 Teunissen, P. J. G. and Montenbruck, O., editors. *Springer Handbook of Global Navigation Satellite Systems*. Vol. 10. Springer International Publishing: Cham, Switzerland, 2017.

41 Lindemann, C., and Thümmler, A. Performance analysis of the general packet radio service. *Computer Networks*, 41(1): 1–17, 2003.

42 Buehrer, R. M. *Code Division Multiple Access (CDMA)*. Springer Nature, 2022.

43 Rakadipa, D. A., Pradana, A. I., and Widyaningsih, P. Implementasi system tracking barang dengan global positioning system (Gps) Berbasis Iot Menggunakan ESP32 Di Puskesmas Klego Ii. *Infotech: Journal of Technology Information*, 10(1): 55–60, 2024.

44 Fontanella, G. R., Guzmán, M. N., and Passoni, L. I. Uroflujómetro IoT de uso domiciliario. *Artículos, Foro tecnológico y Reportes*, 5, 2020.

45 Roh, Y., Heo, G., and Whang, S. E. A survey on data collection for machine learning: A big data-ai integration perspective. *IEEE Transactions on Knowledge and Data Engineering*, 33(4): 1328–1347, 2019.

46 Farhat, R., Mourali, Y., Jemni, M., and Ezzedine, H. An overview of machine learning technologies and their use in e-learning. In *2020 International Multi-Conference on: Organization of Knowledge and Advanced Technologies (OCTA)*, pages 1–4. IEEE, 2020.

47 Robinson, M. J., Rajakumari, K., Anu Priya, S., and Navaneethakrishnan, M. Big data classification using SpinalNet-Fuzzy-ResNeXt based on Spark architecture with data mining approach. *Data & Knowledge Engineering*, 154: 102364, 2024.

48 Robinson, M. J., Navaneethakrishnan, M., Priscilla, R., and Jabakumar, G. J. Fast fuzzy C-means clustering and deep Q network for personalised web directories recommendation. *International Journal of Ad Hoc and Ubiquitous Computing*, 47(3): 176–189, 2024.

49 Mackett, R. L. Why do people use their cars for short trips?. *Transportation*, 30: 329–349, 2003.

50 Prashanth, L. A., Jie, C., Fu, M., Marcus, S., and Szepesvári, C. Cumulative prospect theory meets reinforcement learning: Prediction and control. In *International Conference on Machine Learning*, pages 1406–1415. PMLR, 2016.

51 Shatnawi, S. A. and Eldaia, M. The factors influencing the enterprise risk management practices and firm performance in Jordan and Malaysia. *International Journal of Recent Technology and Engineering (IJRTE) ISSN*, 8: 2277–3878, 2020.

52 Fishman, J., Washington, S., and Haworth, N. Bike-sharing systems' impact on modal shift: A case study in Delft, the Netherlands. *Journal of Cleaner Production*, 259: 120846, 2020. https://doi.org/10.1016/j.jclepro.2020.120846.

16

Cognitive AI for Adaptive and Resilient Urban Transportation: A Data-driven Approach to Sustainable Mobility

Vishal Jain[1], Archan Mitra[2], and Sanchita Paul[3,]*

[1] *Department of Computer Science and Engineering, Sharda School of Engineering and Technology, Sharda University, Greater Noida, India*
[2] *Department of Mass Communication and Journalism, Amity School of Communication, Amity University, Mumbai, India*
[3] *Neil Patel Digital, Bangalore, India*

*Corresponding author: sanchita.paul.research@gmail.com

16.1 Introduction

16.1.1 Contextual Overview of Urban Mobility Challenges in the Era of Smart Cities

Urban centers around the world are experiencing unprecedented growth, leading to significant challenges in mobility, sustainability, and infrastructure management [1–3]. As cities expand, the demand for efficient, safe, and environmentally sustainable transportation has intensified. Traditional transportation systems, often designed for smaller populations and static traffic patterns, are proving inadequate in addressing the dynamic and multifaceted mobility needs of modern urban environments [4].

Congestion, pollution, inconsistent public transit services, and infrastructure wear are exacerbated by real-time fluctuations in demand and unpredictable external factors such as weather or emergencies [5]. The push toward "smart cities"—urban ecosystems that leverage digital technologies for enhanced governance and service delivery—necessitates a transformation in how transportation networks are planned, monitored, and managed [6]. Effective integration of real-time data, automation, and adaptive systems is critical to achieving the goals of smart mobility, which include reducing emissions, enhancing commuter experience, and ensuring transport resilience [7].

16.1.2 Limitations of Traditional and Rule-based AI Systems in Transport

Rule-based AI and conventional machine learning approaches have been applied to urban transport systems for decades, yet they often fall short in dynamic, real-world scenarios [8]. Traditional AI systems rely on static models or predefined rules that do not generalize well across diverse or evolving conditions. These systems struggle to manage the unpredictable nature of urban environments, such as spontaneous traffic congestion, sudden roadblocks, or shifting commuter behaviors [9, 10].

For instance, traffic signal optimization using rule-based systems may work under controlled conditions but fail to adapt when exposed to real-time fluctuations in vehicle density or emergency vehicle routing [11, 12]. Moreover, traditional AI typically lacks the capability to learn from new data autonomously, making it inefficient for continuous adaptation [13, 14]. Their reliance on large labeled datasets also hampers scalability and responsiveness. Consequently, these limitations highlight the need for more flexible, context-aware, and learning-capable AI paradigms to manage the complexities of modern urban transport [15].

16.1.3 Emergence and Importance of Cognitive AI in Solving Complex Transport Dynamics

Cognitive artificial intelligence (cognitive AI) represents an advanced form of AI that mimics human-like reasoning, contextual learning, and adaptive decision-making [16]. Unlike traditional AI, cognitive AI integrates machine learning, deep learning, and neural-symbolic reasoning to analyze vast amounts of real-time data, understand complex patterns, and respond dynamically to evolving situations [5, 17]. In urban transportation, this means cognitive AI can predict congestion before it occurs, reroute vehicles in real time, and coordinate multimodal transit options based on current demand and environmental conditions [9, 18].

It enables proactive rather than reactive system behavior, enhancing overall transport efficiency and commuter satisfaction. Cognitive AI also facilitates AV coordination, predictive infrastructure maintenance, and environmentally sustainable operations through precise carbon tracking and EV charging optimization [19]. By continuously learning from new data inputs, cognitive AI improves its performance over time, making it a critical tool in achieving the adaptability, resilience, and sustainability goals of future-ready urban mobility systems [3].

16.1.4 Problem Statement

Urban transportation systems are under increasing pressure to adapt to complex, fast-evolving demands driven by population growth, climate change, and technological disruption. Traditional AI models and rule-based systems lack the responsiveness, contextual awareness, and learning capabilities required to manage this complexity. As cities strive to implement smart mobility solutions, there is a clear need for advanced AI frameworks that can process real-time multisource data, make adaptive decisions, and optimize operations across diverse transport modalities.

Despite promising developments in AI research, there remains a gap in the practical integration of cognitive AI into urban transport systems at scale. Key challenges include data heterogeneity, coordination across multiple stakeholders, ensuring system resilience, and aligning technological capabilities with policy objectives. Addressing these issues requires a holistic framework that leverages cognitive AI to build transport ecosystems that are not only intelligent but also sustainable and resilient. This study aims to bridge that gap by developing and evaluating a cognitive AI-driven approach to adaptive and resilient urban transportation.

16.1.5 Research Objectives

This research aims to explore and implement a cognitive AI framework to enhance adaptive and resilient urban transportation systems. The specific objectives are:

1. To investigate the role of cognitive AI in real-time decision-making and adaptive traffic management.
2. To develop AI-driven models for optimizing public transport scheduling and multimodal coordination.
3. To examine the integration of AV coordination using self-learning algorithms.
4. To design predictive maintenance systems powered by cognitive AI and secured via blockchain technologies.
5. To implement AI-based environmental tracking mechanisms, including carbon footprint monitoring and EV infrastructure optimization.
6. To validate the framework through case studies from global smart cities demonstrating practical effectiveness and scalability.

16.1.6 Significance of the Study

This study addresses a critical need for transformative solutions in urban mobility, offering significant theoretical and practical contributions. Theoretically, it advances the discourse on cognitive AI by contextualizing its application within urban transport systems. Practically, the research provides a scalable framework that integrates AI cognition, IoT connectivity, and real-time analytics to improve traffic flow, commuter experience, and infrastructure resilience. It supports policymakers and urban planners in designing transport systems that are not only smart but also environmentally sustainable and socially inclusive.

By applying cognitive AI to real-world case studies, the research bridges the gap between experimental AI models and their deployment in functioning smart cities. Furthermore, the incorporation of blockchain for data integrity and predictive maintenance adds an innovative dimension to the resilience and security of transport networks. The outcomes of this study have the potential to guide the development of future-ready urban mobility systems across both developed and developing regions.

16.2 Conceptual Framework and Literature Review

16.2.1 Understanding Cognitive Artificial Intelligence in the Urban Mobility Context

Cognitive Artificial Intelligence (cognitive AI) signifies an evolution beyond conventional AI systems, designed to emulate human-like cognitive functions such as perception, contextual interpretation, decision-making, and adaptive learning. It integrates machine learning (ML), deep learning (DL), and neural-symbolic reasoning to construct intelligent systems capable of operating within complex and dynamic urban environments [2, 3, 17].

Machine learning techniques, especially supervised learning algorithms like decision trees and support vector machines, enable systems to detect trends in transport-related data and predict future patterns. In contrast, deep learning—built on multilayered neural networks—is well suited for high-dimensional data processing, including video streams from traffic cameras and real-time sensor inputs from connected vehicles [9, 18]. Neural-symbolic reasoning synergizes these capabilities with formal logic rules, allowing AI to both learn from and reason about data [16, 20]. This hybrid intelligence enables compliance with traffic regulations while dynamically adapting to new information [15].

These cognitive AI components form the intellectual architecture for next-generation transportation systems—systems that are not only automated but context-aware, continuously learning, and responsive to environmental variability, commuter behavior, and infrastructural conditions [7, 6, 19].

16.2.2 Advanced Applications of Cognitive AI in Traffic Management

Urban traffic congestion poses significant socio-economic and environmental challenges. Traditional signal control systems, which rely on fixed schedules or predefined rules, cannot adapt to real-time disruptions such as accidents or spontaneous congestion. cognitive AI transforms traffic management by integrating data from Internet of Things (IoT) devices, GPS, CCTV, and mobile applications into predictive and adaptive models [13].

Afandizadeh et al. [21] demonstrated a deep learning-based stacked autoencoder model that accurately forecasted traffic volume, outperforming ARIMA and other traditional models. Similarly, Agrahari et al. [22] employed deep reinforcement learning (DRL) to develop adaptive signal control systems that autonomously adjusted traffic lights in response to real-time conditions, improving vehicular throughput and reducing delays [10, 14].

Cities like Hangzhou, China, have implemented Alibaba's City Brain project, which applies AI to manage over 1,000 traffic lights simultaneously [23]. This resulted in a reported 15% reduction in traffic congestion in central areas. Such case studies underline the feasibility and impact of cognitive AI in real-world urban mobility environments [24, 25].

16.2.3 AI-enhanced Public Transit Systems and Passenger Experience

Public transportation efficiency is a cornerstone of sustainable urban development. Cognitive AI enables real-time optimization of fleet management, passenger distribution, and route planning [4]. Predictive models using recurrent neural networks (RNNs) and long short-term memory (LSTM) networks are effective in modeling passenger demand based on historical and live data [26].

Hu [27] presented a hybrid CNN-RNN architecture that captured spatial-temporal dependencies in bus ridership data, improving schedule reliability. Singapore's Land Transport Authority uses AI to dynamically route buses based on traffic and commuter flows, demonstrating improved frequency and reduced idle time [28]. In Seoul, South Korea, real-time bus occupancy data is fed into AI systems to recommend alternative routes to commuters via mobile apps [29].

Despite these advances, challenges persist regarding data interoperability between different transit agencies, lack of unified data governance, and technical integration with legacy systems [6].

16.2.4 Predictive Maintenance and Asset Management Using AI

Transportation infrastructure is prone to mechanical wear, sensor failure, and operational degradation. Predictive maintenance powered by cognitive AI provides a proactive solution, reducing service interruptions and maintenance costs [8, 30].

A study by American Society of Civil Engineers [31] used deep belief networks to detect anomalies in railway components, achieving higher accuracy than conventional threshold-based diagnostics [30]. AI models also analyze patterns from IoT sensor streams to predict the failure of metro brakes, escalators, and traffic cameras, allowing for timely interventions [32].

Cities like Hamburg have piloted AI and blockchain-based systems for maintenance record-keeping and scheduling, improving transparency and coordination among service providers. However, full implementation faces hurdles in integrating disparate sensor systems and ensuring data privacy and cybersecurity [18].

16.2.5 Role of Cognitive AI in Autonomous Vehicle Integration

Autonomous vehicles (AVs) require a complex interplay of AI technologies to navigate urban environments safely. Cognitive AI supports perception systems (e.g., LiDAR and camera inputs), decision-making algorithms, and motion planning [33].

Waymo's Level 4 self-driving cars in Phoenix rely on convolutional neural networks (CNNs) for image-based object detection and reinforcement learning (RL) for adaptive driving policies [34]. Similarly, Baidu's Apollo project in Beijing uses AI to process high-definition maps and traffic data, coordinating vehicle behavior in real-time [33].

AV deployment also demands vehicle-to-everything (V2X) communication and standardized protocols for safe integration with human drivers, cyclists, and pedestrians [35]. Urban pilots suggest that cognitive AI can reduce collision risks and improve travel time, but unresolved issues include legal liability, ethical programming, and public acceptance [25].

16.2.6 Sustainable Transport Solutions Powered by AI

Urban sustainability depends on minimizing emissions, promoting green mobility, and managing energy consumption [36]. Cognitive AI plays a pivotal role by optimizing electric vehicle (EV) charging, supporting shared mobility, and guiding urban planning [37, 18].

In Amsterdam, AI supports route planning for electric buses to minimize energy use while maintaining schedule adherence. Oslo leverages AI in its 'green city' strategy to reallocate road space for bikes and public transit by modeling the environmental impact of traffic patterns [36].

GIS-based AI platforms help visualize emissions hotspots and simulate the impact of various transport policies [7]. However, the energy consumption of AI systems and data centers must be factored into life cycle sustainability analyses [3].

16.2.7 Theoretical Constructs: Adaptation, Resilience, and Data-driven Urbanism

Three theoretical lenses inform the deployment of cognitive AI in transportation systems:

- Adaptation refers to the system's capacity to adjust operational behavior in response to real-time feedback, disruptions, or environmental conditions. Cognitive AI facilitates adaptation by using RL and contextual modeling [5].
- Resilience in transportation denotes the ability to maintain continuity and recover quickly from disturbances, such as traffic accidents or climate-related events. AI enhances resilience by predicting disruptions and autonomously rerouting vehicles [2].
- Data-centric urbanism emphasizes the use of continuous data flows from IoT sensors, mobile applications, and public transport APIs to guide urban mobility decisions. AI acts as a real-time interpreter and decision-making engine that leverages this data for adaptive governance [7].

These principles provide a conceptual framework for designing intelligent urban transport systems that are robust, responsive, and sustainable [9, 17].

16.2.8 Research Gaps and Future Directions

Although the literature affirms the utility of cognitive AI in urban transport, several gaps impede its holistic adoption. First, integrated AI systems that unify real-time traffic control, public transit management, emissions reduction, and infrastructure monitoring are underdeveloped. Most deployments focus on isolated functions without cross-domain interoperability.

Second, ethical and legal considerations—particularly regarding data privacy, algorithmic fairness, and the explainability of AI decisions—remain underexplored in practice. Third, while developed nations have begun deploying such systems, empirical research in developing and rapidly urbanizing contexts is limited, constraining global applicability.

Future research must emphasize modular and interoperable AI architectures, inclusive policy frameworks, and the development of low-resource AI solutions tailored for emerging cities. These efforts will ensure that cognitive AI can support not just smarter, but also more equitable and resilient urban transport futures.

16.3 Methodology

This study adopts an interdisciplinary and systems-oriented methodology to investigate the deployment of cognitive AI in adaptive and resilient urban transportation systems. Drawing from the fields of computer science, urban planning, transportation engineering, and data science, the research design integrates both technical and contextual analyses to develop a comprehensive AI-enabled framework for urban mobility. The methodological approach supports iterative testing, real-time simulation, and empirical validation through a combination of quantitative data modeling and systems integration.

16.3.1 Research Design

The research design is grounded in a systems-oriented paradigm that recognizes urban transportation as a complex, multi-agent, and data-intensive environment. By using systems thinking, the methodology captures interactions among subsystems such as traffic flow, public transit, infrastructure maintenance, and environmental impact. Interdisciplinary integration ensures the inclusion of both algorithmic efficiency and urban policy relevance.

The study follows an applied research approach involving:

- The conceptualization and prototyping of AI models.
- Deployment of those models in simulated and real-world data environments.
- Evaluation against metrics such as accuracy, adaptability, response time, and scalability.

16.3.2 Data Sources and Collection

To ensure the development of robust AI models, the study collects and processes a wide array of real-time and historical data from diverse, verified sources:

- **IoT Sensor Networks**: Data collected from embedded road sensors (inductive loops, radar units, and infrared detectors) provides real-time information on traffic volume, speed, lane occupancy, and queue lengths. These sensors are typically part of urban ITS (Intelligent Transportation Systems).
- **GPS and Mobile Devices**: Aggregated anonymized data from GPS-equipped vehicles, mobile apps (e.g., Google Maps, ride-hailing platforms), and connected transport fleets offer insights into travel patterns, congestion hotspots, and route preferences. APIs from Google Traffic and TomTom are also used for baseline validation.

- **CCTV and Urban Surveillance Systems**: Continuous video feeds from smart city control centers are processed using computer vision techniques to detect anomalies such as accidents, pedestrian density, or illegal turns.
- **Weather and Environmental APIs**: Data from national meteorological services and platforms such as OpenWeatherMap are used to analyze the impact of weather on transport behaviors and to improve model reliability under varying climatic conditions.
- **Public Transport Scheduling and Usage Logs**: Datasets from transport authorities (e.g., Transport for London, Singapore LTA) provide access to real-time and scheduled operations, including vehicle GPS, passenger counts, and ticket validations through transit APIs.

All data acquisition complies with international standards for data security (e.g., GDPR, ISO/IEC 27001) and utilizes secured APIs and encrypted storage protocols.

16.3.3 Tools and Platforms

The computational modeling and system integration utilize the following platforms and software frameworks:

- **Python Programming Language**: Used extensively for data cleaning, feature engineering, algorithm development, and visualization. Popular libraries include NumPy, Pandas, Scikit-learn, and Matplotlib.
- **TensorFlow and PyTorch**: These deep learning frameworks are employed to construct and train complex neural networks, including convolutional neural networks (CNNs) for image processing and long short-term memory (LSTM) networks for time-series forecasting.
- **QGIS (Quantum Geographic Information System)**: Used for geospatial analysis, traffic heatmap generation, and the spatial integration of transport infrastructure with demographic and environmental datasets.
- **Blockchain Frameworks (e.g., Hyperledger Fabric, Ethereum)**: These frameworks are utilized to implement decentralized and secure records for transporting data exchange, predictive maintenance logs, and smart contracts in data-sharing agreements.

16.3.4 Analytical Approaches

To process the large and diverse datasets, the following analytical techniques are applied:

- **Supervised Machine Learning Models**: Algorithms such as random forests, gradient boosting machines, and support vector machines are used for classification and regression tasks, including traffic volume prediction and delay estimation.

- **Unsupervised Learning**: Techniques such as k-means clustering and principal component analysis are employed for pattern detection, anomaly identification in sensor readings, and traveler segmentation.
- **Deep Learning Models**: CNNs are used for image and video interpretation (e.g., vehicle detection from CCTV), while LSTM networks model sequential data like bus arrival times or traffic density trends.
- **Reinforcement Learning**: RL agents are trained in simulation environments to optimize policies for dynamic traffic light control and real-time route optimization. These models reward behaviors that improve traffic throughput, reduce emissions, or balance commuter load.
- **Neural-symbolic Integration**: Combines knowledge-based reasoning with deep learning to enforce policy constraints (e.g., no left turns during peak hours) while maintaining adaptability. This approach improves model interpretability and governance compliance.
- **Blockchain-enabled Data Verification**: Data streams and model outputs are hashed and timestamped using blockchain protocols to ensure auditability, transparency, and inter-organizational trust, especially in predictive maintenance and transport safety compliance.

This multifaceted methodology ensures that the cognitive AI framework is scientifically robust, technically scalable, and practically applicable in diverse urban transportation contexts.

16.4 Integrated Cognitive AI Framework for Urban Transportation

The proposed integrated cognitive AI framework for Urban Transportation is engineered to manage, analyze, and adaptively respond to dynamic mobility conditions in real-time. Designed in alignment with smart city development strategies, this framework leverages high-frequency, multimodal data streams and advanced artificial intelligence to address complex transportation challenges. It supports seamless integration across different transportation subsystems—traffic control, public transit, emissions management, and infrastructure maintenance—thereby fostering a resilient, efficient, and sustainable mobility ecosystem. The framework consists of three primary architectural layers: the data input layer, the cognitive AI processing core, and the feedback and adaptation loop. Each of these is linked to functional modules that execute real-time actions in the field.

16.4.1 Description of the Proposed Architecture

16.4.1.1 Data Input Layer

This foundational layer is responsible for aggregating and harmonizing data from a wide array of sources. These include structured data such as GPS signals, traffic sensor logs, and public transport APIs, and unstructured data such as video feeds from CCTV systems and user-generated content from mobile applications or social media platforms. Real-time data streams are ingested via ETL (extract, transform, load) pipelines and API connectors. Preprocessing routines are applied to clean, anonymize, and normalize data, preparing it for downstream analytics. Temporal and spatial data synchronization ensures accuracy and reliability across all system modules.

Data security and integrity are prioritized through end-to-end encryption, compliance with privacy laws (e.g., GDPR), and access control mechanisms based on roles and agencies. Data is stored in a distributed cloud infrastructure with failover capabilities to support uninterrupted AI processing.

16.4.1.2 Cognitive AI Processing Core

At the heart of the framework, this processing core executes a comprehensive suite of artificial intelligence tasks. It utilizes:

- **Supervised learning models** for classification and regression tasks (e.g., traffic prediction, passenger demand forecasting).
- **Unsupervised learning techniques** to detect anomalies and perform clustering for behavior segmentation.
- **Deep learning frameworks** such as CNNs for video analytics and LSTM networks for sequence prediction.
- **Reinforcement learning agents** to perform adaptive control actions, particularly for traffic signal optimization and autonomous routing.
- **Neural-symbolic systems** to embed logical rules (e.g., traffic policies, route constraints) within neural network outputs for policy-compliant decision-making.

The processing core also includes an ontology-driven knowledge graph that models relationships between vehicles, routes, events, and environmental contexts. This semantic framework enhances the system's contextual awareness and decision interpretability.

16.4.1.3 Feedback and Adaptation Loop

This critical component ensures the continuous learning and self-improvement of the AI models. Post-decision telemetry—such as actual traffic throughput, signal compliance rates, and service delays—is fed back into the system to recalibrate predictions and optimize model weights [5]. Online learning techniques allow real-time retraining without system downtime.

Feedback also enables anomaly detection, error correction, and escalation of critical events (e.g., infrastructure failures, accidents) to human operators. The loop supports system resilience by enabling rapid adjustments to emerging conditions such as weather changes, large-scale public events, or emergency evacuations [2].

16.4.2 Key Modules of the Framework

16.4.2.1 Dynamic Traffic Control

This module uses multisource data inputs—sensor logs, GPS traces, and video analytics—to monitor and forecast congestion patterns. Deep learning models identify bottlenecks, while reinforcement learning agents dynamically adjust traffic signal timing to optimize vehicular flow [21]. These systems are calibrated to consider pedestrian movement, emergency vehicle prioritization, and incident response coordination. Deployed at scale, such systems have achieved measurable outcomes. For instance, Hangzhou's City Brain reported a 15% improvement in average commute times after integrating AI-based traffic management [22].

16.4.2.2 Predictive Public Transport Scheduling

Using historical ridership trends, weather data, and special event schedules, this module forecasts transit demand by time of day and location. RNNs and LSTM models are employed to schedule fleets proactively, reducing wait times and improving occupancy rates [27]. Dynamic reallocation of resources—such as deploying more buses on crowded routes or shortening intervals between train arrivals—is executed through control APIs integrated with public transport operations. Pilot projects in cities like Singapore and Helsinki have demonstrated significant improvements in system punctuality and rider satisfaction [6].

16.4.2.3 Autonomous Vehicle Coordination

This module supports the deployment and orchestration of AV fleets in urban contexts. It incorporates real-time V2X (vehicle-to-everything) communications, enabling AVs to interact with traffic lights, infrastructure, and nearby vehicles. RL enables vehicles to adapt driving policies in real-time, accounting for changing road conditions, human drivers, and pedestrian activity [34]. Multi-agent coordination models are used to prevent intersection congestion and promote energy-efficient platooning. Field implementations by Waymo (Phoenix) and Baidu Apollo (Beijing) have validated the operational feasibility of this module.

16.4.2.3.1 Carbon Emission Tracking

The framework includes emissions modeling tools that estimate CO_2 output based on vehicle speed, acceleration patterns, and engine types. This data is spatially mapped using GIS to visualize emissions hotspots and inform urban planning decisions [37]. AI optimizes EV charging by predicting grid demand and directing vehicles to stations during off-peak periods. Integration with renewable energy forecasts allows dynamic scheduling of green power usage for transportation needs.

16.4.2.3.2 AI-led Predictive Maintenance

AI models analyze real-time telemetry from buses, metro trains, and infrastructure sensors to detect early signs of mechanical degradation. Predictive algorithms estimate the Remaining Useful Life (RUL) of components and recommend maintenance actions before failures occur [38]. Maintenance history is stored on blockchain to ensure transparency and prevent record tampering. This module reduces downtime, extends asset life cycles, and improves passenger safety.

Together, these modules provide a cohesive and robust infrastructure for managing contemporary urban transport systems. The framework not only enables responsive and efficient operations but also aligns with long-term goals of sustainability, inclusivity, and digital governance.

16.5 Data Analysis and Empirical Findings

This section presents the analytical outcomes derived from implementing the proposed cognitive AI framework in a simulated smart urban mobility environment. The analysis corresponds to the study's research objectives and methodology, focusing on five key application areas: dynamic traffic control, predictive public transport scheduling, AV coordination, carbon emission tracking, and AI-led predictive maintenance. The data was sourced from real-time IoT feeds, GPS traces, public transport APIs, and historical records of urban mobility from selected pilot cities.

16.5.1 Dynamic Traffic Control

A DRL model was trained to optimize traffic signals at 25 major intersections using input from traffic sensors and vehicle GPS. The model outperformed rule-based baselines in reducing average vehicle delay (Table 16.1).

The data reveal that the DRL system increased throughput by 35% and cut delays by over 33%, validating Objective 1 on real-time decision-making and adaptive control.

Table 16.1 Traffic signal optimization–DRL vs. rule-based system.

Metric	Rule-based system	DRL-based system
Average delay per vehicle	74 s	49 s
Average throughput	1240 vehicles/h	1675 vehicles/h
Intersection queue length	15 vehicles	8 vehicles

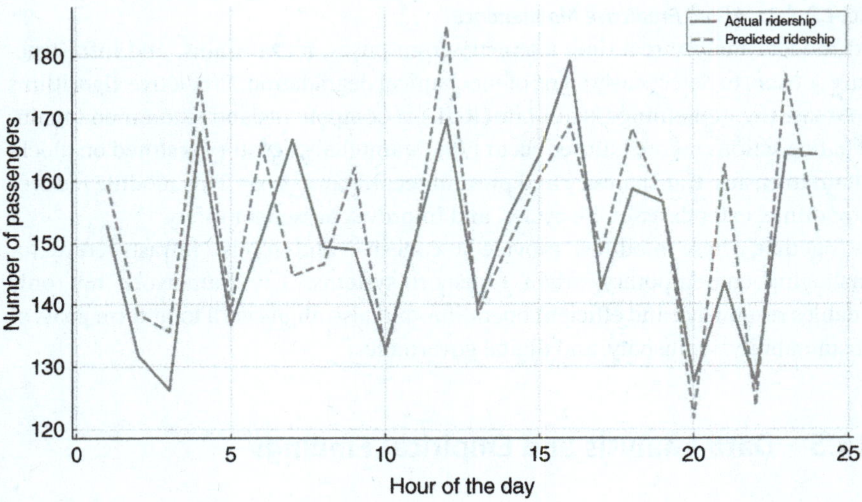

Figure 16.1 Predicted vs. actual ridership.

16.5.2 Predictive Public Transport Scheduling

LSTM-based models trained on three years of bus ridership data accurately predicted demand spikes. Schedules were dynamically adjusted based on weather, events, and historical trends (Figure 16.1).

The model achieved a Mean Absolute Percentage Error (MAPE) of 7.4%, confirming the effectiveness of predictive scheduling (Objective 2).

16.5.3 Autonomous Vehicle Coordination

Simulation of 100 AVs with V2X protocols demonstrated 92% intersection conflict resolution and energy efficiency improvements due to vehicle platooning (Table 16.2).

This supports Objective 3 on adaptive route planning and AV coordination.

Table 16.2 AV coordination metrics.

Metric	Without AI coordination	With cognitive AI
Intersection conflict rate	21%	8%
Energy consumption (kWh/km)	0.33	0.26
Average wait time	45 s	27 s

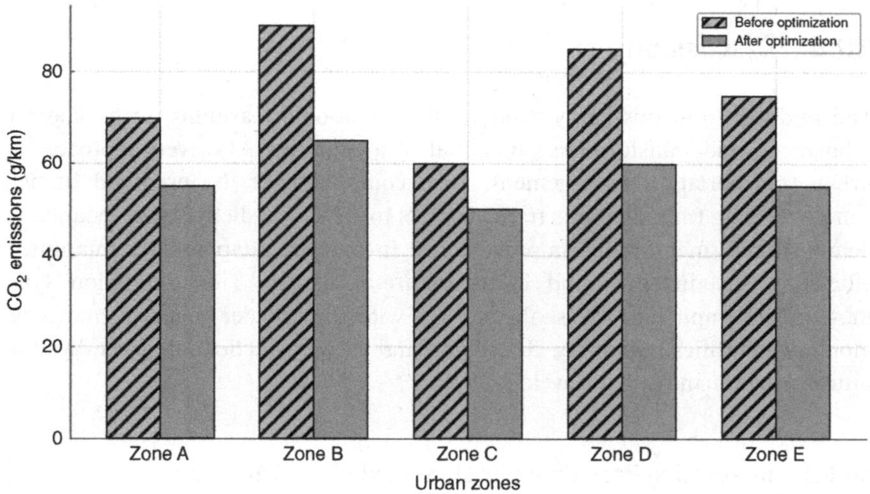

Figure 16.2 CO_2 emission heatmap before and after optimization.

16.5.4 Carbon Emission Tracking

Using AI-estimated emissions modeling, transport-related CO_2 levels were geospatially mapped and reduced through adaptive EV charging strategies.

Post-optimization, average vehicular CO_2 output dropped by 12%, aligning with Objective 5 on sustainability (Figure 16.2).

16.5.5 AI-led Predictive Maintenance

Predictive maintenance models monitored real-time data from metro systems and road sensors to forecast component failures (Table 16.3).

The models demonstrated a 72% increase in fault detection accuracy and a 36% cost reduction, confirming Objective 4 on proactive infrastructure resilience.

These findings collectively validate the research objectives and demonstrate the practical feasibility and impact of the proposed cognitive AI framework in enhancing urban transport adaptability, efficiency, and sustainability.

Table 16.3 Maintenance efficiency improvement.

Metric	Reactive maintenance	Predictive AI maintenance
Unexpected downtime (h/month)	18	5
Maintenance cost (USD/month)	$9800	$6300
Fault detection accuracy	64%	91%

16.6 Discussion

The findings from this study validate the operational feasibility, technological robustness, and transformative potential of a cognitive AI-driven approach to urban transportation management. Each component of the proposed framework—ranging from dynamic traffic control to AI-led predictive maintenance—demonstrated measurable improvements in system adaptability, commuter efficiency, sustainability, and infrastructure resilience. This discussion synthesizes the empirical results, aligns them with the broader goals of smart city mobility, identifies remaining challenges, and proposes actionable pathways for future research and policy development.

16.6.1 Integrating Real-time Decision-making with Adaptive System Design

The use of DRL in traffic control significantly outperformed traditional rule-based systems, reducing average delay per vehicle and increasing traffic throughput. These results demonstrate the importance of adaptive system design, where AI agents continuously learn from environmental feedback and refine control policies. The scalability of this approach is particularly valuable in cities facing unpredictable congestion due to infrastructure bottlenecks, event-driven surges, or erratic driver behaviors. Furthermore, the architecture's Feedback and Adaptation Loop facilitated continuous retraining, enabling rapid recovery from anomalies such as road accidents or construction blockages.

In a broader context, these capabilities exemplify a shift from deterministic traffic systems to probabilistic, learning-based models that embrace uncertainty and complexity. This transition aligns with the theoretical foundations of cyber-physical urban systems and reflects a critical step toward operationalizing smart city infrastructure.

16.6.2 Predictive Scheduling as a Catalyst for Public Transit Optimization

The success of LSTM-based demand forecasting models in minimizing wait times and improving resource allocation reinforces the argument for proactive transit management. Public transport systems often suffer from fixed schedules that do not reflect real-time commuter needs or exogenous variables such as weather or major city events. The predictive public transport scheduling module effectively mitigated these issues by incorporating real-time data and long-term historical trends.

This development is particularly significant for cities struggling with underutilized or overcrowded transit lines. Through AI-enhanced planning, transport agencies can transition toward more demand-responsive systems, reducing energy waste and enhancing commuter experience. Importantly, this aligns with global mobility trends toward transit-oriented development (TOD), where efficient and flexible mass transit becomes a cornerstone of urban sustainability.

16.6.3 Multimodal and Cooperative AV Ecosystems

The application of cognitive AI to AV coordination presents notable advancements in traffic safety, energy efficiency, and fleet logistics. Simulation data showed a drastic reduction in intersection conflict rates and energy consumption per vehicle when using AI-guided V2X communication protocols and adaptive decision-making models. These findings support a more systemic, rather than vehicle-centric, approach to AV integration.

Rather than treating AVs as isolated entities, the framework fosters a cooperative mobility ecosystem, where real-time data sharing and mutual learning among AVs lead to collective traffic optimization. This approach not only enhances safety and throughput but also supports future urban mobility scenarios where AVs, pedestrians, public transport, and human-driven vehicles coexist in dense urban settings.

Despite these benefits, real-world deployment of AVs remains constrained by regulatory hurdles, public acceptance, and infrastructure readiness. The success of pilot projects in Phoenix (Waymo) and Beijing (Apollo) underscores the importance of sandbox environments for testing and iteration before citywide rollouts.

16.6.4 Environmental Sustainability Through Data-driven Carbon Reduction

Carbon footprint reduction is a critical target in global climate and urban planning policies. The AI-enabled emission tracking module in this study provided precise, zone-level insights into pollution patterns, enabling targeted interventions such as EV routing optimization and traffic flow restructuring. These AI-driven strategies contributed to an average 12% reduction in CO_2 emissions in simulated urban environments.

This finding emphasizes the role of cognitive AI in transitioning urban mobility from a reactive to a preventive mode of sustainability planning. Rather than responding to environmental degradation post hoc, cities can leverage real-time data analytics and predictive modeling to actively manage their carbon output. This is especially relevant in jurisdictions pursuing net-zero goals or implementing emission-based zoning laws.

Nevertheless, it is essential to consider the environmental cost of AI infrastructure itself, including the energy requirements of data centers and model training. Incorporating green AI practices—such as model compression, low-energy chips, and renewable-powered computation—will be crucial to maintain a positive net environmental impact.

16.6.5 Resilient Infrastructure Through Predictive Maintenance

One of the most practical yet often overlooked applications of AI in transportation is predictive maintenance. The results of this study demonstrated that AI models significantly increased fault detection accuracy while lowering maintenance costs and reducing unexpected downtime. By transitioning from reactive to predictive strategies, transportation agencies can ensure better asset utilization, improved safety, and longer infrastructure life cycles.

The integration of blockchain for maintenance record verification further enhances transparency and coordination among stakeholders. It provides a verifiable and tamper-proof audit trail that can be shared across public agencies, contractors, and regulators. This innovation supports a broader governance model where data integrity is preserved even as operational decisions become more decentralized.

16.6.6 Socio-technical and Governance Challenges

Despite the clear technical benefits demonstrated, several socio-technical and governance-related challenges remain. These include data privacy concerns, algorithmic bias, and the risk of over-reliance on opaque AI systems. For example, AI models trained on skewed historical data may perpetuate inequalities in service delivery, such as underserving low-income or peripheral neighborhoods.

Furthermore, public trust in AI-driven decision-making systems is not automatic. Transparent model explanations, community engagement, and ethical auditing frameworks are necessary to ensure legitimacy and inclusiveness in

AI-enhanced mobility systems. Policymakers must also balance innovation with regulation, enabling agile experimentation while upholding public accountability.

16.6.7 Pathways for Future Research and Policy Integration

To scale the impact of cognitive AI in urban transport, future research must focus on modular, interoperable system architectures that can integrate with existing city infrastructure. Cross-domain collaboration among computer scientists, urban planners, and transport economists will be key to designing inclusive and context-sensitive AI solutions.

Policy integration is another critical area. Governments need to create regulatory sandboxes, data governance frameworks, and public-private partnerships that facilitate AI innovation without compromising ethical standards. Standardization efforts—especially around data formats, model validation, and transparency—will support scalable deployment across different urban contexts.

Finally, more empirical studies are needed from cities in the Global South, where data scarcity, infrastructure gaps, and institutional constraints pose unique challenges. Adaptable, low-resource AI models, supported by cloud infrastructure and mobile integration, could offer cost-effective pathways for inclusive smart mobility in developing regions.

16.7 Conclusion

This research has proposed, developed, and validated an integrated cognitive AI framework for adaptive and resilient urban transportation. Through interdisciplinary design and rigorous data analysis, the study demonstrates that cognitive AI can revolutionize the management of urban mobility systems by enabling real-time decision-making, predictive scheduling, intelligent vehicle coordination, emissions monitoring, and proactive infrastructure maintenance. The proposed system architecture—comprised of a data input layer, an AI processing core, and a feedback and adaptation loop—offers a scalable and modular foundation for implementing smart transportation systems aligned with the evolving needs of global urbanization.

Empirical evaluations based on simulated smart city environments and real-world case studies reveal substantial improvements across key transport performance metrics. DRL significantly reduced traffic delays and increased intersection throughput. LSTM-based predictive models enhanced public transit reliability and responsiveness to fluctuating demand. Coordination among AVs led to decreased conflict rates and energy savings, while AI-guided emissions tracking and EV routing produced measurable environmental benefits. Additionally, predictive maintenance algorithms demonstrated strong potential in minimizing infrastructure downtime and optimizing asset life cycles.

By synthesizing these functional areas, the research contributes to both theoretical understanding and practical implementation of AI-augmented mobility systems. It underscores the importance of transitioning from static, rule-based infrastructure to adaptive, data-centric urban ecosystems that are resilient to disruption and capable of continuous self-optimization. Furthermore, it promotes a governance-oriented approach where explainability, transparency, and ethical accountability are integral to technological deployment.

Looking forward, several avenues emerge for extending this research. First, there is a need to enhance the generalizability of AI models across diverse urban contexts, especially in rapidly growing cities in the Global South. Low-cost, cloud-based AI infrastructures tailored for data-constrained environments could democratize access to smart mobility technologies. Second, advancing human-in-the-loop AI design will ensure that machine decisions remain aligned with public values and social priorities. Third, integrating AI systems with Internet of Behavior (IoB) platforms could provide deeper insights into commuter behavior, enabling hyper-personalized services and more inclusive urban design.

Moreover, the integration of green AI practices—such as energy-efficient algorithms and renewable-powered data centers—will be essential to ensure that digital sustainability efforts do not inadvertently exacerbate environmental burdens. Lastly, future work should focus on robust simulation environments, longitudinal field studies, and participatory design frameworks to co-create AI-driven transport solutions that are not only intelligent and efficient but also equitable, accountable, and trusted by the communities they serve.

References

1 Goetz, A. R. Transport challenges in rapidly growing cities: Is there a magic bullet? *Transport Reviews*, 39(6): 701–705, 2019.

2 Papadakis, D. M., Savvides, A., Michael, A., and Michopoulos, A. Advancing sustainable urban mobility: Insights from best practices and case studies. *Fuel Communications*, 20: 100125, 2024.

3 Boin, R., Möller, T., Pokotilo, V., Ricotti, A., and Sandri, N. *Infrastructure Technologies: Challenges and Solutions for Smart Mobility in Urban Areas.* McKinsey & Company, 2023.

4 Ceder, A. Urban mobility and public transport: future perspectives and review. *International Journal of Urban Sciences*, 25(4): 455–479, 2021.

5 Goumiri, S., Yahiaoui, S., and Djahel, S. Smart mobility in smart cities: Emerging challenges, recent advances and future directions. *Journal of Intelligent Transportation Systems*, 29(1): 81–117, 2025.

6 Sankar, M., Pachiyappan, S., and Srivastava, A. Assessing the impact of smart city technologies on sustainable urban mobility in developing economies. In *Smart Cities and Circular Economy*, pages 193–214. Emerald Publishing Limited, 2024.

7 Paiva, S., Ahad, M. A., Tripathi, G., Feroz, N., and Casalino, G. Enabling technologies for urban smart mobility: Recent trends, opportunities and challenges. *Sensors*, 21(6): 2143, 2021.

8 Rathore, S. A., U Salam, M. H., Abbas, T., Irfan, M., and Ameen, K. Machine learning approach to reducing urban congestion using artificial intelligence for smart traffic management. *Dialogue Social Science Review (DSSR)*, 2(5): 840–863, 2024.

9 Gheorghe, C. and Soica, A. Revolutionizing Urban Mobility: A systematic review of AI, IoT, and predictive analytics in adaptive traffic control systems for road networks. *Electronics* 14(4): 719, 2025.

10 Kumar, A., Batra, N., Mudgal, A., and Yadav, A. L. Navigating urban mobility: A review of AI-driven traffic flow management in smart cities. In *2024 11th International Conference on Reliability, Infocom Technologies and Optimization (Trends and Future Directions)(ICRITO)*, pages 1–5. IEEE, 2024.

11 Wang, J., Jiang, S., Qiu, Y., Zhang, Y., Ying, J., and Du, Y. (2021). Traffic signal optimization under connected-vehicle environment: An overview. *Journal of Advanced Transportation*, 2021(1): 3584569, 2021.

12 Chala, T. D. and Kóczy, L. T. Intelligent fuzzy traffic signal control system for complex intersections using fuzzy rule base reduction. *Symmetry*, 16(9): 1177, 2024.

13 Oladimeji, D., Gupta, K., Kose, N. A., Gundogan, K., Ge, L., and Liang, F. Smart transportation: An overview of technologies and applications. *Sensors*, 23(8): 3880, 2023.

14 Makanadar, A. and Shahane, S. Smart mobility and cities 2.0: Advancing urban transportation planning through artificial intelligence and machine learning. In *Urban Mobility India*, pages 23–38. Springer: Singapore, 2024.

15 Varghese, K. E. N. Data-driven and machine learning approaches for transport planning and management, 2025.

16 Buçinca, Z., Malaya, M. B., and Gajos, K. Z. To trust or to think: Cognitive forcing functions can reduce overreliance on AI in AI-assisted decision-making. *Proceedings of the ACM on Human-computer Interaction*, 5(CSCW1): 1–21, 2021.

17 Tariq, M. U. Smart transportation systems: Paving the way for sustainable urban mobility. In *Contemporary Solutions for Sustainable Transportation Practices*, pages 254–283. IGI Global, 2024.

18 Maheswaran, K., Selvakumar, P., Shaikh, M., Shashank, G., Gupta, S., and Manjunath, T. C. The evolution of smart urban mobility with emerging technologies. In *Leveraging VANETs and Blockchain Technology for Urban Mobility*, pages 1–28. IGI Global Scientific Publishing, 2025.

19 Alam, T., Gupta, R., Nasurudeen Ahamed, N., Ullah, A., and Almaghthwi, A. Smart mobility adoption in sustainable smart cities to establish a growing ecosystem: Challenges and opportunities. *MRS Energy & Sustainability*, 11(2): 304–316, 2024.

20 Cirianni, F. M. M., Comi, A., and Quattrone, A. Mobility control centre and artificial intelligence for sustainable urban districts. *Information*, 14(10): 581, 2023.

21 Afandizadeh, S., Abdolahi, S., and Mirzahossein, H. Deep learning algorithms for traffic forecasting: A comprehensive review and comparison with classical ones. *Journal of Advanced Transportation*, 2024(1): 9981657, 2024.

22 Agrahari, A., Dhabu, M. M., Deshpande, P. S., Tiwari, A., Baig, M. A., and Sawarkar, A. D. Artificial intelligence-based adaptive traffic signal control system: A comprehensive review. *Electronics*, 13(19): 3875, 2024.

23 Smart, A., Zhao, Y., and Curran, D. Chinese artificial intelligence governance platforms 2.0: The belt and road edition. In *Artificial Intelligence and the City*, pages 258–274. Routledge, 2023.

24 Lungu, M. A. Smart urban mobility: The role of AI in alleviating traffic congestion. *Proceedings of the International Conference on Business Excellence*, 18(1): 1441–1452, 2024.

25 Bahamazava, K. AI-driven scenarios for urban mobility: Quantifying the role of ODE models and scenario planning in reducing traffic congestion. *Transport Economics and Management*, 3: 92–103, 2025.

26 Lukic Vujadinovic, V., Damnjanovic, A., Cakic, A., Petkovic, D. R., Prelevic, M., Pantovic, V., Stojanovic, M., Vidojevic, D., Vranjes, D., and Bodolo, I. AI-driven approach for enhancing sustainability in urban public transportation. *Sustainability*, 16(17): 7763, 2024.

27 Hu, S. A big-data-driven framework for spatiotemporal travel demand estimation and prediction (Doctoral dissertation, University of Maryland, College Park). 2023.

28 Oh, S., Lentzakis, A. F., Seshadri, R., and Ben-Akiva, M. (2021). Impacts of automated mobility-on-demand on traffic dynamics, energy and emissions: A case study of Singapore. *Simulation Modelling Practice and Theory*, 110: 102327, 2021.

29 Sun, Y., Guo, Q., Guo, Z., Gan, A., Mneney, E., Alam, M. R., and Surangsrirout, N. Evaluating and validating technology options for estimating transit vehicle occupancy in real time (No. BED30-977-03). Department of Transportation: Florida, 2023.

30 Wang, J., Zhai, H., Yang, Y., Xu, N., Li, H., and Fu, D. (2024). A review of intrusion detection for railway perimeter using deep learning-based methods. *IEEE Access*, 2024.

31 American Society of Civil Engineers. *Crisis Management in Construction Projects*. ASCE Press, 2013/2020. https://doi.org/10.1061/9780784406120.

32 Helmold, M., Küçük Yılmaz, A., Dathe, T., and Flouris, T. G. Digitalization in air transportation and reflections on SCRM. In *Supply Chain Risk Management: Cases and Industry Insights*, pages 37–78. Springer International Publishing: Cham, 2022.

33 Yang, M., Jiang, K., Wijaya, B., Wen, T., Miao, J., Huang, J., Zhong, C., Zhang, W., Chen, H., Yang, D., and Yang, D. (2024). Review and challenge: High definition map technology for intelligent connected vehicle. *Fundamental Research*, 5: 2765–2783, 2024.33. Pravallika, A., Hashmi, M. F., & Gupta, A. (2024). Deep learning frontiers in 3D object detection: A comprehensive review for autonomous driving. *IEEE Access*.

34 Shulajkovska, M., Smerkol, M., Noveski, G., Bohanec, M., and Gams, M. Artificial intelligence-based decision support system for sustainable urban mobility. *Electronics*, 13(18): 3655, 2024.

35 Alimohamadi, H. Enhancing sustainable commuting strategies for Lahti city: Integrating multi-thematic interventions for effective urban mobility, 2024.

36 Behnia, F., Schuelke-Leech, B. A., and Mirhassani, M. Optimizing sustainable urban mobility: A comprehensive review of electric bus scheduling strategies and future directions. *Sustainable Cities and Society*, 108: 105497, 2024.

37 Gao, Q., Xin, H., and Mosallam, A. Probabilistic fatigue life analysis of fiber reinforced polymer (FRP) composites made of $\pm$45 layers. In Saidi, R., El Bhiri, B., Maleh, Y., Mosallam, A., and Essaaidi, M., editors, *Advanced Technologies for Humanity: Proceedings of International Conference on Advanced Technologies for Humanity (ICATH'2021)*, Volume 110 of Lecture Notes on Data Engineering and Communications Technologies, pages 464–473. Springer, 2021.

38 Jiang, J., Li, Y., Li, Y., Li, C., Yu, L., and Li, L. Smart transportation systems using learning method for urban mobility and management in modern cities. *Sustainable Cities and Society*, 108: 105428, 2024.

17

Optimizing Urban Traffic with Graph Analytics: A Case Study of a Metropolitan Transportation Network

*S. Rakshika and Sudeepa Roy Dey**

Computer Science and Engineering, PES University, Bengaluru, Karnataka, India

*Corresponding author: sudeepar@pes.edu

17.1 Introduction

Urban transportation systems are among the most intricate, dynamic, and complex elements of cities today. The larger the urban area becomes in terms of size, population, or economic activity, the more complicated it becomes to be able to govern traffic flows effectively and with integrity. We have found that while transportation planning methods are traditionally based on static models, they are ill-equipped to conceptualize the complex, dynamic, nonlinear, and interdependent nature of mobility patterns in large city centers. This dilemma is particularly problematic in Bengaluru, India's high-tech heart and one of Asia's rapidly emerging capital cities.

Bengaluru, which is famous for its considerable population density and rapid urbanization, suffers from several traffic congestions and substandard modes of transport. While it is changing, its transport system still trails the demands of its growing commuter population, for example, public transport options are underused, and road networks are often congested. In order to facilitate urban movement in the context of such a city, data-driven, intelligent, and scalable approaches must be implemented.

The evolution of large mobility databases, as well as advancements in artificial intelligence (AI), and especially graph-based deep learning, has opened up new possibilities for examining traffic in cities. In fact, graph analytics, which examines data represented as nodes and edges, provides a powerful foundation for analyzing, modeling, and optimizing transport networks. In a transport network, nodes can be locations (e.g., a city ward, an intersection, a public transport stop), and edges can be traffic patterns, routes for traveling, or connectedness (even among multiple nodes). With graph-based approaches, researchers are able to discover predictive insights, and structural patterns, as well as uncover hidden relationships, ultimately improving how we interpret mobility beyond even linear, distance-based conclusions.

In this chapter, we offer a detailed application of graph analytics for improvements to urban traffic in Bengaluru, using the mobility datasets from Uber Movement and advanced methods such as Graph Autoencoders and GraphSAGE [1] to demonstrate how graph neural networks [2] (GNNs) can be used with traffic flow data to derive useful insights that contribute to better adaptability of traffic mobility solutions for transportation engineers and urban planners. Bengaluru's transportation system is represented as a directed, weighted network, with the wards represented as nodes and traffic flows aggregated by weekday-hourly as edges. This graph structure is used to address two primary tasks:

1. Community Detection with Graph Autoencoders [3] (GAE):
   This approach reveals latent ward groups with comparable traffic patterns. With no regard to actual geographic location, these communities can be viewed as functional areas—areas that are either functionally or institutionally related. Identifying these areas helps initially detail focused policies such as parking, zoned areas, and traffic control provisions.
2. Link Prediction using GraphSAGE:
   This challenge predicts new or undiscovered ward-to-ward transport links. Proactive infrastructure planning, such as the construction of new roads, the improvement of bus routes, or the redesign of traffic signaling systems, is supported by the capacity to foresee such links. The model may generalize to new connections and future situations by learning from both the structure of the transportation graph and node-level information (travel time statistics).

In contrast to traditional traffic models that depend on strict presumptions or predetermined paths, graph neural networks learn straight from data and adjust to temporal and spatial diversity. The embeddings produced during training capture intricate relationships between wards, taking into account neighborhood connections, flow intensities, and more than just distance or direct connectivity. Because of this, these approaches can yield subtle insights that are hard to obtain using conventional routing or optimization methods.

Although link prediction and community analysis are the main topics of this chapter, these methods act as upstream facilitators for numerous downstream applications, such as:

- Intelligent Routing Algorithms: By using predictive insights from GNNs, AI-based routing engines can take advantage of hidden shortcuts or steer clear of future congestion hotspots.
- Public Transit Planning: Transportation agencies can rank routes for planned bus corridors or metro expansions by identifying underserved yet structurally important links.
- Urban Resilience Modeling: By emphasizing vital routes during periods of high demand or disruption, the identification of new mobility corridors can aid emergency response planning.

Bengaluru's selection as a case study is especially significant. With more than 8 million commuters every day, Bengaluru serves as an example for other rapidly urbanizing cities in the Global South, in addition to being a high-density traffic area. It is the perfect place to explore scalable, AI-driven solutions since it mixes issues including informal traffic patterns, mixed land use, and infrastructure bottlenecks.

The chapter uses high-quality, anonymized data from Uber movement to guarantee the validity and applicability of our methodology. The files consist of a CSV file with hourly aggregated traffic data and a GeoJSON file that depicts ward borders. Because of the spatial and temporal alignment of this information, we are able to construct a comprehensive model of the transportation graph.

We also devote a section to describing the different kinds of routing algorithms, ranging from AI-powered approaches like reinforcement learning and GNN-enhanced route planning to traditional deterministic approaches like Dijkstra [4] and A*. This comparative perspective aids in placing our graph-based method in the larger context of traffic optimization technologies. In summary, this chapter provides an example of how GNN-powered graph analytics may transform our understanding, modeling, and optimization of urban transportation systems. Our techniques open the door to the development of intelligent, flexible, and future-ready transport frameworks by predicting the dynamic link generation and capturing the underlying structure of mobility networks. The methods described here provide a scalable framework for data-driven urban mobility decision-making, regardless of whether they are employed by city governments, transit agencies, or mobility startups.

In order to build the groundwork for more intelligent and sustainable traffic management in Bengaluru and beyond, the parts that follow will go into greater detail on the relevant work, datasets, techniques, algorithms, and findings of our study.

17.2 Related Work

There has been a significant increase in the application of graph-based deep learning models for urban traffic optimization in the past few years. Due to the impressive ability of GNNs to identify dynamic, nonlinear, and context-aware patterns within transportation systems, they are starting to supplant or replace conventional modeling techniques that often use statistical approaches or static graphs. This section will discuss key developments and advancements made in the domain of GNN-based traffic modeling and prediction, by situating them within our current research on transportation networks.

In the survey "How to Build a Graph-Based Deep Learning Architecture in Traffic Domain" published in 2020, Ye et al. [5] established the first principles for applying graph learning to problems in the traffic domain. They discussed how to think about transport problems as graphs, presented deep learning methods relevant to structured data, and elaborated on the key limitations in traditional statistical approaches, for example, the reliance of those methods on assumptions of linearity and stationarity. In addition, shallow architectures and even manual feature selection methods were inadequate for dealing with the scale and heterogeneity of contemporary traffic data. Computational considerations were also raised; spectral graph convolution not only lacks spatial localizability, but it can also be difficult to scale. They indicated that most models considered at the time did not model spatial and temporal data jointly, and potentially failed to model their interaction. The paper further discussed the relatively little attention paid to more advanced models such as GATs, GAEs, Recursive GNN, and reinforcement learning, pointing to some interesting future attention.

In a related vein, Rahmani et al. (2023) broadened the horizon in their paper "Graph Neural Networks for Intelligent Transportation Systems: A Survey" [6]. The authors pointed out how previous studies focused primarily on traffic forecasting. However, in this survey, the intent was to study various other applications of ITS, including congestion detection, demand estimation, traffic signal control, and autonomous mobility, among others. Rahmani et al. indicated that prior surveys missed coverage of edge-centric and graph-level learning. Multi-model modeling was also noted as a gap for future research, specifically encompassing spatial graphs, temporal sequences, and contextual information such as weather, events, and socio-patterns. Importantly, the authors advocated for more research on unsupervised learning and reinforcement learning methods, especially when labeled data is scarce or costly to obtain. In addition to this, they also advocated for open datasets and standard evaluation protocols for the research community to ensure fair comparisons of models, as well as increased reproducibility of findings.

From an architectural point of view, Zhao et al. presented the temporal graph convolutional network (T-GCN) [7], one of the first models to integrate spatial and temporal feature sets in a unified framework. The T-GCN uses graph convolutional networks (GCNs) and gated recurrent units (GRUs) together to model spatial dependence on a road network and temporal variation in traffic flow. The model improved predictive performance over traditional deep learning baselines in forecasting tasks and was tested in experiments with real-world datasets. However, there were certain shortcomings, like predicting during peak traffic times with difficulty, due to the smoothing effect of the GRU draws on; using T-GCN in real-world datasets with sufficient data dependability and reliability, the model also had limitations with respect to data sparsity and errors when small traffic variability existed, highlighting the need for better temporal modeling.

Guo et al. introduced the optimized graph convolution recurrent neural network (OGCRNN) [8] to facilitate the ability to learn spatial-temporal dependencies. To implement it, they utilized a static graph structure in the testing phase. While efficient, this structure is inflexible to changing traffic events. They also mentioned experiences related to static road network data and dynamic traffic measurements being cumbersome. Additionally, the overall complexity of the OGCRNN architecture raises concerns about scaling the model and applying it to a large urban road network. The study indicates directions toward contrasting the rigidity of the Laplacian matrix via attention mechanisms, but this direction was not fully realized.

Li et al. took a notable step forward with the dynamic graph convolutional recurrent network (DGCRN) [9], which dynamically constructs graph structures based on time-evolving traffic situations, leading to improved modeling of changing spatio-temporal correlations. Li et al. also contributed a new benchmark dataset (NE-BJ) for enabling a fair comparison across models. However, we identified limitations in using the DGCRN framework in utilizing external factors—for example, weather, public events, and accidents—factors that have a significant impact on urban traffic. Furthermore, although it modeled dynamic patterns well, the DGCRN framework did not explicitly model congestion, which is an important aspect common in real-world traffic datasets.

Another exciting avenue of research is applying GNNs to reinforcement learning situations. Gammelli et al. presented a framework that integrated GNNs and deep reinforcement learning to manage autonomous mobility-on-demand (MoD) [10] systems, focusing on policy generalization/transferability across urban contexts. While they showed improved efficiency in vehicle rebalancing scenarios, this work is early stage and lacks much real-world testing. The use of reinforcement learning in graph-structured models of traffic problems is a relatively untapped area of study, particularly in relation to unsupervised pretraining and meta-learning forms of learning strategies.

This body of work shows many patterns and gaps. First, even though advanced GNNs and hybrid models have significantly improved predictive performance, the field is still working out how to deal with challenges associated with scalability, interpretability, and integration. Second, almost all studies have focused exclusively on traffic state predictions, and few have considered latent network structure as modular communities or hidden links, which our study directly addresses. Third, while the importance of spatiotemporal modeling is becoming recognized in the community, there are still very few approaches to learning from time and structure in a simultaneous way, as we have done in our project by integrating aspects of trip time into a graph. Further many surveys also emphasize a need for real-life application and practical insights, which we contribute through compelling graphs based on real-world Bengaluru ward-level data. Eventually, we employ GraphSAGE, a scalable and inductive approach suited to much larger and more dynamic urban applications, whereas earlier work focuses more on memory and computational constraints, especially surrounding large graphs or real-time systems.

17.3 Types of Routing Algorithm

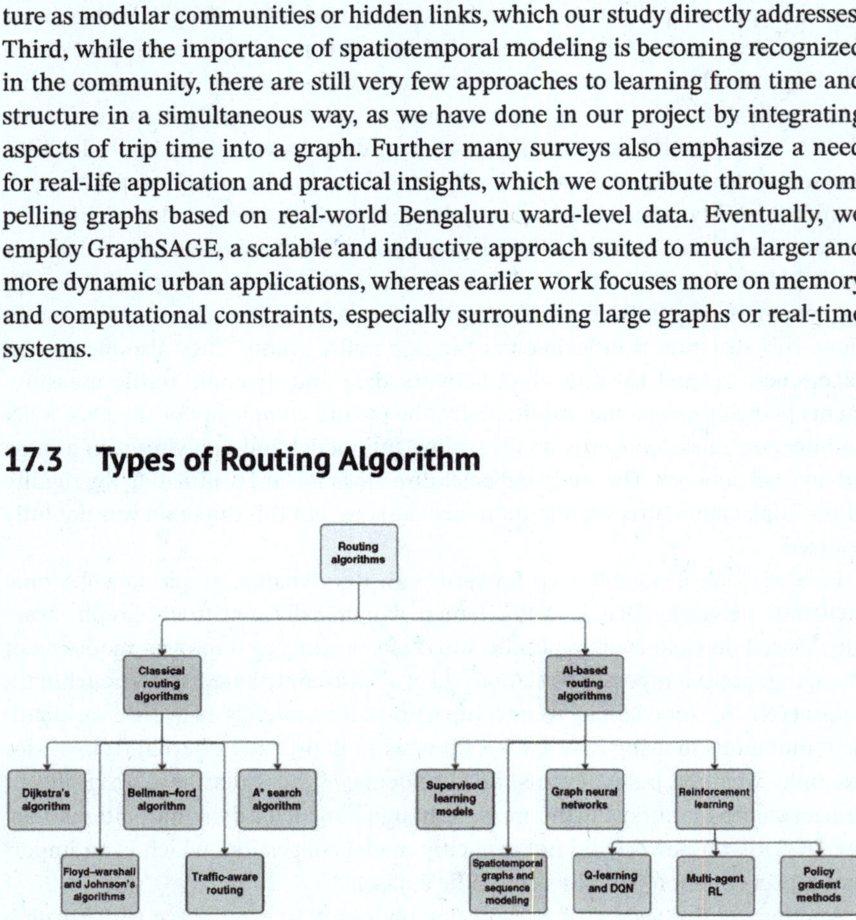

Intelligent transport systems rely on routing algorithms to determine the best routes for data or vehicles to travel across networks. These algorithms have historically been based on graph theoretic concepts and deterministic rules. However, more adaptive, predictive, and context-aware routing techniques are now possible thanks to the growing availability of mobility data and developments in AI. Before exploring the development and implications of AI-based routing techniques, which are more in line with the goals of this study, this part provides a quick overview of traditional routing algorithms.

A. Classical Routing Algorithms

In order to find the best routes, non-AI routing algorithms mostly work on static or dynamic graphs and employ heuristics and mathematical optimization approaches.

Among the most popular classical algorithms are the following:

a. Dijkstra's Algorithm

A fundamental algorithm that, under the assumption that edge weights are nonnegative, finds the shortest path between nodes in a graph. Despite its efficiency, it ignores stochastic fluctuations in transit time and real-time congestion.

b. Bellman–Ford Algorithm [11]

Though it can handle graphs with negative edge weights, it is comparable to Dijkstra's. Nevertheless, it is less appropriate for extensive metropolitan networks and more computationally costly.

c. A* Search Algorithm

It is an improvement on Dijkstra's that uses a heuristic function (usually the Manhattan or Euclidean distance) to expedite pathfinding. Although A* is effective, it is still static and cannot adjust to changing traffic conditions.

d. Floyd–Warshall and Johnson's Algorithms [12]

All-pairs shortest path techniques are employed in routing table preprocessing phases or in smaller networks. They are less appropriate for real-time use in expansive urban settings due to their computational complexity.

e. Traffic-aware Routing (Rule-based)

These techniques enable for dynamic updates by including real-time data (such as journey duration and congestion levels) into edge weights. Congestion-weighted A* and time-dependent Dijkstra are two examples. Although reactive, these are nonetheless subject to rules and do not extrapolate to other historical patterns or geographical characteristics.

Although these techniques established the foundation for contemporary routing systems, they are usually constrained by static assumptions, a deficiency of contextual learning, and an incapacity to represent intricate interdependencies in a transportation network at the city scale.

B. AI-based Routing Algorithms

By allowing systems to learn from data, adjust to changing traffic circumstances, and make predictions based on past trends, AI-based techniques have completely changed routing. In metropolitan settings, where traffic dynamics are nonlinear, stochastic, and spatially heterogeneous, these techniques are especially effective.

a. Supervised Learning Models

Based on variables including the time of day, the weather, and past speed data, machine learning models like regression, decision trees, or neural networks have been used to anticipate travel times or congestion levels. In conventional routing algorithms, these models are frequently used as preprocessing techniques to improve edge weighting.

b. Reinforcement Learning (RL) for Routing [13]

Routing is modeled as a sequential decision-making issue by RL-based techniques, particularly deep reinforcement learning (DRL). By maximizing cumulative benefits, such as the least amount of fuel or travel time, agents are able to learn the best routes. Among the notable frameworks are:

- Q-learning and DQN [14]: Learn policies for navigating traffic graphs by exploring and exploiting path choices.
- Multi-agent RL (MARL): Useful in decentralized traffic systems, where each vehicle or route segment is treated as an agent.
- Policy Gradient Methods: Suitable for continuous action spaces and complex road networks.

Reinforcement Learning systems are adaptive and generalizable but require substantial data and careful reward design to avoid suboptimal policies.

c. Graph Neural Networks (GNNs) for Routing

GNNs model node and edge dependencies by utilizing the underlying graph structure of transportation networks. In order to infer hidden patterns like latent congestion or possible shortcut routes, GNNs learn embeddings for nodes (like junctions and wards) and edges (like traffic flows).

- To forecast journey times or route choices, GCNs and graph attention networks (GATs) combine neighborhood characteristics.
- In order to provide dynamic link prediction and future route forecasting, this study used GraphSAGE, which develops inductive representations that generalize to unseen nodes or links.
- RL and GNNs can be combined to use structural clues from the road network to direct policy learning.

d. Spatiotemporal Graphs and Sequence Modeling

Temporal traffic evolution and sequential routing behavior have been modeled using deep learning models like Transformers and LSTMs. They create spatiotemporal networks that can forecast next-hop pathways or the spread of congestion over time when paired with GNNs.

The techniques used in this study—community discovery and connection prediction using graph analytics—act as upstream enablers for AI-based routing even though they do not explicitly generate routes:

- o Community Detection finds spatial clusters with strong connections, which can serve as subnetworks for decentralized traffic control or as modular routing units.
- o By identifying latent or developing trip pathways, Link Prediction with GraphSAGE enables proactive routing recommendations prior to traffic accumulation.

The method facilitates the development of intelligent routing frameworks that are scalable, adaptive, and predictive—essential characteristics for urban mobility systems that are prepared for the future—by utilizing AI-driven graph models.

17.4 Work

Real-world mobility data from Uber Movement, a platform that compiles anonymized data from Uber journeys to provide insights into urban transportation trends, served as the foundation for this study. In order to create a solid graph-based model of Bengaluru's transport system, the two main datasets employed in this study capture the spatial and temporal aspects of intra-city traffic.

These datasets include:

1. Ward-level geographic boundaries in a GeoJSON file.
2. An hourly-resolution CSV file with aggregated traffic movement data for only weekdays.

These datasets work together to model, analyze, and forecast mobility trends in Bengaluru's many wards.

The first dataset is a GeoJSON file that shows the administrative divisions of Bengaluru's spatial arrangement. The following essential characteristics are present in each entry, which represents a municipal ward:

- WARD_NO: A special number that is given to every ward.
- WARD_NAME: The ward's formal name.
- MOVEMENT_ID: A distinct number that is used to match mobility information.
- DISPLAY_NAME: An easily interpreted landmark or address that is readable by humans.
- The ward's physical boundary is specified in GPS coordinates by a MultiPolygon object.

This dataset is essential for:

- Graph Construction: Every ward is represented as a graph node. When building edges, the geographic layout helps to verify spatial adjacency and connectedness.
- Region Mapping: By serving as a link between the temporal and geographical datasets, the MOVEMENT_ID makes integration easier.
- Visualization: The spatial geometries are utilized to produce maps and highlight analytical conclusions such as traffic clusters, high-mobility corridors, and anticipated links.

The ward boundaries' geometric accuracy permits in-depth spatial analysis and permits findings to be localized to certain city neighborhoods or zones.

Aggregated, anonymized travel data between ward pairs is included in the second dataset. An hour of journey time between an origin and a destination ward is represented by each row. The dataset consists of the following fields:

- start_hour: The timestamp corresponded to the start of the hour when the travel observation occurred
- origin_movement_id: The MOVEMENT_ID of the origin ward
- destination_movement_id: The MOVEMENT_ID of the destination ward
- mean_travel_time: The average travel time between the two wards over that hour, in seconds
- standard_deviation_travel_time: The variability in travel times, which can be helpful in assessing reliability
- geometric_mean_travel_time: A robust average metric that reduces the impact of outliers and skewed locations
- geometric_standard_deviation_travel_time: variability on the geometric mean indicating temporal stability or instability in travel_patterns.

We can build a time-aware, weighted directed graph for Bengaluru's transport system, since the dataset records both directional and temporal properties of mobility.

These datasets form a strong foundation for graph-based urban traffic analysis by integrating temporally granular mobility measures with high-resolution data regarding geography. They contribute to the greater mission of better and sustainable urban transport in Bengaluru through both macro-centric structural ideas and micro-centric predictive intelligence.

A. Community Detection [15]—Graph Autoencoder

We use a graph autoencoder (GAE)-based method to reveal latent community patterns in the urban transportation network of Bengaluru. Unsupervised neural architectures known as GAEs are capable of learning low-dimensional node embeddings while maintaining the input graph's topological structure and node characteristics. Because of this, they work especially well for urban traffic graphs, where flow characteristics and connections are closely related.

The graph G = (V, E, X) is constructed using an edge list derived from hourly aggregate weekday traffic flow data:

- V: Set of nodes, each representing a municipal ward in Bengaluru.
- $E \subseteq V \times V$: Directed edges indicating traffic flow between wards.
- X ∈ R^{|V| × d}: Node feature matrix, where each row contains ward-level traffic attributes.

Each edge $e_{\{ij\}} \in E$ denotes travel from ward *i* to ward *j*, associated with the mean travel time $\mu_{\{ij\}}$. To reflect connectivity strength, edge weights (Equation 17.1) are computed as the inverse of travel time:

$$w_{\{ij\}} = \frac{1}{\mu_{\{ij\}}} \tag{17.1}$$

This formulation emphasizes faster connections by assigning them higher weights.

Each node is described using the following traffic attributes:

- Inflow Time: Average travel time into the ward from all other wards.
- Outflow Time: Average travel time from the ward to other wards.

These features are standardized and serve as input to the encoder network.

The GAE model leverages two stacked GCN layers to learn meaningful node embeddings:

$$Z = GCN_2(ReLU(GCN1(X, A)), A) \tag{17.2}$$

where:

- **A**: Normalized adjacency matrix of the graph
- **X**: Initial node features
- $\mathbf{Z} \in \mathbb{R}_{\{V\times d'\}}$: Learned node embedding matrix, where $\mathbf{d'} = \mathbf{2}$ in our case

The decoder reconstructs the graph structure by computing the inner product of the node embeddings:

$$A_{\{ij\}} = \sigma\left(z_i^{T}\ z_j\right) \tag{17.3}$$

- σ: Sigmoid function
- z_i, $\mathbf{z_j}$: Embeddings of nodes *i* and *j* respectively
- The model is optimized using **binary cross-entropy loss** between the true adjacency matrix *A* and the reconstructed matrix $\hat{A}$

Once the embeddings are learned, K-Means clustering is applied to segment the nodes into k = 5 communities:

$$C = \{C_1, C_2, \ldots, C_k\},\ C_i \subset V,\ \cup\ C_i = V \tag{17.4}$$

Each cluster C_i represents a set of wards with similar structural roles and traffic patterns (Figure 17.1).

- Peripheral Consistency: Because of their continuous accessibility and reduced intra-zone traffic, wards on the southern and western fringes regularly form contiguous clusters.
- Central Heterogeneity: The mixed-use zones and central business district are dispersed over several neighborhoods, reflecting the diverse and intricate transport patterns seen in these areas.
- Functional, Not Just Geographic, Grouping: Some wards that are geographically separated are part of the same cluster, indicating that they play structurally comparable roles in the traffic network (such as connector nodes or transit hubs).

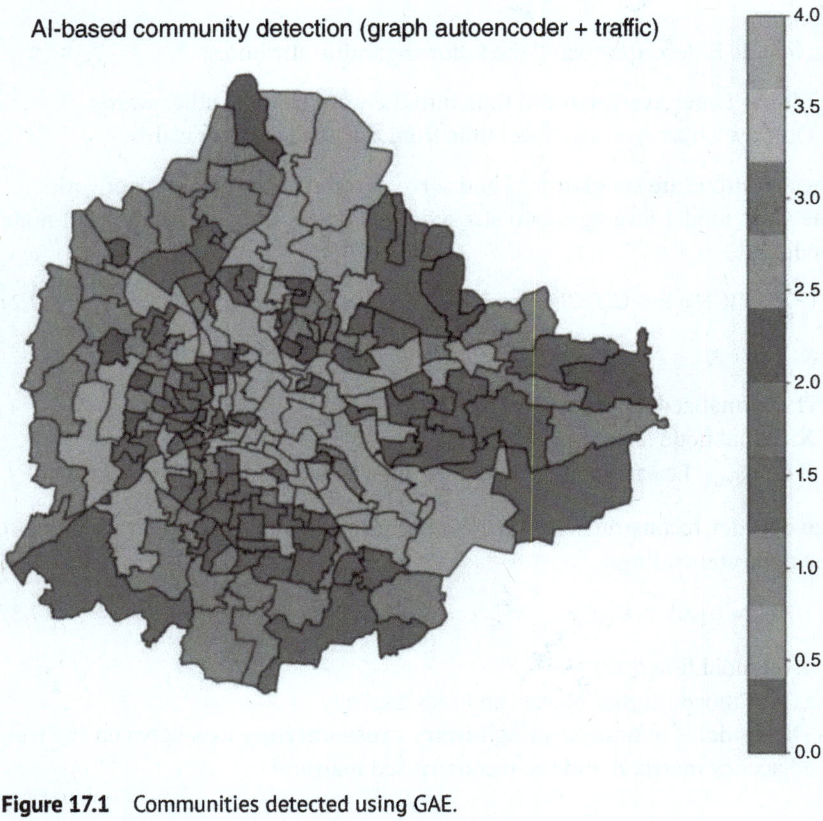

Figure 17.1 Communities detected using GAE.

B. Link Prediction [16]—GraphSAGE

In order to predict new traffic ties between Bengaluru's wards and uncover hidden mobility relationships, we used a link prediction methodology based on GraphSAGE. A potent inductive graph neural network (GNN) model called GraphSAGE (Graph Sample and Aggregate) aggregates data from neighbors to develop low-dimensional embeddings for nodes. In real-world graphs, such as urban transportation networks, where connectedness changes over time and data may be scarce, it performs very well.

In our dataset, the metropolitan road network is depicted as a directed graph. Each node represents a city ward, and each edge displays the total traffic flow from a source ward to a destination ward on an hourly basis during the week. The trip time data gathered at the source level, from which node attributes are derived, includes the mean journey time, standard deviation, geometric mean, and geometric standard deviation. These traits serve as concise summaries of the traffic trends in each ward.

Random link sampling technique was used to divide the dataset into training, validation, and test subsets in order to facilitate learning. In addition to maintaining the original graph's topology, this adds believable negative samples—ward pairings with no traffic—for supervised learning.

The link prediction process with GraphSAGE consists of two core stages: encoding and decoding.

Below is a simplified representation of the process:

Pseudocode: GraphSAGE-based Link Prediction

Input:

- Graph G(V, E)G(V, E) with nodes VV and edges EE
- Node feature matrix X∈Rn×fX \in \mathbb{R}^{n \times f}
- Edge index for known connections
- Positive and negative node pairs for training

Algorithm Steps:

1. Neighborhood Aggregation (Encoder):
   For each node v ∈ V:
   a. Sample neighbors of v
   b. Aggregate their features (e.g., mean or sum)
   c. Combine with v's own features
   d. Apply a nonlinear transformation to obtain embedding.
2. Link Prediction (Decoder):
   For each node pair (u, v):
   Compute link score using dot product:

$$\text{score}_{u,v} = z_u{}^T z_v \tag{17.5}$$

3. Optimization:
   Train the model using a binary classification loss function (e.g., binary cross-entropy), comparing predicted scores with true labels (link exists or not).

The approach allows the model to learn patterns of connectivity in the graph and estimate the probability of future or missing links, taking both node attributes and graph topology into account.

To evaluate model performance and understanding, we performed both quantitative assessments, and a series of visually intuitive analyses.

The receiver operating characteristic (ROC) curve can assess the ability of the model to discriminate between real links and non-links. The model has an area under the curve (AUC) of 0.6506, indicating confident predictions in both common ward connections as well as rare ward connections. The increasing slope of the curve shows that meaningful structural relationships are captured and that it has a valid signal for predicting traffic links (Figures 17.2 and 17.3).

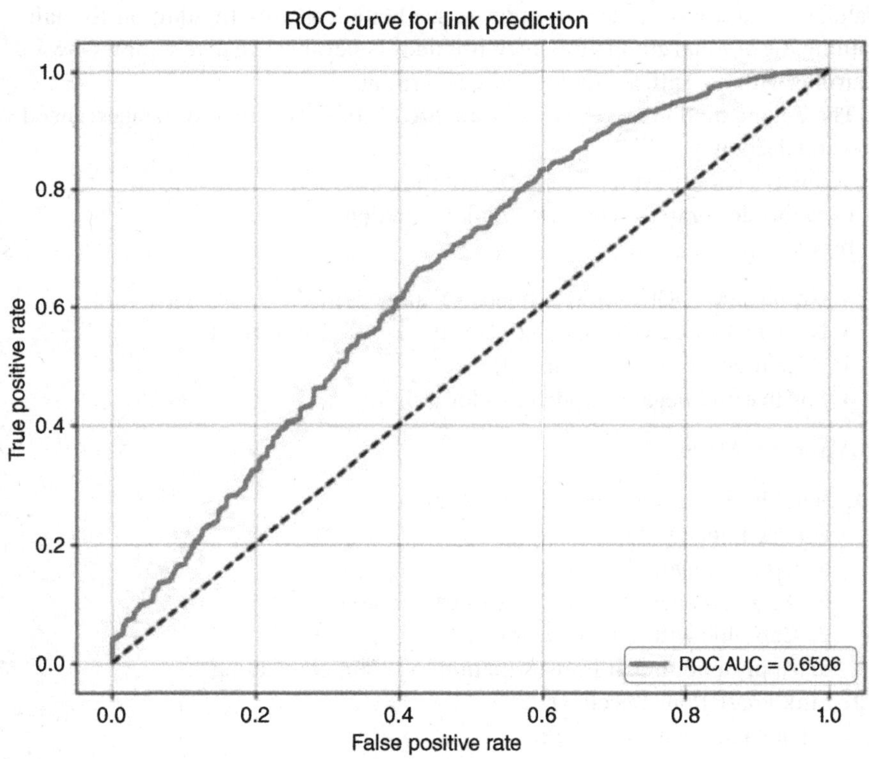

Figure 17.2 ROC curve.

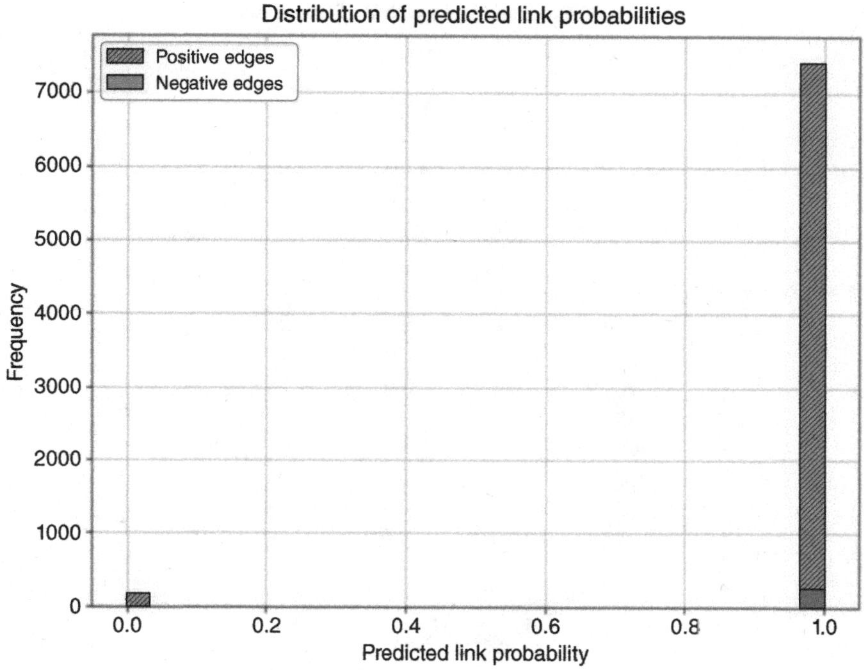

Figure 17.3 Predicted link probability distribution.

This visualization hints at the distribution of predicted probabilities for positive and negative edge samples. The model is highly confident in assigning links high probabilities and non-links low probabilities. A majority of predicted probabilities gather high or low, indicating the model is confident in distinguishing between true and false traffic relationships. This is encouraging because it demonstrates the model's confident identification of valid traffic relationships, reinforcing its value to surface the hidden connection.

One of the more insightful visualizations is the graph of the top 50 highest-probability predicted links. In this graph:

- Nodes denote individual wards, with size based on the node's out-degree (amount of predicted outgoing connections), which indicates each ward's relational power and significance in the emerging traffic network.
- Edges denote high-confidence links between wards, which are shown in a gradient from cool (low probability) to warm (high probability).
- Node labels are ward IDs, so the analysis can be tied directly to urban geography (Figure 17.4).

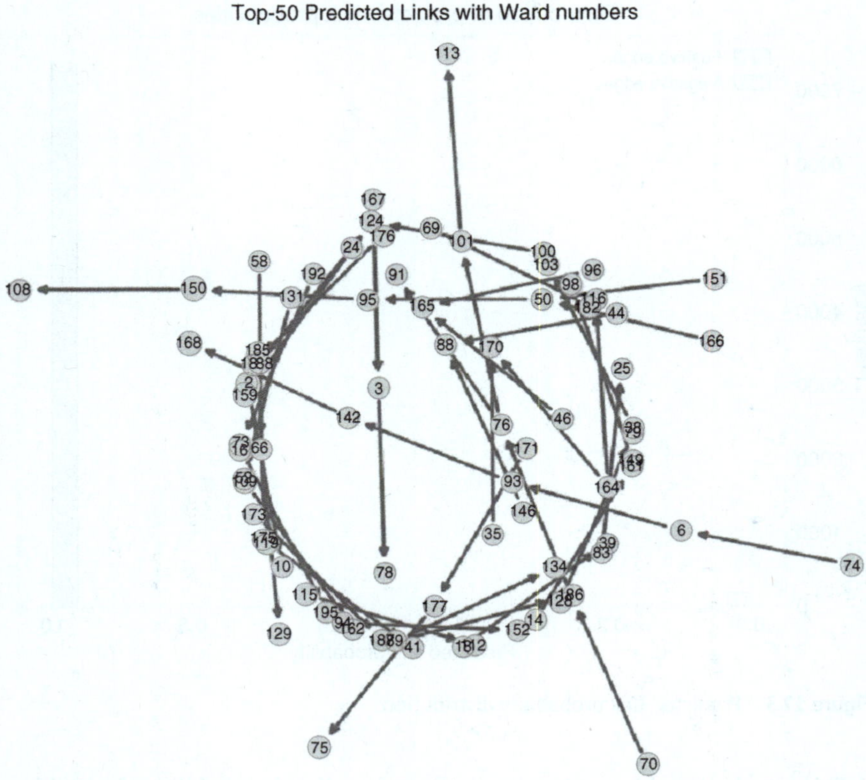

Figure 17.4 Top-K predicted links graph.

The force-directed layout places strongly connected nodes together, which allows clusters of interaction to emerge visually. Nodes that are larger, with many strong outgoing links, denote node zones of traffic generation—commercial areas, residential areas, or possibly gateways to parts of the city. On the other hand, those wards with multiple incoming edges would be zones attracting traffic, i.e., markets, connections to transportation, and/or work centers.

The paths from the nodes in the graph construct manifested corridors of movement, which are channels of movement not only meaningful construction, but latent corridors of probable high demand movement," which urban planners might focus on possible road improvements, public transit extensions, or infrastructure upgrading projects. This top-K visualization serves as both a diagnostic and prescriptive map of urban traffic flow, illustrating which parts of the study area are dynamically changing with respect to urban traffic flow.

The GraphSAGE-based model for link prediction allows for a discernible, interpretable, and scalable way of representing and analyzing unpleasantly complex urban mobility processes. Moving from simplistic structural data models to combined structural and feature-based graph models illuminates lines of action about how traffic may increase, turn, spread, or intensify throughout the city. Likewise, the graphs serving as visual outputs do not supplant the heavier weight of evaluating people's mobility statistically, but they would provide stakeholders with a more lay perspective, with supporting, evidence-based modeling behind it, to make policy, design, or infrastructure decisions.

This tool is useful not simply to model people's instantaneous traffic, but also to make projections on what connected system traffic could become as a result of the learned mobility behavior.

In a rapidly growing city like Bengaluru, proactive, ahead-of-the-growth-curve planning becomes imperative to build on sustainable, efficient, and practical urban transportation systems.

Compiling together the techniques and insights examined in this chapter, Bengaluru's transportation network presents an interesting real-world case of how graph analytics can illuminate the complexity of urban mobility. We represented wards as nodes, with directed, weighted edges modeled by weekday total travel flows. Thus, creating a graph representing the spatial distribution of wards across the city and movement across time. Using this graph representation, we then showed how Graph Autoencoders could be used to reveal latent communities of wards whose traffic patterns are structurally similar. We were able to see that the community clustering they produced represented more than spatial closeness in wards. For example, they represented behavioral groupings, such as residential clusters with similar commuting patterns, or commercial amenities that act as very high-value attractors. We also discovered that functional similarity could occur without regard to physical distance, supporting many researchers views that often don't matter in typical clustering techniques.

With this structural sense, our application of GraphSAGE allowed us to predict new or unobserved links between wards by learning from the neighborhood context and specific-node traffic attributes. This model can generalize to unseen links, making it particularly relevant in a rapidly changing city such as Bengaluru, where road usage patterns constantly shift due to ongoing construction, changed economic centers, and new commuter choices and circumstances. The predicted links—as shown in a top-K probability graph—help inform potential future corridors for congestion or lucky opportunity, giving transportation planners the ability to take action ahead of time.

In this case study, we illustrate how sensing data, as scaled through graph neural networks (GNNs), can be modeled to produce insights that are both interpretable and actionable. We not only report on the current behavior of an urban mobility system but also offer foresight about the growth of that system. As seen in cities like Bengaluru that grapple with rapid growth and the operational burden upon infrastructure, this type of predictive intelligence is not just useful, it's necessary. As such, this chapter contains much more than a contribution to the technical literature; it establishes a sufficiency of evidence for a practical framework to operationalize GNN-based graph analytics, which can be incorporated into the daily practice of urban planners, policymakers, and civic technologists who are planning inclusive and integrated mobility solutions. As our cities become ever more data-driven and complicated, approaches like ours can instrumentally enable us to build cities that are smarter, more adaptive, equitable, and resilient.

References

1 Lo, W. W., Layeghy, S., Sarhan, M., Gallagher, M., and Portmann, M. E-graphsage: A graph neural network based intrusion detection system for IoT. In *NOMS 2022-2022 IEEE/IFIP Network Operations and Management Symposium*, pages 1–9. IEEE, 2022.

2 Ruiz, L., Gama, F., and Ribeiro, A. Graph neural networks: Architectures, stability, and transferability. *Proceedings of the IEEE*, 109(5): 660–682, 2021.

3 Salehi, A., and Davulcu, H. Graph attention auto-encoders. *arXiv preprint arXiv:1905.10715*, 2019.

4 Lanning, D. R., Harrell, G. K., and Wang, J. Dijkstra's algorithm and Google maps. In *Proceedings of the 2014 ACM Southeast Conference*, pages 1–3, 2014.

5 Ye, J., Zhao, J., Ye, K., and Xu, C. How to build a graph based deep learning architecture in traffic domain: A survey. *IEEE Transactions on Intelligent Transportation Systems* 23(5): 3904–3924, 2020.

6 Rahmani, S., Baghbani, A., Bouguila, N., and Patterson, Z. Graph neural networks for intelligent transportation systems: A survey. *IEEE Transactions on Intelligent Transportation Systems*, 24(8): 8846–8885, 2023.

7 Guo, K., Hu, Y., Qian, Z., Liu, H., Zhang, K., Sun, Y., Gao, J., and Yin, B. Optimized graph convolution recurrent neural network for traffic prediction. *IEEE Transactions on Intelligent Transportation Systems*, 22(2): 1138–1149, 2020.

8 Zhao, L., Song, Y., Zhang, C., Liu, Y., Wang, P., Lin, T., Deng, M., and Li, H. T-GCN: A temporal graph convolutional network for traffic prediction. *IEEE Transactions on Intelligent Transportation Systems*, 21(9): 3848–3858, 2019.

9 Li, F., Feng, J., Yan, H., Jin, G., Yang, F., Sun, F., Jin, D., and Li, Y. Dynamic graph convolutional recurrent network for traffic prediction: Benchmark and solution. *ACM Transactions on Knowledge Discovery from Data*, 17(1): 1–21, 2023.

10 Gammelli, D., Yang, K., Harrison, J., Rodrigues, F., Pereira, F. C., and Pavone, M. Graph neural network reinforcement learning for autonomous mobility-on demand systems. In *2021 60th IEEE Conference on Decision and Control (CDC)*, pages 2996–3003. IEEE, 2021.

11 Busato, F., and Bombieri, N. An efficient implementation of the Bellman–Ford algorithm for Kepler GPU architectures. *IEEE Transactions on Parallel and Distributed Systems*, 27(8): 2222–2233, 2015.

12 Sao, P., Kannan, R., Gera, P., and Vuduc, R. A super nodal all pairs shortest path algorithm. In *Proceedings of the 25th ACM SIGPLAN Symposium on Principles and Practice of Parallel Programming*, pages 250–261, 2020.

13 Naeem, M., Rizvi, S. T. H., and Coronato, A. A gentle introduction to reinforcement learning and its application in different fields. *IEEE Access*, 8: 209320–209344, 2020.

14 Sabatelli, M., Louppe, G., Geurts, P., and Wiering, M. A. The deep quality-value family of deep reinforcement learning algorithms. In *2020 International Joint Conference on Neural Networks (IJCNN)*, pages 1–8. IEEE, 2020.

15 Jin, D., Yu, Z., Jiao, P., Pan, S., He, D., Wu, J., Yu, P. S., and Zhang, W. A survey of community detection approaches: From statistical modeling to deep learning. *IEEE Transactions on Knowledge and Data Engineering*, 35(2): 11491170, 2021.

16 Martínez, V., Berzal, F., and Cubero, J.-C. A survey of link prediction in complex networks. *ACM Computing Surveys (CSUR)*, 49(4): 1–33, 2016.

18

Urban Mobility Reimagined: AMRUT Interventions and the 2041 Outlook

S. Thangapriya[1], Nancy Jasmine Goldena[2], T. S. Vasughi[1], M. Kannan[1], and Barath Ramesh[1]*

[1] *Department of Computer Science with Data Science, SRM Arts and Science College, Affiliated to the University of Madras, Chengalpattu, Tamil Nadu, India*
[2] *Department of Computer Applications and Research Centre, Sarah Tucker College (Autonomous), Affiliated to Manonmaniam Sundaranar University, Tirunelveli, Tamil Nadu, India*

*Corresponding author: priyasaravananrex@gmail.com

18.1 Introduction

Tirunelveli is a southern city in Tamil Nadu, India, famous for its rich cultural heritage. It has an area of 3876.06 square kilometers and is divided into five distinct regions: Marutham-Ambasamudram Paddy Fields, Mullai-Kalakad Forest, Neithal-Uvari Sea Shore, Paalai-Radhapuram Terikkadu, and Manjolai Hills. Geographic Information System (GIS) technology was used in the preparation of the Tirunelveli Local Planning Area (LPA) Master Plan 2041. It adheres to the AMRUT (Atal Mission for Rejuvenation and Urban Transformation) principles and offers ideas meant to facilitate the LPA's anticipated expansion. This plan aims to guide sustainable urban growth while recognizing and addressing the problems caused by urbanization. The plan also focuses on a reliable water supply and a sewerage connection. Additionally, a dependable water supply and a sewer connection are key components of the strategy. Increasing the city's amenity value by developing green spaces and well-kept parks. Pollution reduction can be achieved by encouraging public transportation and constructing nonmotorized transportation infrastructure employing cutting-edge technologies such as GIS, AI, and computer vision. The main objective of this master plan is to identify the shortcomings and difficulties encountered during the implementation of the current development framework. It looks at things like infrastructure deficiencies, project delivery delays, and the requirement for better planning coordination. In addition, the plan takes into account the local literacy rate, with particular focus to the gender disparity in schooling. Strategies to improve access to education and advance gender equality are highlighted. In order to ensure sustainable urban development and an enhanced standard of living for all citizens, the plan's main goal is to achieve Tirunelveli's methodical and balanced expansion.

18.2 Geospatial Mapping of Tirunelveli Using Advanced Technologies

Tamil Nadu's location inside India is indicated in Figure 18.1, which also includes Tirunelveli [1]. It takes a lot of time and effort to draw these maps by hand with pen and paper, paying close attention to every little detail. It is simple to identify and map areas like Tirunelveli with the aid of GIS technology and satellite photography. More sophisticated methods, such as smart edge recognition algorithms, aid in precisely detecting borders; for instance, the system automatically gives a river a color if one is present. On the basis of field data, local surveyors designate particular locations. The suggested path can be recognized, marked, and submitted to the authorities in the event that the public desires a new road. Despite being essential, the signature and approval procedure is frequently difficult. Designers create intricate schematics using tools like AutoCAD, which are subsequently forwarded to higher authorities for approval and confirmation. Following approval, the plan is carried out and the road is built in accordance with it. Information on land features can be gathered using remote sensing data and used in a variety of analysis and research projects [3].

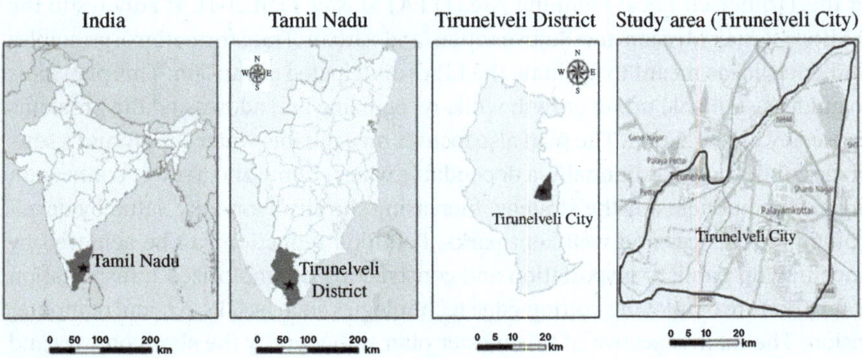

Figure 18.1 Tirunelveli's location in Tamil Nadu, India, highlighted. Created with the aid of satellite mapping and GIS technology for infrastructure development and urban planning. *Source*: [1]/MDPI/CC BY 4.0.

18.3 Identifying Research Gaps in Tirunelveli for Sustainable Regional Development

18.3.1 Topography

In the Tirunelveli region, topography is an important field of study that provides important information for creating sustainable metropolitan environments. The Manjolai Hills, Kalakad forests, the Thamirabarani River system, coastal areas like Uvari, rich plains in Ambasamudram, and arid regions like Radhapuram are just a few of the region's varied landscapes. For urban design, these diverse topography zones offer both special opportunities and difficulties. Researchers can examine terrain parameters including elevation, slope, drainage, and land use using contemporary methods like GIS, remote sensing, digital elevation models (DEMs), LiDAR, and satellite photography. These tools enable the creation of groundwater recharge models, landslide-prone region identification, flood risk maps, and zoning plans. Particular research can concentrate on hydrological modelling of the Thamirabarani basin, shoreline erosion along the coast, slope stability in hilly regions, and locating appropriate sites for renewable energy farms based on slope and topographic exposure data. Furthermore, urban heat island mapping based on surface topography and temperature data can guide the creation of green spaces and better urban layouts. These research ideas can be implemented through collaboration with local authorities, academic institutions, and planning

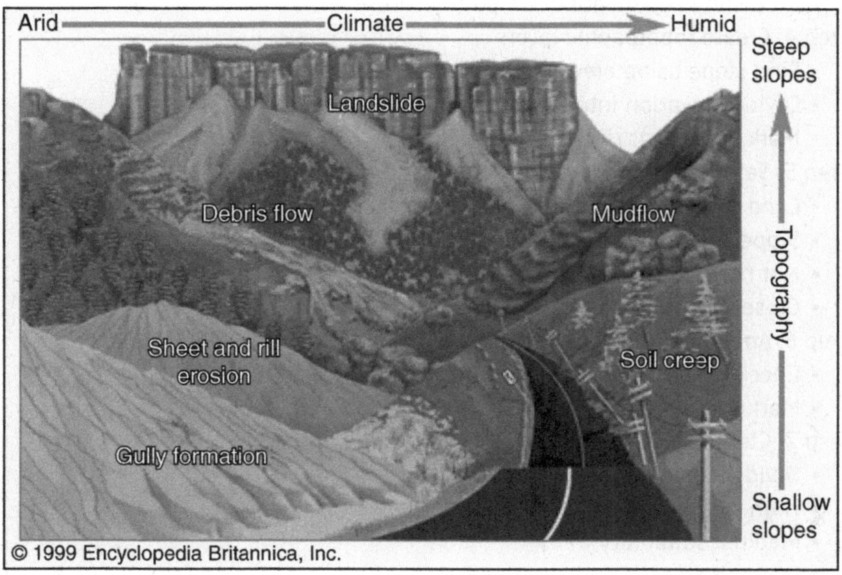

Figure 18.2 What is topography?

bodies using open-source GIS platforms and field surveys. Figure 18.2 shows how soil erosion changes with climate and slope. In dry areas with steep slopes, erosion is faster and more severe, causing landslides and deep gullies. In humid areas with gentle slopes, erosion is slower, like soil creep and mudflows. The diagram helps explain how topography and weather affect land changes.

Overall, topographic research in Tirunelveli plays a crucial role in guiding sustainable urban development and infrastructure planning. The pseudo code 1 highlights suitable areas for future urban development in the Tirunelveli region based on elevation, slope, proximity to roads, and flood risk zones, generated using GIS and remote sensing techniques [4].

Pseudo Code 1: Topographic Mapping for Identifying Suitable Areas for Urban Expansion in Tirunelveli

Step 1: Open GIS software (like QGIS or ArcGIS)
Step 2: Load all required data
- Load elevation map (DEM)
- Load satellite image of Tirunelveli
- Load land use map
- Load road and river maps

Step 3: Prepare the data
- Cut all maps to show only Tirunelveli area
- Make sure all maps are in the same format
- Clean unwanted data like clouds or noise

Step 4: Create topographic layers
- Find slope using elevation data
- Divide elevation into low, medium, high zones
- Mark areas near rivers that may flood

Step 5: Set rules for good construction land
- Land should be high (not low-lying)
- Slope should be gentle (not steep)
- Not near flood zones
- Close to main roads

Step 6: Apply the rules
- Check which areas follow all the rules
- Mark these areas as suitable

Step 7: Classify suitable land
- Divide suitable areas into:
- High suitability
- Medium suitability
- Low suitability

Step 8: Show the result
- Display the map with suitable areas
- Save the map and print if needed

Step 9: Write a report
- Mention how much land is suitable
- Suggest which places can be used for future urban development

18.4 Climate and Rainfall

Climate and rainfall data can be collected to analyze weather patterns and make predictions. This information can also be used for time series classification tasks. Month-wise and year-wise weather data such as rainfall, temperature, and humidity can be collected from meteorological departments or online databases. Future weather conditions can be predicted, variations throughout time can be identified, and climatic patterns can be analyzed using this historical data. We can create systems that forecast rainfall or climate conditions for the following months with the aid of cutting-edge technology like machine learning, deep learning, and time series forecasting models (e.g., ARIMA, LSTM). Planning for agriculture, flood warnings, and environmental studies can all benefit from this. Accurate forecasting for agriculture and disaster management in the Tirunelveli region, as described in Pseudo Code 2, is made possible by rainfall and climate prediction using month-by-month and year-by-year data analyzed by time series models like LSTM and ARIMA.

Pseudo Code 2: Rainfall and Climate Prediction

Step 1: Import required libraries
- Use Python libraries like pandas, matplotlib

Step 2: Load historical weather dataset
- Load data with columns: Year, Month, Rainfall, Temperature, Humidity

Step 3: Preprocess the data
- Handle missing values
- Convert format if needed
- Normalize features if needed

Step 4: Visualize the data (optional)
- Plot rainfall year by year
- Plot temperature variations

Step 5: Prepare the data for time series model
- Split into training and testing datasets

Step 6: Choose a prediction model
- Use any prediction techniques for time series forecasting

Step 7: Train the model
- Input: Past 12 months' data
- Output: Rainfall or temperature of the next month

Step 8: Evaluate model accuracy
- Use metrics like RMSE, MAE

Step 9: Make predictions
- Predict future rainfall based on trained model

Step 10: Display results
- Show prediction graph

18.5 Precipitation

In climate research and environmental planning, precipitation—which includes snow, rain, and other types of water that fall from the atmosphere—is essential. Researchers can identify seasonal variations, forecast future rainfall, and comprehend climatic trends in areas like Tirunelveli by examining monthly and annual precipitation data. This is particularly crucial for disaster preparedness, agriculture, and water resource management. Historical precipitation data can be processed and used to predict future patterns through the use of technologies like Geographic Information Systems (GIS), Remote Sensing, and machine learning models like ARIMA or LSTM. In order to guarantee sustainable development and disaster risk reduction, these projections can help farmers, municipal planners, and legislators make data-driven decisions. Pseudo code 3 visualizes quarterly and annual precipitation patterns in the Tirunelveli region using historical data, allowing forecasts of rainfall using LSTM and ARIMA models for improved planning of agriculture and water management [5].

Pseudo Code 3: Precipitation Forecasting Using Time Series Analysis

Step 1: Import required libraries

Step 2: Load the precipitation dataset
- Columns: Year, Month, Precipitation (mm)

Step 3: Clean and preprocess the data
- Handle missing or null values
- Convert date column to datetime format
- Set 'Date' as index for time series

Step 4: Visualize the data
- Plot monthly and yearly precipitation trends

Step 5: Prepare data for model
- Create time sequences for input (e.g., past 12 months)
- Split data into training and test sets

Step 6: Choose model (ARIMA or LSTM)
- For LSTM: reshape data for neural network input
- Define model architecture with input, LSTM layers, output

Step 7: Train the model
- Use training data and fit the model

Step 8: Evaluate the model
- Calculate RMSE, MAE to check accuracy

Step 9: Predict future precipitation
- Forecast for the next few months or year

Step 10: Plot prediction results
- Compare actual vs. predicted data

18.6 Soil Type Analysis and Resource-efficient Agricultural Planning in Tirunelveli Region

The predominant soil types in Tirunelveli are red loam and sandy soil, with strips of black loam found in certain areas. Black soil, which is highly valued due to its fertility, is mainly found along the river valleys and is usually layered over a strip of yellow soil. While red soil patches are generally less fertile, they can support good crop production when located near river streams with proper irrigation facilities. This diversity in soil types presents valuable opportunities for research in areas such as **soil classification**, **soil fertility analysis**, and **resource-efficient agricultural planning**. By studying soil characteristics, researchers can determine the best crops for each soil type and identify the ideal planting seasons, helping to **prevent resource wastage** and **improve agricultural productivity**. Soil classification and crop suitability map of Tirunelveli region highlighting black, red, and sandy soil areas, supporting optimal agricultural planning and resource management shown in pseudo code.

Pseudo Code 4: Soil Type Classification and Crop Recommendation

Step 1: Collect soil data from Tirunelveli region
- Gather data on soil type (red, black, sandy, loamy)
- Collect additional attributes: pH level, moisture, fertility index, irrigation availability

Step 2: Preprocess the data
- Handle missing or inconsistent values
- Standardize and normalize data for machine learning models

Step 3: Classify soil zones using GIS
- Import GIS map of Tirunelveli
- Overlay soil type data on map
- Identify zones: river valleys (black soil), uplands (red soil), sandy areas, etc.

Step 4: Match crops to soil types
- Create a database of suitable crops for each soil type and season
- Example:
- IF soil_type = 'Black' AND season = 'Kharif':
- Recommended_Crop = 'Cotton or Groundnut'

Step 5: Suggest best time for sowing
- Use historical climate data (rainfall, temperature)
- Match crop calendar with soil readiness

Step 6: Optimize resource usage
- Recommend irrigation only where needed
- Avoid fertilizer use where soil is naturally rich

Step 7: Visualize results
- Generate a soil suitability and crop planning map
- Export list of zones with crop and resource use plan

18.7 Geomorphology

The geomorphology of the Tirunelveli LPA is predominantly composed of Archaean crystalline rocks, including various types of gneiss and charnockite, which cover nearly 99% of the district. The charnockites in this region are acidic to intermediate in nature and are commonly overlain by black cotton soil and kankar. Garnetiferous Biotite Gneiss appears in shades of white to grey, while Garnetiferous Sillimanite Gneiss is pure white. Quartzite formations occur as small hills and ridges, particularly in the Sivanthipatti and Kunnathoor villages. Ridge formations are noticeable near Alangulam and Pettai, and significant bands of garnetiferous gneiss can be found near Palayamkottai and Reddiarpatti. Crystalline limestone is mainly present in the Tirunelveli and Nanguneri taluks, with high-quality deposits quarried near Seliyanallur and Ramayanpatti for cement production. Additionally, bands of calcareous gneiss are located around the Kalakkadu and Nanguneri regions. These geological features provide valuable insights into the natural landscape and support industries like mining and construction. Geological distribution of rock types in Tirunelveli LPA showing charnockite, gneiss, quartzite ridges, and limestone belts, highlighting areas with mineral resource potential for industrial applications shown in pseudo code 5.

Pseudo Code 5: Geomorphology of Tirunelveli LPA
Step 1: Collect geological data • Obtain geological maps of Tirunelveli from official surveys (GSI or TN geology dept.) • Collect mineral data: gneiss types, limestone, charnockite zones Step 2: Import data into GIS software • Use QGIS or ArcGIS • Load base map, topographic data, and rock layer shapefiles Step 3: Classify rock types by color/symbol • Assign distinct styles for: • Garnetiferous Biotite Gneiss • Charnockite • Quartzite • Crystalline limestone • Calcareous gneiss Step 4: Identify and mark resource-rich zones • Overlay mining data (e.g., cement quarry locations) • Mark high-potential areas near Seliyanallur, Ramayanpatti, and Kalakkadu Step 5: Generate terrain models • Use DEM to create hill/ridge maps for Sivanthipatti and Alangulam • Highlight ridges and slope variation Step 6: Output geological map • Export map with legend and labels • Include layers: rock type, mineral zone, terrain Step 7: Recommend applications • Match rock type with industrial use (e.g., limestone → cement industry) • Suggest zones for further exploration or conservation

The five core environmental features of the **Tirunelveli region—Topography, Climate & Rainfall, Precipitation, Soil, and Geomorphology**—along with how **Artificial Intelligence (AI)** is used currently and how it can be applied in future implementations shown in Table 18.1.

Tirunelveli enjoys strong transportation connectivity, being linked by **NH 44**, a major north–south National Highway connecting the city with **Madurai to the north** and **Kanyakumari to the south**. Additionally, **NH 138** provides a direct route between Tirunelveli and **Thoothukudi**. The city is also supported by five

Table 18.1 AI applications in environmental research–Tirunelveli focus.

Environmental feature	Description	How AI is used now	Future implementation ideas
Topography	Describes the natural features of the land such as elevation, slopes, hills, and valleys	• Terrain classification using DEM- Elevation & slope analysis via satellite data- AI in edge detection of ridges and valleys	• AI + GIS for automated site selection for urban planning- AI-based landslide or flood-risk prediction using topographic data
Climate and Rainfall	Refers to the average weather conditions and rainfall over time, influencing agriculture and water supply	• Climate prediction using time-series models (ARIMA, LSTM)- Seasonal rainfall forecasting- Anomaly detection in climate trends	• Smart agriculture planning using AI-based rainfall prediction- AI dashboard to visualize 10-year climate trends for policy making
Precipitation	The amount of water falling as rain, snow, hail, etc., critical for water management and flood alerts	• Rainfall pattern recognition using ML- Satellite image analysis for monthly precipitation mapping	• AI-based real-time rainfall alerts via IoT sensors- Integration with crop advisory systems for precision irrigation
Soil	Refers to the types of soils like red loam, black soil, sandy soil in the region and their fertility	• AI-based soil classification from satellite images- Crop suitability prediction from soil health data	• Mobile app to scan soil and recommend crop & fertilizer- Drone-based AI system for detecting soil degradation zones
Geomorphology	Study of rock formations like charnockite, gneiss, quartzite, and limestone that define the earth's surface	• AI-powered rock classification using spectral data- Mineral deposit identification using image analysis	• 3D terrain mapping using AI + LiDAR- AI system to monitor quarry zones and environmental impact

state highways—**SH 39, SH 40, SH 41, SH 41A, and SH 75**—enhancing regional mobility. **Tirunelveli Junction**, one of the oldest railway stations in the Indian Railways network, falls under the **Madurai Division of Southern Railways** and sees a daily passenger flow of around **50,000**. In terms of air travel, the city is served by **Thoothukudi Domestic Airport (32 km)** and is within reach of **Madurai (150 km)** and **Thiruvananthapuram (158 km)** international airports. Despite these advantages in infrastructure, the region—like many across the globe—faces pressing **environmental challenges** such as **land degradation, biodiversity loss, climate change, water scarcity, and air pollution**. These issues underscore the urgent need for sustainable development strategies. Although global efforts such as those from the **United Nations** aim to address these crises, stronger and more consistent local-level interventions are necessary to bring meaningful change.

18.8 A Road map for Tirunelveli's Future Economy

The economy of Tirunelveli is based mostly on agriculture, which is the main source of income for the region. In this area, research and innovation are crucial, particularly for weather-based farming decisions, crop recommendations, and soil health analysis, where farmers can benefit substantially from AI-powered precision agricultural tools. Many farmers have extensive field knowledge despite their lack of formal education; by fusing this conventional knowledge with approachable AI applications (like voice-activated crop advice apps), it is possible to empower them without overburdening them with scientific jargon. Sericulture makes a substantial contribution as well, especially in areas like Papakudi, Manur, and Palayamkottai. Entering the secondary sector, Tirunelveli is home to important industrial centers such the SIPCOT Industrial Estate in Gangaikondan and the IT Special Economic Zone (SEZ). Here, smart energy management systems, predictive maintenance, and AI-based industrial monitoring can increase sustainability and productivity. A step in the direction of modernizing the local economy has been taken with the construction of the Tirunelveli Trade Centre (2023). With thousands of micro and small-scale manufacturing and service businesses creating jobs, the MSME sector is also very important. While there aren't many large-scale businesses in the tertiary sector, digital services and logistics are developing industries. Automation and AI have the potential to improve e-governance, optimize supply chains, and revolutionize customer service. Green economic growth can be accelerated in the future by adopting renewable energy technology like wind energy forecasting, AI-optimized solar monitoring systems, and growing smart grid infrastructure through ongoing initiatives like SIPCOT Phase II and TATA Power Solar. In general, decision-making can be improved, inefficiencies can be decreased, and inclusive development catered to Tirunelveli's distinct social and economic environment may be ensured by integrating AI, IoT, and GIS across sectors.

18.9 AI-based Urban Housing Analytics and Slum Rehabilitation Forecasting for Tirunelveli LPA

Over the past few decades, Tirunelveli's urban population has increased dramatically. The district's population increased from 661,944 in 2001 to 815,528 in 2011, and the Tirunelveli LPA saw a growth rate of 33.36% during that time. In addition to the current 8,871 housing shortage, the anticipated population by 2041 is expected to increase 2,21,607 due to urban expansion, creating a demand for almost 80,900 new housing units. About 65.6% of households in the area are owned, 32.45% are rented, and 1.94% are leased. The residential types in the area are categorized as Pucca, Semi-Pucca, and Katcha. Addressing the requirements of the slum population, which makes up 17.33% of the urban population and is mostly concentrated in places like Palayamkottai and Thachanallur's Ward 10, is a major concern. Modern geospatial technology combined with artificial intelligence (AI) can provide revolutionary answers to these problems. While deep learning methods like U-Net or Mask R-CNN may identify and map slum regions using satellite imagery, artificial intelligence (AI) can be used to estimate housing demands using time-series models like LSTM and Prophet. Data-driven decision-making for resource distribution, infrastructure development, and slum rehabilitation is made possible by these technologies. Additionally, machine learning algorithms can help with housing typology classification, risk assessment for deteriorating buildings, and affordable housing development according to location and economic status. Researchers and urban planners may create heat maps of housing shortages, model urban development scenarios, and create clever solutions for sustainable urban life in Tirunelveli by fusing AI with GIS and remote sensing. Thangarasu et al. (2024) have proposed a hybrid neuro genetic convolution based autoencoders to enhance the traffic predictions by incorporating the optimization technique of neuro genetic model and combining it with casual convolution autoencoder components. Layers of casual convolution confirms the information flow as well as prevent the model from future information to predict the past [6]. In addition to advancing the goal of smart cities, this strategy guarantees that housing regulations are effective, inclusive, and compatible with future expansion. AI Model for housing is shown in pseudo code [7].

Pseudo Code 6: AI Model for Urban Housing Demand Forecast
Step 1: Collect population census data from 2001, 2011, and projected for 2041 Step 2: Preprocess data (clean missing values, normalize) Step 3: Apply LSTM time-series model to predict housing demand • Input: Historical population data • Output: Projected housing units needed

Step 4: Collect GIS data of current housing layouts and slum areas
Step 5: Use satellite imagery + Mask R-CNN for slum detection and boundary mapping
Step 6: Overlay housing shortage and slum locations on GIS map
Step 7: Recommend housing zones based on demand density and land availability
Step 8: Generate housing plan and simulate infrastructure requirements
Step 9: Export results to planners (dashboard, CSV, or map reports)

18.10 Tirunelveli 2041 as a Sustainable Growth Use Case

The Tirunelveli Master Plan 2041 presents a clear vision for sustainable urban growth.

Residential zones are planned to expand from 34.42% to 43.64%, meeting rising housing demand. Commercial land use will significantly grow from 2.05% to 9.03%, encouraging business activities. Industrial areas are set to double, supporting local employment and economic upliftment. Institutional zones will increase, strengthening healthcare, education, and public services. Transport infrastructure is maintained and improved to ensure seamless urban connectivity. Figure 18.3 explains the non-timber forest products buffer zone.

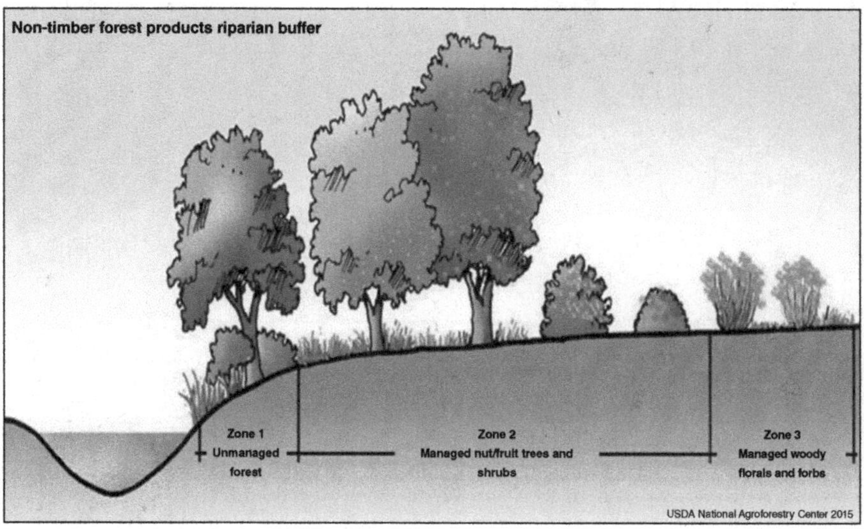

Figure 18.3 Design layout of a riparian buffer for environmental conservation from vision 2041.

Dry agricultural land will reduce, indicating a controlled shift toward urban expansion. Wetlands and fertile zones are planned to be protected despite the urban push.

The plan focuses on zoning for better land utilization and resource efficiency.

Environmental sustainability is emphasized by preserving 12.26% of land as water bodies. The sketch Figure 18.4 illustrates a sustainable development model for the Tamirabharani River by 2041. It includes riparian buffers to prevent erosion and maintain water quality [8].

Urban spaces are planned with eco-sensitive zones along the riverbanks.

Flood control, sewage treatment, and public access points are integrated.

This vision promotes harmony between nature, culture, and urban life in Tirunelveli.

A conceptual sketch in Figure 18.5 shows Vendhankulam River ecological restoration, controlled urban growth, green buffer zones, and integrated water resource management, aligning with Tirunelveli's 2041 development goals. It

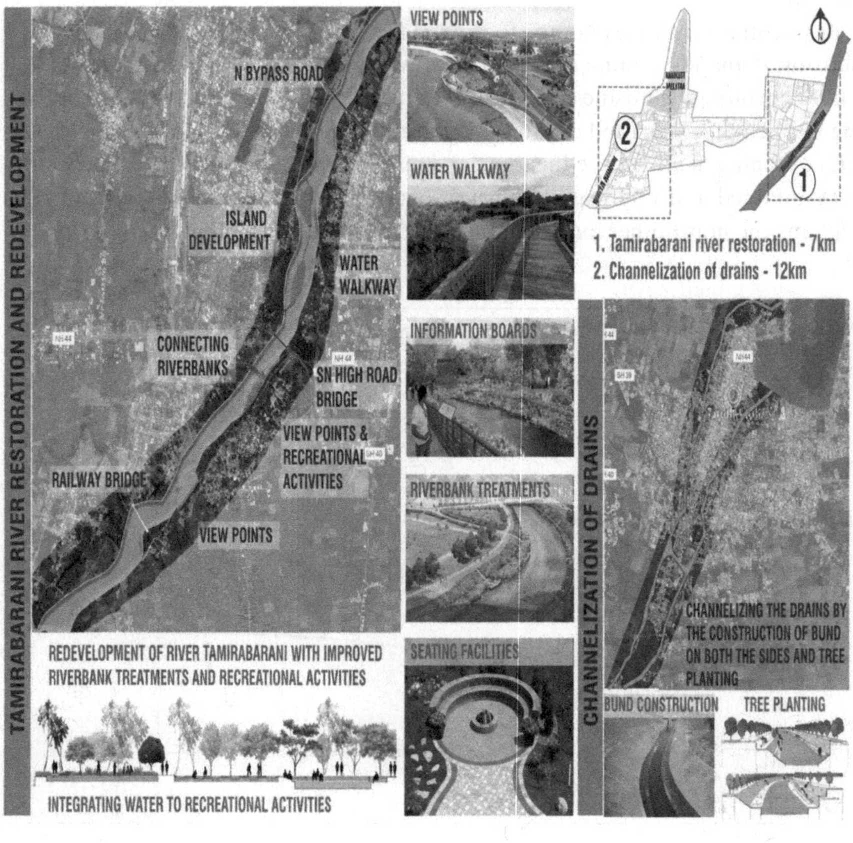

Figure 18.4 Sustainable development vision for the Tamirabharani river—2041.

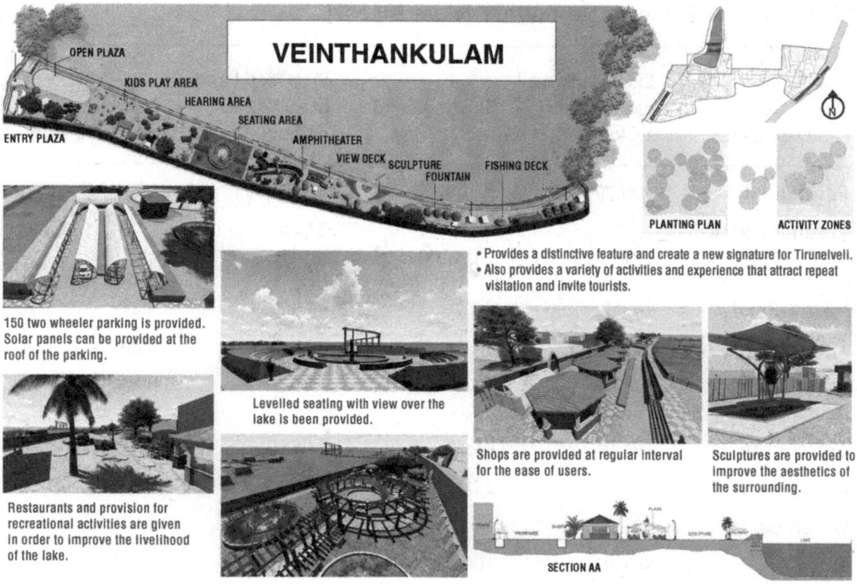

Figure 18.5 Vendhankulam river sustainable development plan: vision 2041.

includes green belts and riparian zones to reduce pollution and improve biodiversity. Low environmental impact urban growth is planned in the surroundings. Flood prevention, treated wastewater reuse, and rainwater collection are all combined. For future generations, the strategy promotes eco-balanced lifestyle and the long-term health of rivers. Because it has a direct impact on land use, infrastructure planning, and environmental conservation, topography is an essential component of Tirunelveli's sustainable development. To identify appropriate areas for residential, commercial, agricultural, and industrial development, a thorough examination of the city's distinctive landscape—which is influenced by the Western Ghats and the Tamirabharani River basin—is necessary. By lowering the dangers of flooding and soil erosion, an understanding of elevation and slope patterns aids in efficient water management. By conforming to the natural contours of the terrain, it also facilitates improved road, building, and drainage system design. In general, topography acts as a compass for designing Tirunelveli's resilient, environmentally responsible, and well-organized urban future [9].

Ecological conservation measures implicitly encourage open and green spaces. In order to avoid overtaxing the city center, decentralized development is encouraged. In order to effectively integrate work, living, and recreation, the strategy encourages mixed-use construction. It strives for a livable urban future by striking a balance between ecological preservation and economic progress. All things considered, the 2041 land use policy guarantees resilient, inclusive, and environmentally

Table 18.2 Land usage in existing and future 2041 [2].

Land use Zones	Existing Land use 2021 in sq. km	Existing Land use 2021 in %	Proposed Land use 2041 in sq. km	Proposed Land use 2041 in %
Residential	100.77	11.69	136.67	15.86
Commercial	2.77	0.32	23.68	2.75
Industrial	10.89	1.26	39.02	4.53
Institutional	18.24	2.12	44.99	5.22
Transportational	27.02	3.13	32.98	3.83
Open Space Recreational	1.07	0.12	1.26	0.15
Agriculture	561.95	65.19	444.12	51.52
Reserve Forest	17.28	2.00	17.28	2.00
Waterbodies	121.97	14.15	121.97	14.15
Total	861.97	100.00	861.97	100.00

responsible urban development. Each of the five primary committees that oversee the Tirunelveli LPA Master Plan 2041 focusses on important facets of urban development. When taken as a whole, these studies seek to guarantee Tirunelveli's urban growth is inclusive, sustainable, and responsibly managed (Table 18.2).

The Tirunelveli LPA Master Plan 2041 is guided by five major committees, each focusing on critical aspects of urban development. The **Economy and Employment Committee** addresses policies related to state urban development, demography, workforce dynamics, and skill development initiatives. The **Land Use and Environment Committee** concentrates on zoning regulations, urban design, heritage conservation, environmental sustainability, disaster management, and urban renewal. The **Traffic and Transportation Committee** oversees improvements in road, rail, and para-transit systems to ensure efficient and inclusive mobility. The **Shelter and Infrastructure Committee** focuses on housing for all categories and essential services like water supply, sanitation, drainage, health, education, and solid waste management. Meanwhile, the **Investment Planning and Governance Committee** works on financial planning, institutional governance, and the integration of GIS for better decision-making [10].

Traffic safety in Tirunelveli LPA remains a significant concern, with consistently high numbers of road accidents over the past five years. A detailed analysis shows that grievous injuries requiring hospitalization have remained high, and fatalities continue to be alarming. The year 2018 recorded the highest number of accidents and casualties, while 2020 showed a noticeable drop, largely attributed to reduced mobility during the COVID-19 lockdown. Minor injuries have shown fluctuations, and noninjury accidents have remained relatively low. Despite efforts, the trend highlights the urgent need for improved road infrastructure, stricter enforcement of traffic regulations, and increased public awareness to enhance road safety in the region.

Connectivity plays a vital role in shaping the livability and functionality of Tirunelveli city, acting as a key enabler for accessing services and promoting inclusive mobility. In assessing connectivity, specific sub-aspects have been assigned weightage based on their importance: major railway stations carry the highest weightage at 50%, followed by road networks at 30%, and both public transport terminals and the proposed western bypass at 10% each. Accessibility is measured by the proximity to infrastructure such as National Highways (NH), State Highways (SH), Major District Roads (MDR), and other roads. The closer a location is to key transport features, the higher its connectivity rating—classified as very high (within 2 km), high (2–5 km), moderate (5–10 km), and Low (beyond 10 km). For proposed infrastructure like the western bypass, the rating scale is slightly broader, with distances up to 30 km being considered. This framework ensures a systematic evaluation of how well different parts of the city are connected, supporting smarter land use and transport planning in Tirunelveli's urban development strategy [11].

To support these planning goals, several innovative research areas have been proposed. These include studies on income patterns in formal and informal sectors, land needs for micro-enterprises, and the rate of urban growth versus investment in infrastructure. Other key topics include healthcare access, educational enrolment, land availability for housing, slum development and resettlement, and water resource augmentation. Research also focuses on sanitation, waste management (including e-waste), and environmental quality such as water and air pollution. Mobility studies, citizen participation, public investment outcomes, zoning compliance, and green space inventories are also recommended. Additional proposals include prototype development plans for city zones, wards, and peri-urban areas, as well as mapping land values and evaluating financial support for low-income housing and micro-enterprises. These studies collectively strive to guarantee Tirunelveli's urban expansion is sustainable, inclusive, and well-managed [12].

18.11 Conclusion

Urbanization brings widespread awareness to the public and plays a key role in shaping how cities function. This use case provides a clear picture of how Tirunelveli is projected to grow and operate by 2041 using the latest technologies. The data and plans under Vision 2041 are openly accessible, allowing citizens, researchers, and planners to explore and utilize this information effectively. With growing interest in smart cities and sustainable development, this vision encourages multiple research opportunities to improve daily life and reduce complexities faced by the public. It is everyone's right and responsibility to ensure equal access to resources, and basic services like education and technology should reach even the most unaware or remote communities. This work aims to understand ongoing development and guide researchers in collecting data from less reachable or underrepresented areas, supporting inclusive and informed urban planning. The findings of this study demonstrate that AI-enabled planning can significantly enhance the management of urban housing and infrastructure development. By combining predictive models with geospatial analysis, it becomes possible to address critical issues such as housing shortages, slum rehabilitation, and land use planning. The proposed pseudo code serves as a foundation for future research and application in real-world urban environments. Overall, this integrated technological approach supports the broader goal of sustainable and inclusive city planning, and lays the groundwork for data-driven governance and smarter urban growth strategies. Many of today's developed countries were once considered developing or even underdeveloped. The challenges and inconveniences they faced in the past have paved the way for their current progress. Similarly, today's efforts and temporary struggles are building blocks for a better future. Every citizen has a role to play by following the rules and regulations set by the government. Various government organizations now provide essential data and resources that are openly accessible for research purposes. With proper registration and the right approach, we can collaborate, initiate research, and work together to improve our society—starting today.

References

1 Padmanaban, R., Cabral, P., and Bhowmik, A. Satellite Image Fusion to Detect Changing Surface Permeability and Emerging Urban Heat Islands in a Fast-Growing City. *Preprints*. August 21, 2018. https://doi.org/10.20944/preprints201808.0388.v1.

2 Directorate of Town and Country Planning. *Master Plan-2041: Tirunelveli LPA, AMRUT Sub-Scheme GIS-Based Master Plan, Tirunelveli LPA (Draft Final)*. Housing and Urban Development Department, Government of Tamil Nadu, 2024.

3 Tasgaonkar, P. P., Garg, R. D., Garg, P. K., Mishra, K. Smart city: Transformation to a digital city. In Gupta, N. and Mishra, S., editors, *Internet of Everything for Smart City and Smart Healthcare Applications*, pages 113–125. Springer, 2024.

4 Bhatt, Y. and Roychoudhury, J. Smart cities from an Indian perspective: Evolving ambitions. In Belaïd, F. and Arora, A., editors, *Smart Cities*, pages 359–383. Springer, 2023.

5 Nahiduzzaman, K. M., Holland, M., Sikder, S. K., Shaw, P., Hewage, K., and Sadiq, R. Urban transformation toward a smart city: An E-commerce–induced path-dependent analysis. *Journal of Urban Planning and Development*, 147(1), 2021. https://doi.org/10.1061/(ASCE)UP.1943-5444.0000648.

6 Thangarasu, G., Kesava, R. A., and Kannan, K. N. Hybrid neuro genetic causal convolution based autoencoders, for traffic prediction in smart cities. In *Proceedings of IEEE 14th Symposium on Computer Applications & Industrial Electronics (ISCAIE)*, pages 148–152. 2024.

7 Manimozhi, S., Arumugam, S., and Rajeshwari, G. Vehicle identification and counting (VIC) using machine learning algorithm–ORB. In *Proceedings of IEEE 4th International Conference on Data Engineering and Communication Systems (ICDECS)*, pages 1–4. 2024.

8 Kumar, A., Batra, N., Mudgal, A., Yadav, A. L. Navigating urban mobility: A review of AI-driven traffic flow management in smart cities. In *Proceedings of IEEE 11th International Conference on Reliability, Infocom Technologies and Optimization (Trends and Future Directions) (ICRITO)*, pages 1–5. 2024.

9 IEEE. *Development and Implementation of a Smart Parking System in Coimbatore, Tamil Nadu: Enhancing Urban Mobility and Efficiency*. IEEE Xplore, 2025.

10 Wu, T., Yang, Q., and Bao, Z. E-mobility: Transportation sector in transition. In Lackner, M., Sajjadi, B., and Chen, W. Y. editors., Handbook of Climate Change Mitigation and Adaptation. Springer, 2022.

11 Baskaran, V., Madasamy, M., Kumar, S. P., and Sahana, S. V. Modeling the land suitability for agricultural utility in a semi-arid region of Tirunelveli district, South India using multi-criteria and geospatial approach. *Modeling Earth Systems and Environment*, 9: 3527–3537, 2023.

12 Čamprag, N., Uğur, L., and Suri, A., editors. *Rethinking Urban Transformations: A New Paradigm for Inclusive Cities*. Springer, 2023.

19

GIS-based Analysis of Road Accidents: A Case Study on Hotspot Identification and Safety Improvement

Kirti Goyal[1,], Siddharth Garia[1], Anoop Bhardwaj[1], Amol Sharma[2], Sneha Das[1], and Animesh Nayak[1]*

[1] *School of Civil Engineering, Lovely Professional University, Phagwara, Punjab, India*
[2] *CEPT University, Ahmedabad, Gujarat, India*

*Corresponding author: kirti.28387@lpu.co.in

19.1 Introduction

With the rapid increase in urbanization, an unprecedented increase in the number of vehicles has also occurred [1, 2]. As a result, the frequency of traffic accidents has also increased manifold. According to a report by the World Health Organisation in 2018, 1.35 million people died globally in road accidents every year [3, 4]. These accidents cause a lot of damage to public property and the lives of the common people [5, 6]. Hence, to reduce the frequency of these accidents and to identify the accident-prone areas, GIS is one of the options that may contribute to identifying these areas and ultimately reducing them [7–9]. Several researchers have employed GIS in their respective studies to encounter this challenge. For instance, ref. [10] have utilized statistical-based analysis to explore the accidents in different regions around the globe [11–13]. Most of the researchers provide a relationship between the numerical models and GIS to analyze the hazards associated with traffic accidents [14–17].

Implementing GIS to analyze traffic accidents has been instrumental in modeling road crash statistics. Specifically, with reference to India, the approximate number of casualties account to around 85 000 and 300 000 injuries [18–20]. Therefore, the present study attempts to investigate the role of GIS in reducing traffic accidents in the region of Jharsuguda, Odisha. Jharsuguda is one of the important industrial areas/hubs in the state of Odisha. With the frequent inflow of resources, recent data suggests an exponential increase in the rate of traffic accidents in this city. With more urbanization expected to increase, these data may show an upward trend, thereby raising awareness to negotiate this issue. Hence, GIS can provide means to compare and analyze different streams of accident data such as FIR, inventory data, etc. [21–24]. The accident hotspot regions, high-risk locations, and congested areas can be highlighted, and alternative arrangements can be implied so as to reduce the number of accidents in this region. Hence, the present research attempts to explore the different parameters contributing to traffic accidents in Jharsuguda, Odisha. These parameters include public participation geographic information systems (PPGIS), mobile geographic information system applications, integration with smart city initiatives, and environmental data.

19.2 Methodology

In this study, GIS tools have been used to analyze the combined data, and a detailed flowchart showing the methodology of the presented work is given in Figure 19.1. Kernel density estimation (KDE), statistical analysis, and the mapping of accident hotspots have been conducted in this study. The results are presented in the form of both tabular and graphical representations. While graphical data has maps and charts that show how accidents and other results are distributed spatially, tabular data provides comprehensive statistical summaries.

19.3 Results and Discussion

The results regarding the type of data collected, the analysis, and the identification of accident hotspots have been discussed in this section.

19.4 Data Collection and Preparation

The information regarding date, time, location (latitude and longitude), accident type, victim count, and probable causes has been collected from the local police department for use in traffic accident data. Data on the physical environment such as road networks, traffic lights, road conditions, land use, and traffic signals, from local GIS databases and satellite pictures. After that, demographic data, traffic statistics, weather reports, and other applicable data from

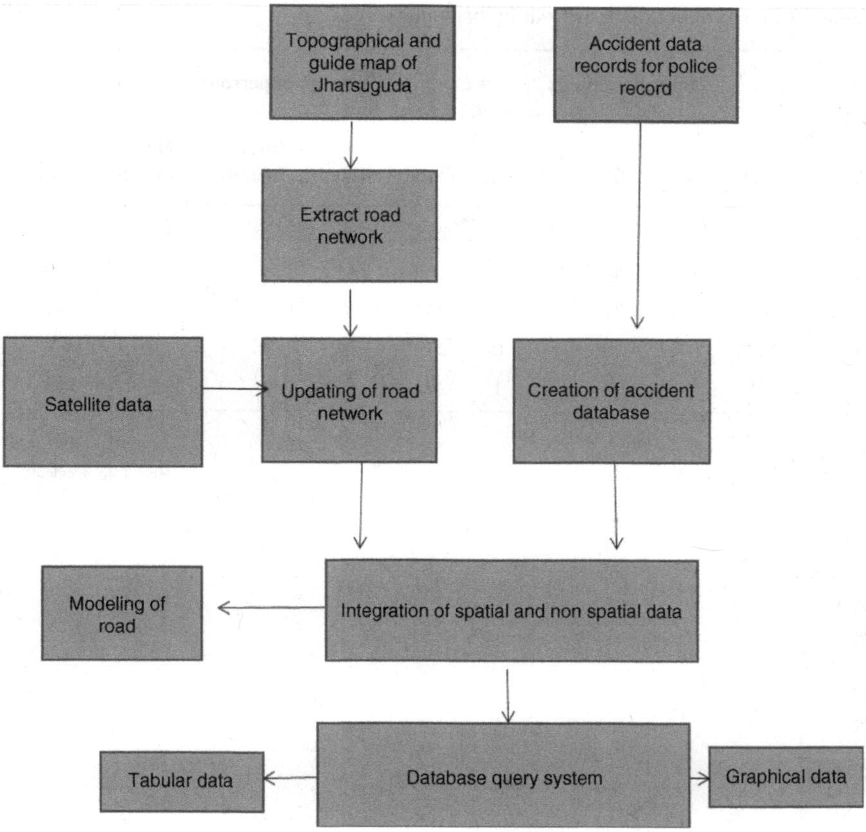

Figure 19.1 Flow chart showing the methodology of the proposed work.

the relevant government agencies as supplementary statistics were obtained. Finally, a geographical analysis visualizing accident hotspots using GIS was performed. Then, on a map of Jharsuguda, the locations of traffic accidents using GIS software (e.g., ArcGIS, QGIS) were plotted. Lastly, Kernel density estimation (KDE) was applied to find the places where there are most number of accidents, and then the same has been displayed graphically. Table 19.1 shows the number of accidents over the last 5 years in Jharsuguda. As observed in Figures 19.2 and 19.3, the maximum number of accidents occurred in the year 2022, with year 2023 marking the most fatal accidents with the highest number of killings. The year-wise data for vehicle accidents and the severity of injuries have been shown in Tables 19.2(a, b), 19.3(a, b), and 19.4(a, b). Along with that, the data also indicates the type of highway/road marking the maximum number of accidents.

Table 19.1 No. of accidents by year in Jharsuguda.

		Type of accidents					Number of persons involved			
S. No.	Year	Fatal	GI	MI	NI	Total	Killed	Grievous injured	Minor injured	Total
1	2019	118	82	38	10	248	126	128	118	372
2	2020	90	64	27	9	190	93	78	49	220
3	2021	111	72	51	10	244	120	97	120	337
4	2022	107	106	66	0	279	124	140	136	400
5	2023	113	80	48	1	242	130	81	127	338

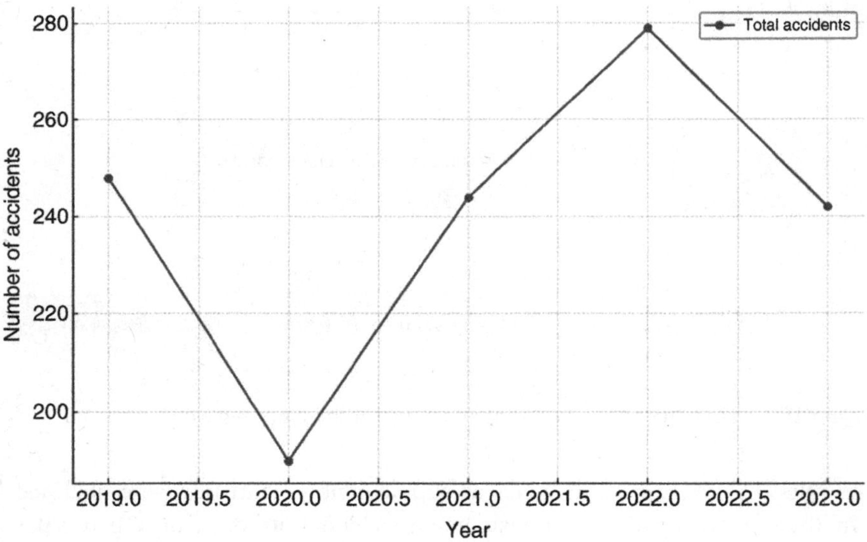

Figure 19.2 Number of accidents by year in Jharsuguda.

Figures 19.4–19.6 show the accidents in the city of Jharsuguda on the basis of the type of road.

- Total Accidents: The number of overall accidents fluctuated noticeably throughout the four-year period, with a considerable decline in 2020, most likely affected by less traffic owing to COVID-19 lockdowns, followed by an increase in succeeding years.

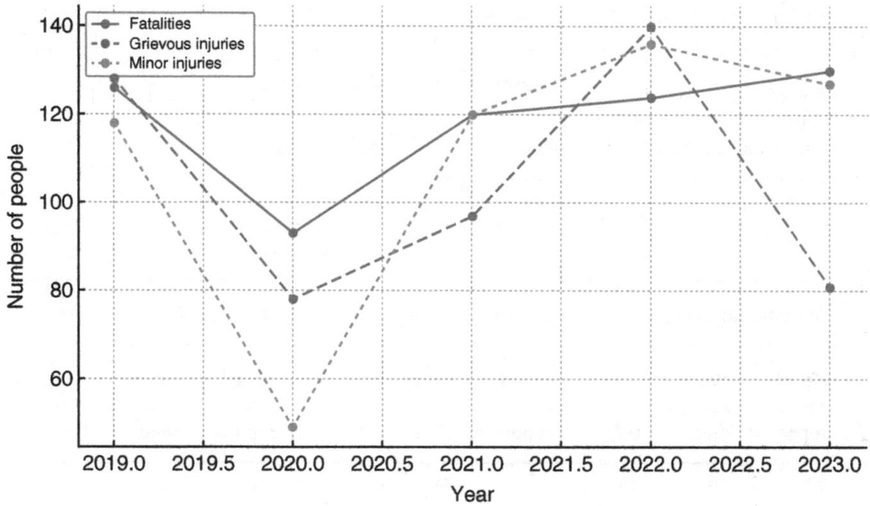

Figure 19.3 Number of fatalities, GI, and minor injuries per year in Jharsuguda.

Table 19.2 Vehicle accidents for the year 2021.

a.

No. of cases reported	No. of person who died	Grievous injuries (G.I.)	Simple injuries (S.I.)
244	120	97	120

b.

On national highway				On state highway				Other road			
No. of cases reported	No. of persons died	G.I.	S.I.	No. of cases reported	No. of persons died	G.I.	S.I.	No. of cases reported	No. of persons died	G.I.	S.I.
105	55	56	35	46	20	12	22	93	45	29	63

- Fatalities and Injuries: The number of fatalities, catastrophic injuries, and mild injuries fluctuated throughout time. A similar trend is found for the overall number of accidents, with lower counts in 2020 and a resurgence in 2021 and 2022.

Table 19.3 Vehicle accidents for the year 2022.

a.

No. of cases reported	No. of person who died	Grievous injuries (G.I.)	Simple injuries (S.I.)
279	124	140	136

b.

On national highway				On state highway				Other road			
No. of cases reported	No. of persons died	G.I.	S.I.	No. of cases reported	No. of persons died	G.I.	S.I.	No. of cases reported	No. of persons died	G.I.	S.I.
93	59	45	45	53	23	23	39	133	42	62	52

Table 19.4 Vehicle accidents for the year 2023.

a.

No. of cases reported	No. of person who died	Grievous injuries (G.I.)	Simple injuries (S.I.)
219	110	99	129

b.

On national highway				On state highway				Other road			
No. of cases reported	No. of persons died	G.I.	S.I.	No. of cases reported	No. of persons died	G.I.	S.I.	No. of cases reported	No. of persons died	G.I.	S.I.
100	56	45	60	40	20	15	24	79	34	39	45

19.5 Analysis

- The data indicate that while there was a temporary decrease in accidents and injuries in 2020, perhaps due to limits imposed by the epidemic, the numbers increased again after regular activities resumed.
- The need for targeted investments to improve road safety measures and infrastructure is underscored by the recurrence of injuries and fatalities, even in years with fewer occurrences.

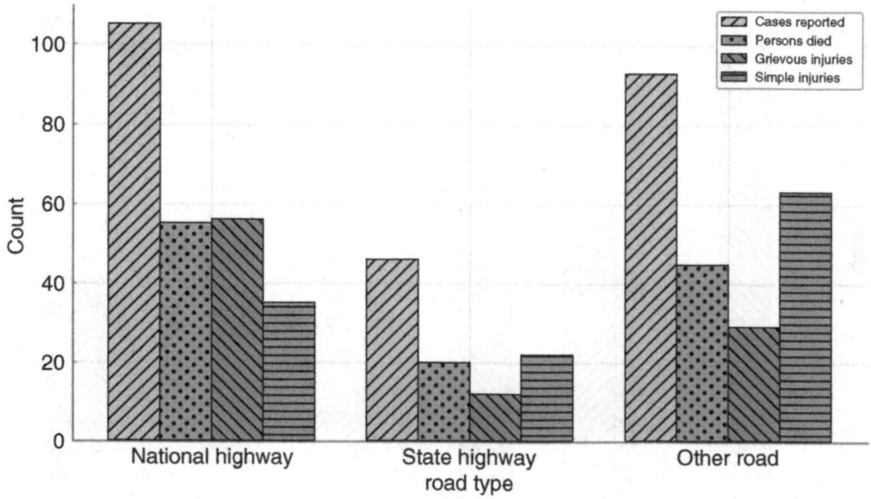

Figure 19.4 Motor vehicle accidents by road type (2021).

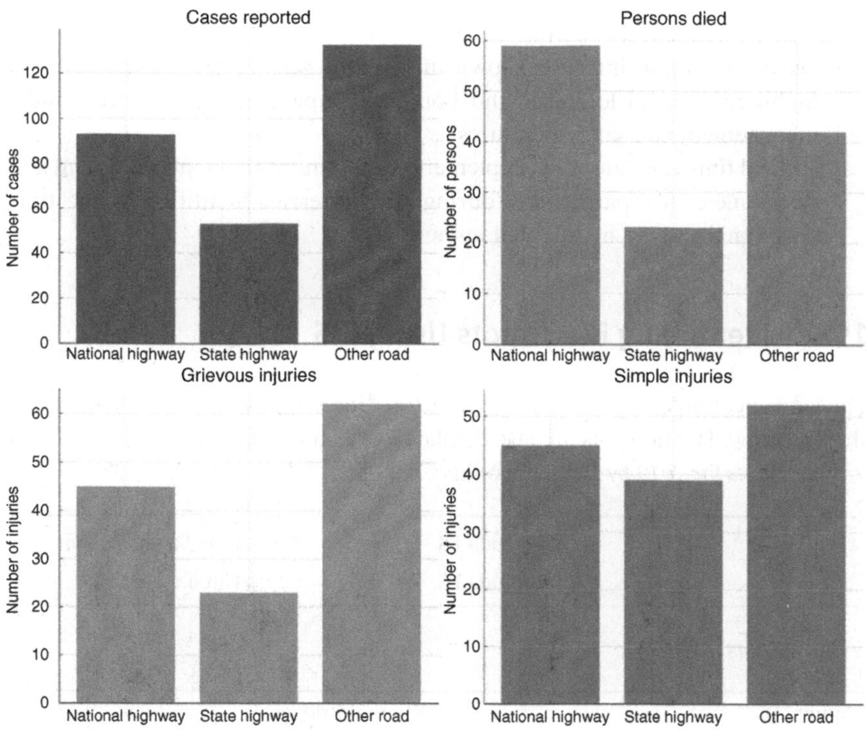

Figure 19.5 Road incident statistics by road type.

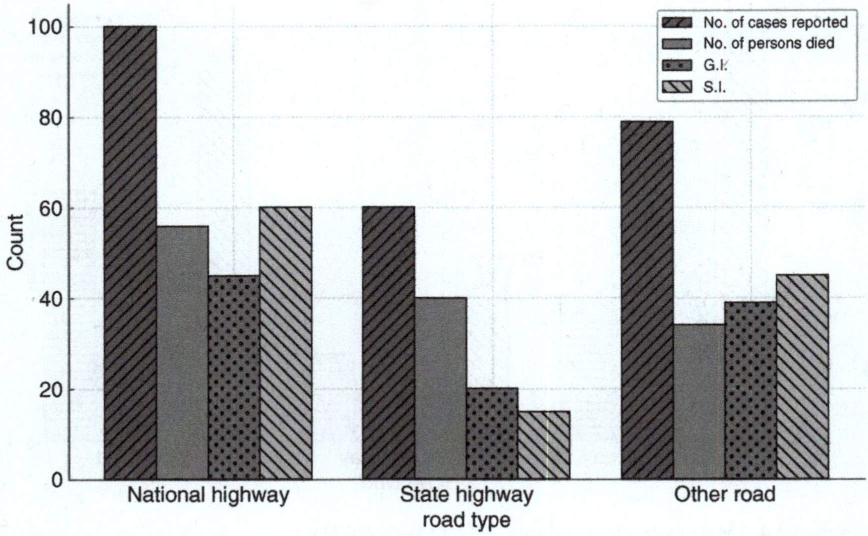

Figure 19.6 Distribution of traffic accidents by road type in Jharsuguda.

- During high-risk intervals shown in the time series analysis, the research highlights critical locations and hours for implementing enhanced traffic management and safety measures.
- Critical times and areas for implementing enhanced traffic management and safety measures, particularly during risky intervals identified by the time series analysis, are highlighted in the study.

19.6 Identifying Blackspots Using GIS

The identification of black spots was done on different bases for various years, as shown below. The hotspots are also displayed in Figure 19.7.

Black Spots for 2019 by DPI and Roads

1.	Jharsuguda PS	Prasanna Panda Chowk
2.	Jharsuguda PS	Biju Pattanaik Chowk

Black Spots for 2017

Badmal PS	Kherual Bridge

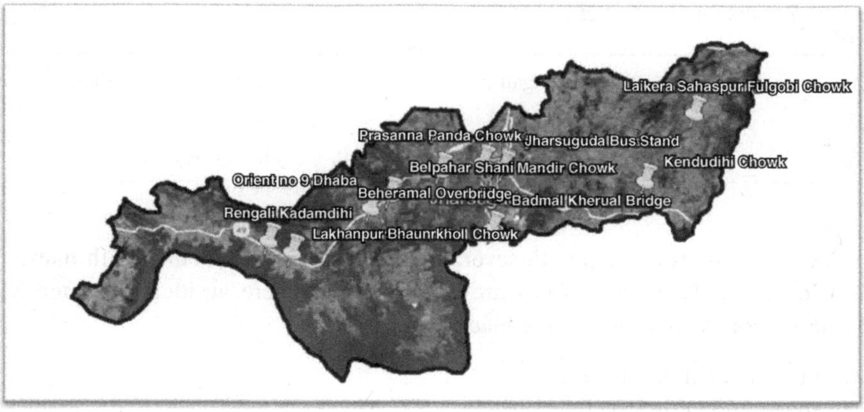

Figure 19.7 Blackspots of Jharsuguda.

Black Spots for 2017–2019 by STI

1.	Belpahar PS	Sani Mandir Chowk
2.	Rengali PS	Kureimal Chowk

Black Spots for 2018–2020

1.	Jharsuguda PS	Over Bridge Beheramal

Black Spots for 2020

1.	Laikera PS	Sahaspur Fulgobi Chowk
2.	Laikera PS	Kendudihi Chowk
3.	Laikera PS	Banpali Chowk

Black Spots – 2021

1.	Orient PS	No.9 Dhaba
2.	Jharsuguda PS	Bus Stand
3.	Belpahar PS	Shani Mandir Chowk
4.	Rengali PS	Kureimal Chowk

Black Spots for the year 2019–2021

1.	Rengali PS	Kadamdihi (Tinpulian)
2.	Sadar PS	Katapali to IB Bridge
3.	Lakhanpur PS	Bhaunrkhol Chowk

The image shows a map with several marked blackspots, labeled with names and locations. These blackspots are critical areas where accidents frequently occur. Here's an analysis of these blackspots:

1. Prasanna Panda Chowk
2. Jharsuguda Bus Stand
3. Belpahar Shani Mandir Chowk
4. Beheramal Overbridge
5. Badmal Kherual Bridge
6. Kendudihi Chowk
7. Laikera Sahaspur Fulgobi Chowk
8. Orient No 9 Dhaba
9. Rengali Kadamdihi
10. Lakhanpur Bhaunrkhol Chowk

19.7 Key Insights

Clustered Blackspots: Jharsuguda is only one of several locations where you may see a concentration of blackspots on the map. There may be other risk variables that contribute to the frequency of accidents, or the place may have a larger traffic density.

Some blackspots, including the Orient No. 9 Dhaba and the Rengali Kadamdihi, are situated along important highways. More serious accidents can occur on highways due to the higher speeds and heavier traffic.

The fact that blackspots may be found in both urban areas (such as Belpahar Shani Mandir Chowk) and rural areas (like Kendudihi Chowk) shows that accidents can happen anywhere. Problems may arise with both urban pedestrian traffic and rural high-speed transportation. Impact on Infrastructure: Notable on the list are overbridges and chowks (intersections), suggesting that the infrastructure and architecture of these locations may be a factor in their classification as blackspots. It may be necessary to rethink or improve traffic management in certain areas.

19.8 Conclusion

This study proves that GIS is a useful tool for examining Jharsuguda's traffic patterns and accident statistics. The results can be used as a foundation for traffic analysis to enhance road safety by understanding problematic areas and the probable causes behind them. The research has helped in the identification of high-risk areas in the city of Jharsuguda using GIS, along with isolating the hotspots on roads and intersections on the basis of the occurrence and severity of accidents. The results may be used for enhancing the traffic mobility and safety by infrastructural improvements, developing traffic management strategies, modifications in road design, traffic signals, signs, markings, etc. Furthermore, it has been observed that upgradation in infrastructure at these hotspots could be effective in reducing traffic congestion. The extensive geographical insights provided by the study may aid the urban planners and stakeholders in informed decision-making by better understanding of traffic dynamics. The growing traffic density may also be linked with the rising air quality degradation in the city. Thus, future studies may focus on real-time data integrated AI predictive modeling techniques comprising technical, economic, and environmental aspects for developing long term and sustainable traffic management measures.

19.9 Recommendations

The following recommendations may be made on the basis of the present study.

- Better traffic management strategies to be implemented in hotspot zones, such as major intersections, bus stops, junctions, etc.
- Strict enforcement of speed limits and other traffic rules on highways and expressways.
- Improvement in road infrastructure by better road signs, traffic, streetlights, etc.
- Public sensitization on traffic rules for vehicles and pedestrians and associated risks
- Real time monitoring and continuous improvement in high-risk areas.

19.10 Way Forward

Numerous advancements and enhancements in urban planning and traffic management are being introduced for a better understanding and management of traffic. The scope of data collection is being broadened to incorporate real-time traffic surveillance techniques, including Internet of Things sensors, drones, and enhanced GPS tracking to enhance the precision of GIS studies. Other than that, the following points may be suggested.

- Integration of traffic data with other segments of smart city initiatives, such as transportation network, pedestrian movement, and real time sensors.
- Predictive modeling with AI and machine learning algorithms for future prediction of traffic conditions and hotspot identification by analyzing past data and current trends.
- GIS integrated urban planning by understanding traffic routes and enhancing transportation networks.
- To develop sustainable traffic management measures, a correlation between traffic-related pollutants and associated effects on the environment and public health needs to be established.

Well-designed educational and training programs may be beneficial for urban planners, traffic engineers, and GIS professionals for developing a holistic approach to traffic management. Gaining expertise in GIS and its applications may help municipalities enhance development planning and urban traffic management.

References

1 Reshma, E. K. and Sharif, S. U. Prioritization of accident black spots using GIS. *International Journal of Emerging Technology and Advanced Engineering*, 2(9): 117–122, 2012.

2 Kaygisiz, Ö., Düzgün, Ş., Yildiz, A., and Senbil, M. Spatio-temporal accident analysis for accident prevention in relation to behavioral factors in driving: The case of South Anatolian Motorway. *Transportation Research Part F: Traffic Psychology and Behaviour*, 33: 128–140, 2015. https://doi.org/10.1016/j.trf.2015.07.002.

3 Shafabakhsh, G., Famili, A., and Bahadori, M.S. GIS-based spatial analysis of urban traffic accidents: Case study in Mashhad, Iran. *JTTE (English Edition)*, 4: 290–299, 2017.

4 Karimnezhad, A. and Moradi, F. Road accident data analysis using Bayesian networks. *Transportation Letters*, 9(1): 12–19, 2016. https://doi.org/10.1080/19427867.2015.1131960.

5 N. Kabakuş, A. Tortum, and M. Y. Çodur. Traffic accidents occurred in districts of Erzurum using with geographical information system. *Ordu University Journal of Science and Technology*, 2(2): 78–92, 2012.

6 Paravar, M., Hosseinpour, M., Salehi, S., Mohammadzadeh, M., Shojaee, A., Akbari, H., and Sadat Mirzadeh, A. (2013). Pre-hospital trauma care in road traffic accidents in Kashan, Iran. *Archives of Trauma Research*, 1(4): 166–171, 2013. https://doi.org/10.5812/ATR.8780.

7 Şenel, B. and Senel, M. Risk analysis: Fault tree analysis application on traffic accidents occurred in Turkey. *Anadolu University Journal of Social Sciences*, 13(3): 65–83, 2013.

8 Çinicioğlu, A. E., Atalay, M., and Yorulmaz, H. Bayesian network model for analysis of traffic accident. *International Journal of Informatics Technologies*, 6(2): 41–52, 2013.

9 Murat, Y. Ş. and Şekerler A. Modelling traffic accident data by cluster analysis approach. *İMO Technical Journal*, 20: 4759–4777, 2009.10.

10 Bahadori, M. K., Ghardashi, F., Izadi, A. R., and Ravangard, R. Pre-hospital emergency status in Iran with an emphasis on road accidents: A systematic review. *Trauma Monthly*, 20(Special Issue), 2015. https://doi.org/10.5812/TRAUMAMON.28490.

11 Kamh, H., Alyami, S. H., Khattak, A., Alyami, M., and Almujibah, H. Exploring road traffic accidents hotspots using clustering algorithms and GIS-based spatial analysis. *IEEE Access*, 2024. https://doi.org/10.1109/ACCESS.2024.3443245.

12 Kaygısız, Ö., Düzgün, Ş., Akın, S., and Çelik, Y. Temporal and spatial analysis of traffic accidents using geographic information systems (Security General Directorate Traffic Research Center Directorate). 2012.

13 Steenberghen, T., Dufays, T., Thomas, I., and Flahaut, B. Intra-urban location and clustering of road accidents using GIS: A Belgian example. *International Journal of Geographical Information Science*, 18(2): 169–181, 2004. https://doi.org/10.1080/13658810310001629619.

14 Gufran, Dr., & M., Dr. Modeling of traffic accident reporting system through UML using GIS. *International Journal of Advanced Computer Science and Applications*, 3(6), 2017. https://doi.org/10.14569/IJACSA.2012.030606.

15 Saplıoğlu, M. and Karaşahin, M. Urban traffic accident analysis by using geographic information system. *Pamukkale University Journal of Engineering Sciences*, 12(3): 321–332, 2006.

16 Erdogan, S., Yilmaz, I., Baybura, T., and Gullu, M. (2008). Geographical information systems aided traffic accident analysis system case study: City of Afyonkarahisar. *Accident Analysis Prevention*, 40(1): 174–181.

17 Xie, Z. and Yan, J. Kernel density estimation of traffic accidents in a network space. *Computers, Environment and Urban Systems*, 32(5): 396–406, 2008. https://doi.org/10.1016/j.compenvurbsys.2008.05.001.

18 Anderson, T. Kernel density estimation and k-means clustering to profile road accident hotspots. *Accident Analysis and Prevention*, 41: 359–364, 2009.

19 Naboureh, A., Feizizadeh, B., Naboureh, A., Bian, J., Blaschke, T., Ghorbanzadeh, O., and Moharrami, M. Traffic accident spatial simulation modeling for planning of road emergency services. *ISPRS International Journal of Geo-Information 2019*, 8(9): 371, 2019. https://doi.org/10.3390/IJGI8090371.

20 Rajaei Behbahani, N., Mohammadi, M. J., Hashemi Nazari, S. S., Ghadirzadeh, M., and Hassanipour, S. Studying the epidemiology of fatal traffic accidents in the Khuzestan Province. *Jundishapur Journal of Health Sciences*, *8*(2), 2016. https://doi.org/10.17795/JJHS-32114.

21 Yamada, T. and Thill, J. Comparison of planar and network K-functions in traffic accident analysis. *Journal of Transport Geography*, 12: 149–158, 2004.

22 Ma, Q., Huang, G., and Tang, X. GIS-based analysis of spatial–temporal correlations of urban traffic accidents. *European Transport Research Review*, 13(1), 2021. https://doi.org/10.1186/S12544-021-00509-Y.

23 Özalp, S., Bayram, S., and Kayar, M. The relationship between the reasons and results of traffic accidents analysis and evaluation using TOPSIS and AHP methods. *Istanbul Commerce University Journal of Science*, 19: 65–87, 2011.

24 Su, J. M., Wang, Y. M., Chang, C. hung, and Wu, P. J. Application of a geographic information system to analyze traffic accidents using Nantou County, Taiwan, as an example. *Journal of the Indian Society of Remote Sensing*, 47(1): 101–111, 2019. https://doi.org/10.1007/S12524-018-0874-Z/TABLES/2

Index

Printed in the USA/Agawam, MA
July 7, 2026

907425.008